ON TO MARS

— *Volume 2* —

Exploring and Settling
A New World

Compiled and Edited by
Dr. Frank Crossman
and Dr. Robert Zubrin
The Mars Society

The proposals and opinions expressed in this
book are those of the contributing authors
and do not necessarily reflect those of either
the Mars Society as a whole or of the publisher.

All rights reserved under article two of the Berne Copyright Convention (1971).
We acknowledge the financial support of the Government of Canada through the
Book Publishing Industry Development Program for our publishing activities.
Published by Collector's Guide Publishing Inc., Box 62034,
Burlington, Ontario, Canada, L7R 4K2
Printed and bound in Canada
ON TO MARS - VOLUME 2
Compiled and Edited by
Dr. Frank Crossman and Dr. Robert Zubrin, The Mars Society
ISBN 1-894959-30-2
ISSN 1496-6921
Apogee Books Space Series No. 55
©2005 Apogee Books

Table of Contents

Introduction – Getting Space Exploration Right
by Robert Zubrin

In early 2004, President George W. Bush delivered a major policy speech charting a new course for the National Aeronautics and Space Administration (NASA). Instead of focusing on perfecting flight to and operations in low-Earth orbit, the space agency would henceforth set its sights on a return to the Moon and then "human missions to Mars and to worlds beyond." The President's move was a direct response to concerted criticism of the nation's space policy following the shuttle *Columbia* accident of February 2003. Numerous members of Congress had decried the fact that the US manned space program had gone adrift, spending huge amounts of money and putting lives at risk without any discernible objective. Accordingly, in a reversal of previous administration pronouncements, the new "Vision for Space Exploration" was created to pose grand goals for America in space.

There is no doubt that a radical policy shift was in order. During the first dozen years of its existence, NASA took the nation from having no space capability to landing humans on the Moon, but since then, the manned space program has been stuck in low-Earth orbit. Clearly, three decades of stagnation are enough. The question is whether the new policy is adequate to remedy the problems that have mired the space program in confusion and impotence, or whether it will amount to nothing. What needs to be done to make the Bush vision real?

To answer this question, we need to examine NASA's fundamental mode of operation, and see how the new policy bears on the organization's pathology. Then, to assess how the proposed cure is working, we need to examine the developments that have occurred since the President's announcement. While there are many hopeful signs, there remain large causes for concern, and radical changes in both the policy itself and its method of implementation will be required for the President's vision to succeed. Finally, we need to understand the deeper significance of this endeavor for both America and the human future. We need to ask: Why should human beings explore space at all, and why us?

But first things first. Before we can present the cure, we need to understand the disease.

Why Has NASA Been Failing?
Over the course of its history, NASA has employed two distinct modes of operation. The first prevailed during the period from 1961 to 1973, and may be called the Apollo Mode. The second has prevailed since 1974, and may be called the Shuttle Mode.

In the Apollo Mode, business is (or was) conducted as follows: First, a destination for human space flight is chosen. Then a plan is developed to achieve this objective. Following this, technologies and designs are developed to implement that plan. These designs are then built and the missions are flown.

The Shuttle Mode operates entirely differently. In this mode, technologies and hardware elements are developed in accord with the wishes of various technical communities. These projects are then justified by arguments that they *might* prove useful at some time in the future when grand flight projects are initiated.

Contrasting these two approaches, we see that the Apollo Mode is *destination-driven*, while the Shuttle Mode pretends to be technology-driven, but is actually *constituency-driven*. In the Apollo Mode, technology development is done for mission-directed *reasons*. In the Shuttle Mode, projects are undertaken on behalf of various pressure groups pushing their own favorite technologies and then defended using *rationales*. In the Apollo Mode, the space agency's efforts are *focused and directed*. In the Shuttle Mode, NASA's efforts are *random and entropic*.

To make this distinction completely clear, a mundane metaphor may be useful. Imagine two couples, each planning to build their own house. The first couple decides what kind of house they want, hires an architect to design it in detail, and then acquires the appropriate materials to build it. That is the Apollo Mode. The second couple polls their neighbors each month for different spare house-parts they would like to sell, and buys them all, hoping eventually to accumulate enough stuff to build a house. When their relatives inquire as to why they are accumulating so much junk, they hire an architect to compose a house design that employs all the knick-knacks they have purchased. The house is never built, but an excuse is generated to justify each purchase, thereby avoiding embarrassment. That is the Shuttle Mode.

In today's dollars, NASA's average budget from 1961 to 1973 was about $17 billion per year – only slightly higher than NASA's current budget. To assess the comparative productivity of the Apollo Mode with the Shuttle Mode, it is therefore useful to compare NASA's accomplishments during the years 1961-1973 and 1990-2003, as the space agency's total expenditures over these two periods are roughly the same.

Between 1961 and 1973, NASA flew the Mercury, Gemini, Apollo, Skylab, Ranger, Surveyor, and Mariner missions, and did all the development for the Pioneer, Viking, and Voyager missions as well. In addition, the space agency developed hydrogen-oxygen rocket engines, multi-staged heavy-lift launch vehicles, nuclear rocket engines, space nuclear reactors, radioisotope power generators, space suits, in-space life support systems, orbital rendezvous techniques, soft landing rocket technologies, interplanetary navigation technology, deep space data transmission techniques, reentry technology, and more. In addition, such valuable institutional infrastructure as the Cape Canaveral launch complex, the Deep Space tracking network, and the Johnson Space Center were all created in more or less their current form.

In contrast, during the period from 1990 to 2003, NASA flew about fourscore shuttle missions, allowing it to launch and repair the Hubble Space Telescope and partially build what is now known as the International Space Station. About half a dozen interplanetary probes were launched (compared to over 40 lunar and planetary probes between 1961 and 1973). Despite innumerable "technology development" programs, no new technologies of any significance were actually developed, and no major operational infrastructure was created.

Comparing these two records, it is difficult to avoid the conclusion that NASA's productivity – both in terms of missions accomplished and technology developed – was vastly greater during its Apollo Mode than during its Shuttle Mode.

The Shuttle Mode is hopelessly inefficient because it involves the expenditure of large sums of money without a clear strategic purpose. It is remarkable that the leader of any technical organization would tolerate such a senile mode of operation, but NASA administrators have come to accept it. Indeed, during his first two years in office, Sean O'Keefe (the NASA administrator from 2001 until early 2005) explicitly endorsed this state of affairs, repeatedly rebutting critics by saying that "NASA should not be destination-driven."

Yet ultimately, the blame for this multi-decade program of waste cannot be placed solely on NASA's leaders, some of whom have attempted to rectify the situation. Rather, the political class must also accept major responsibility for failing to provide any coherent direction for America's space program – and for demanding more than their share of random projects that do not fit together and do not lead anywhere.

Advocates of the Shuttle Mode claim that by avoiding the selection of a destination they are developing the technologies that will allow us to go anywhere, anytime. That claim has proven to be untrue. The Shuttle Mode has not gotten us anywhere, and can never get us anywhere. The Apollo Mode got us to the Moon, and it can get us back, or take us to Mars. But leadership is required – and for the last three decades, there has been almost none.

The New Bush Policy

While a growing chorus of critics has decried overspending and other fiscal inefficiencies at NASA over the years, it was only the *Columbia* accident of February 2003 that provided the impetus for policy makers to examine the fundamental problem of America's manned space program.

In the aftermath of *Columbia*'s destruction, both Congress and the administration initiated inquiries into the affair. These included extensive hearings in both the House and Senate and a special blue-ribbon commission appointed by the President and headed by retired Navy Admiral Harold Gehman, Jr. While much of the attention in these investigations focused on determining the specific causes of the accident itself, both Gehman and many of the congressional and press critics took a broader view, identifying as problems not only the particular management failures that led to the shuttle's loss, but also the overall lack of strategic direction of the space agency.

Columbia was lost on a mission that had no significant scientific objectives, certainly none commensurate with the cost of a shuttle mission, let alone the loss of a multi-billion dollar shuttle and seven crew members. In war, when soldiers are lost attempting a military mission of no value, the fallen are still heroes, but the generals have some explaining to do. The *Columbia* flight program included conducting experiments in mixing paint with urine in zero-gravity, observing ant farms, and other comparable activities – all done at a cost greater than the annual federal budgets for fusion energy research and pancreatic cancer research, combined.

After the Columbia Accident Investigation Board's report was issued in August 2003, this line of criticism became a refrain. In response, the Bush administration initiated an internal deliberative process to try to define strategic goals for the American space program. This process was carried out primarily behind closed doors, although a number of outsiders were invited to present their views. From these discussions and a series of congressional hearings, three distinct factions emerged. First, there were those who supported continuing business as usual at NASA, with appropriate cosmetic adjustments to get past the immediate crisis, but no fundamental changes. Second, there were those who called for making a human return to the Moon the central goal of the manned space flight program. And third, there were those who argued for an initiative to get humans to Mars.

President Bush announced the new policy on January 14, 2004, in a speech at NASA headquarters. As articulated in that speech and an accompanying National Security Presidential Directive, the new policy, dubbed the "Vision for Space Exploration," included something for each faction. The vision called for:

• Implementing a sustained and affordable human and robotic program to explore the solar system and beyond;
• Extending a human presence across the solar system, starting with a human return to the Moon by the year 2020, in preparation for human exploration of Mars and other destinations;

- Developing the innovative technologies, knowledge, and infrastructures both to explore and to support decisions about the destinations for human exploration; and
- Promoting international and commercial participation to further US scientific, security and economic interests.

The directive then lists a series of actions and activities to achieve these stated goals. These include returning the space shuttle fleet to flight, using it to complete construction of the International Space Station, and then retiring the shuttle and moving beyond it by "the end of this decade." The directive also states that NASA should develop "a new crew exploration vehicle to provide crew transportation for missions beyond low-Earth orbit," and should conduct "the initial test flight before the end of this decade in order to provide an operational capability to support human exploration missions no later than 2014." It also says that NASA should "acquire crew transportation to and from the International Space Station, as required, after the space shuttle is retired from service."

Beyond low-Earth orbit, the policy instructs NASA to "undertake lunar exploration activities to enable sustained human and robotic exploration of Mars and more distant destinations in the solar system." By 2008, NASA should begin a series of lunar robotic missions intended to "prepare for and support future human exploration activities." The first human mission is supposed to commence between 2015 and 2020. And unlike the short, three-day stay on the Moon that is the previous record (set by *Apollo 17* in 1972), this would be an *"extended* human expedition."

In addition to studying the Moon itself, these lunar activities are meant to "develop and test new approaches, technologies, and systems . . . to support sustained human space exploration to Mars and other destinations." The plan calls for robotic exploration of the solar system – Mars, asteroids, Jupiter's moons – as well as a search for habitable planets outside our solar system. The knowledge gathered from the robotic exploration of Mars, along with the lessons learned from long-term stays on the Moon, along with new technologies for "power generation, propulsion, life support, and other key capabilities," are aimed at making possible "human expeditions to Mars" at some unspecified date.

The most obvious problem with the Bush plan is its long, slow time-line. The only activities that the Vision for Space Exploration actually mandates before the end of the Bush administration's second term are the return of the shuttle to flight, the use of the shuttle to complete the International Space Station, the flight of one lunar robotic probe, and the initiation of a development program for the Crew Exploration Vehicle. The ten-year schedule for the development of the Crew Exploration Vehicle is especially absurd. Technically, it makes no sense: starting from a much lower technology base, it only took five years to develop the Apollo command module, which served the same functions. Politically, it is unwise: the delay makes the development of the Crew Exploration Vehicle reversible by the next administration. And fiscally, it is foolish: the long timeline only serves to gratify the major aerospace industry contractors, which desire a new long-term, high-cost activity to replace the recently canceled Orbital Space Plane. Stranger still is the decision to set the next manned Moon landing as late as sixteen years into the future – twice as long as it took the United States to reach the Moon back in the 1960s – and to place the Mars mission at some nebulous time in the future. Such a drawn-out timeline is unlikely to serve as a driving force on the activities of this slow-moving bureaucracy.

Still, there are aspects of the new policy that make it a positive step forward. By declaring that Moon-Mars would be the next order of business after the completion of the space station, the Bush vision precludes starting alternative initiatives that would get in the way. More importantly, by declaring that human exploration of the Moon and Mars is the goal of NASA, the new policy makes it legitimate for the space agency to allocate funds for technology development to support this objective. This is very important, since such spending previously could not be justified unless it could be defended as a necessary part of other programs, such as the space station or the robotic planetary exploration program. The mere designation of the Moon-Mars objective broke a formidable dam against the agency's progress, and the administration rapidly showed its bona fides by requesting several hundred million dollars to support such newly permissible research and development. In addition, it was made clear that funds would be available to demonstrate some of these new technologies using subscale units on robotic missions to the Moon and Mars, starting around the end of this decade. But even this positive news must be viewed with caution. For in the absence of an actual Moon-Mars program – one that develops an efficient mission plan that designates the program's technology needs – broad R&D expenditures can be quite inefficient.

Relative to the decisive form of leadership that drove the success of the Apollo program, the Bush policy set forth a large vision without the sense of urgency to make it real. But an uncertain trumpet is still better than none at all. Before President Bush's announcement, the idea of an American program to pioneer the space frontier seemed to many like the stuff of science fiction writers, wistful dreamers, and marginal visionaries. Suddenly, it was a mainstream political idea, and significant social forces began to rally both for and against the plan.

The Hubble Blunder

The new Bush space policy received mixed reviews in the press. But it was nearly derailed two days after its release when Administrator O'Keefe announced his decision to cancel the planned shuttle mission to maintain and upgrade the Hubble Space Telescope, thereby dooming the instrument to destruction. Lacking any scientific or technical background, O'Keefe might be forgiven for not understanding Hubble's value to astronomy. Yet, as an experienced bureaucrat, he should have had some appreciation of the significance of abandoning several billions of dollars of the American tax-payers' property. Apparently, however, he did not, and the affair that ensued produced one of the worst public relations disasters in NASA's history.

Built, launched, repaired, and successively upgraded at a total cost of some $4 billion, the Hubble Space Telescope has made numerous important discoveries about the nature and structure of the universe. It is the most powerful instrument in the history of astronomy, and far and away the most productive spacecraft that NASA has ever launched. Because it orbits above the atmosphere, which both smears light and blocks out major portions of the spectrum, the Hubble can see things that no ground-based telescope will ever see. It took decades of hard work by very dedicated people to create Hubble, and an equivalent space-based replacement remains decades away. In contrast to the general run of meaningless shuttle missions carrying silly science fair experiments, the shuttle flights to Hubble stand as epochal achievements. If one considers the moral significance of the scientific enterprise to our society and culture, Hubble stands out not just as NASA's finest work, but as one of the highest expressions of the human creative spirit in the twentieth century.

At a cost of $167 million, two new instruments, the Wide Field Camera 3 and the Cosmic Origins Spectrometer, had been developed and built which, once installed on Hubble, would together triple the instrument's sensitivity. Accordingly, NASA had scheduled a shuttle mission to the telescope for 2006, both to add these capabilities and perform certain other maintenance tasks that would extend the life of Hubble through at least 2010. Under the new Bush space policy, the shuttles were scheduled to remain operational through 2010, permitting a final shuttle mission to Hubble to occur toward the end of the decade. This would allow one last replacement of the telescope's batteries and gyros and a reboost of its orbit, thereby making it functional beyond 2015. If no missions to Hubble were flown, however, the space observatory's aging gyroscopes would put it out of commission by 2007.

Incredibly, on January 16, 2004, O'Keefe announced that he had decided to allow that to happen. He justified his decision by claiming that shuttle missions to Hubble were unsafe since they offer no alternative safe haven to the crew, in contrast to missions to the International Space Station (under the President's policy, about 25 more such shuttle missions would be flown). This argument was basically nonsense, since there are numerous features of space station missions that make them more dangerous than Hubble flights. For example, Hubble missions depart Cape Canaveral flying east-southeast, which means that in the event of an abort, the crew can ditch in tropical waters where their survival chances would be much better than in the frigid North Atlantic and Arctic oceans overflown by the northeast-flying ISS missions. Hubble missions also take off much more lightly laden than ISS missions, which makes them safer, as less performance is required of the engines to make it to orbit. Moreover, the danger from micrometeorite and orbital debris is estimated by NASA to be about 60 percent greater at the space station's altitude than at Hubble's.

So NASA's own risk analysis did not support O'Keefe's argument that Hubble missions posed too high a risk, and while the administrator declined to include such information in his briefings to congressional committees, outraged NASA personnel quickly leaked the relevant data to the press. O'Keefe countered by ordering high-level NASA officials who were known to be ardent supporters of Hubble to take public stands supporting his decision. The disgusting spectacle of bureaucratic self-humiliation that followed only excited derision in the press.

Mr. O'Keefe then argued that regardless of the actual risk, the recommendations of Admiral Gehman's Columbia Accident Investigation Board precluded a shuttle flight to Hubble. But this claim was rejected by Gehman himself, in a letter to Senator Barbara Mikulski (D.-Md.), a strong Hubble supporter. Almost all the risk in any shuttle mission occurs during the ascent and descent; "where one goes on orbit makes little difference" to overall safety, Gehman wrote. "Only a deep and rich study of the entire gain / risk equation can answer the question of whether an extension of the life of the wonderful Hubble telescope is worth the risks involved."

Admiral Gehman's response provided Mr. O'Keefe an exit opportunity from his policy blunder, but the NASA Administrator chose not to take it. Not only that, but when Senator Mikulski and Senator Sam Brownback (R.-Kans.) ordered a review from the National Research Council, Mr. O'Keefe responded by saying that while he welcomed a review from such a prestigious body, he would not change his decision regardless of anything they said.

As a final dodge, Mr. O'Keefe then announced that he sincerely wanted to save Hubble, but could not bring himself to risk human life to do so. Accordingly, he would request $1.9 billion in new funds to develop robots capable of performing the mission. This proposal was thoroughly disingenuous. A Hubble upgrade mission requires the coordinated efforts of seven highly trained and superbly skilled astronauts using a spacecraft and other equipment that has been specifically designed and extensively tested as suitable for this purpose. In contrast, there isn't a robot on this planet that can change an overhead kitchen lighting fixture. What's more, the robots touted by O'Keefe as candidates for repairing Hubble ranked much too low on the agency's standard system of "technology readiness levels," meaning that to use them would be a complete abandonment of NASA mission planning discipline.

In December 2004, the National Research Council panel reported back, rejecting the robotic repair – such a robotic mission "would require an unprecedented improvement" in technology in the next few months, the panel concluded – and calling for a manned shuttle mission "as early as possible." A few days later, Mr. O'Keefe announced his resignation, but before departing he submitted a NASA budget containing no funds for either a manned or robotic mission to repair Hubble. Instead, he requested $300 million to develop a special spacecraft to deorbit Hubble – that is, to crash it into the ocean in a controlled fashion. Even aside from the rest of the Hubble controversy, this proposal is remarkable for its irrationality. NASA calculates that if Hubble were to reenter Earth's atmosphere without direction, there is a 1 in 10,000 chance that the resulting debris would strike someone. If saving lives is the goal, that $300 million could do a lot more good spent on tsunami relief, body armor for the troops, highway safety barriers, childhood vaccinations, swimming lessons – take your pick.

The fate of Hubble remains undecided at this writing, but the damage done to the new initiative has been substantial, and threatens to become much worse if Mr. O'Keefe's decision is allowed to stand. Effectively, by choosing the most valuable part of the old space program and selecting it for destruction as collateral damage of implementing the new, the former administrator has branded the President's vision with the mark of Cain. Opponents of the new policy, such as the New York Times, have blamed the loss of the space telescope on the Moon-Mars initiative, and indeed, it is difficult to take seriously the claims of scientific purpose of an agency which chooses to abandon its capabilities so flippantly. Why should NASA receive more funds to build new space telescopes when, like a spoiled child bored with a two-hour old toy, it willfully throws away the one it already has? And how can anyone believe that an agency too scared to launch astronauts to Hubble will ever be ready to send humans to Mars? Congress has spent billions of taxpayer dollars to create the hardware needed to implement the Hubble program and the supporting shuttle infrastructure, only to be confronted with a NASA administrator who refuses to use it. If O'Keefe's decision to desert Hubble is not reversed, how can Congress know that after they spend further tens of billions for human flight systems to the Moon and Mars, that the agency leadership won't get cold feet again?

The Aldridge Commission

In order to give the new space policy some blue-ribbon certification – and also to drum up some public support for the plan – the Bush administration launched the President's Commission on Implementation of the United States Exploration Policy. Chaired by former Air Force Secretary Edward "Pete" Aldridge, Jr., the committee was charged with making recommendations for the scientific agenda, technological approach, and organization strategy for the new space initiative. In addition to Aldridge, the committee included two high-level corporate executives, a retired four-star general, a former congressman, three geologists, and an astrophysicist-cum-planetarium director. Some of these people were quite eminent in their chosen fields, but the absence of any astronautical engineer (or indeed anyone who had ever worked as an engineer in any field) or any astrobiologist was striking. The committee thus lacked credentials in two central areas of its charge. Of the committee members, only one, lunar geologist Paul Spudis, had ever participated in studies of human planetary exploration before, and his scientific interests are so narrowly focused on the Moon that he has been known to make extravagant claims in support of his research agenda (such as maintaining that lunar geology is the key to understanding mass extinction processes on Earth).

Between February and May 2004, the committee held hearings in ten American cities. About a hundred witnesses were invited to testify, but it rapidly became clear that the committee was not actually interested in ideas that diverged from a predetermined mantra. This was partially forgivable, since much of the testimony the committee chose to entertain was quite absurd, like the presentation from one crankish invitee arguing that the best place to look for Martian fossils was on the Moon, by searching for ejected Mars rocks landed there. (This idea was strange, to say the least, since there are many more Martian rocks on Earth than on the Moon – and, of course, there are significantly more on Mars itself.) But while the committee was hard-headed enough to set such nonsense aside, it was also impervious to necessary ideas. A very sad example of this was exhibited at the San Francisco hearings, when noted science fiction author Ray Bradbury testified. Bradbury gave an impassioned and eloquent speech in which he said that the American people could be inspired to support the new space policy if it were presented as the first step in the growth of humanity into a multi-planet spacefaring species. After he concluded, Aldridge replied with a question about how we "sell this to the American taxpayer." With great patience and poetic clarity, Bradbury explained his point again. Spudis then responded, saying it would be easier to just tell the American people that space is "a source of virtually unlimited wealth." One has to wonder how a group of people who don't actually believe in a great enterprise can hope to lead it.

On June 4, 2004, the Commission finally released its report. Remarkably, the group managed to get the answers completely wrong in the three central areas of its responsibility: the scientific goals, the technical strategy, and the reform of NASA.

First, the scientific goals. The commission proposed a sixteen-point science agenda that ranged from discovering the origin of the universe to assessing global climate change. Many of these points represented important fields of scientific research, but fourteen of the sixteen had very little to do with human exploration of the Moon and Mars. Rather, the list seemed to be something that had been cut and pasted from prior National Research Council reports on generic scientific priorities in space. Of the two items on the agenda that did have a clear relationship with human exploration, both dealt with planetary geology. While one of these latter points did include "identification and characterization of *environments* potentially suitable" (emphasis added) for past or present biogenic activity as a goal, absent from the list was any search for past or present *life itself*. This is remarkable because the search for life was clearly central to President Bush's new vision for NASA, and because surely the search for life – especially on Mars – is key to understanding the prevalence and diversity of life in the universe. Even as the commission was doing its work, NASA's *Spirit* and *Opportunity* rovers were making headlines identifying the coastal deposits of ancient Martian oceans, and high-level NASA officials were saying things like, "If you have an interest in searching for fossils on Mars, this is the first place you want to go." Astrobiological research conducted on the Martian surface by human explorers provides the most compelling scientific rationale for the new space policy; it is the one really important form of extraterrestrial research that only astronauts can do adequately. Yet the commission did not include it on the agenda. By failing to do so, the commission deprived the human exploration initiative of its strongest rational basis.

Second, the commission identified a list of seventeen technologies that it said need to be developed to enable the new initiative. According to the commission, funds should be spent to create these technologies, after which they should be integrated into the exploration architecture. This is exactly the opposite of the correct way to proceed. Instead of arbitrarily choosing a list of technologies to develop, and then forcing them into the mission plan, NASA should design the mission plan, identify the technologies it requires, and then develop them. To do otherwise is to dissipate resources in random spending. Only about four of the seventeen technologies the commission cited are strictly necessary for human Moon-Mars exploration. Of the rest, about half are generally useful but not necessary mission enhancements, while most of the others are only plausibly useful under certain mission scenarios. Finally, one of the cited technologies is clearly not needed under any circumstances, and one technology that failed to make their list is critically needed. The point is, if you want a system of parts to fit and work together, you design the system first, and then you make the parts. In contrast, the commission approach involves acquiring a bunch of well-marketed items, and then trying to fit them together to make a system – a repeat of the Shuttle Mode approach to spending that has been the primary cause of the past three decades of stagnation.

Third, the commission correctly observed that there is a need for organizational reform in NASA if the new space initiative is to be implemented successfully. They noted that the most effective of the NASA field centers is the Jet Propulsion Laboratory (JPL), and that JPL is not a civil service institution like the other NASA centers but a Federally Funded Research and Development Center (FFRDC). Employee merit can thus be rewarded at JPL with higher pay, or lack of performance punished with dismissal, in a way that is simply not possible in a civil service organization. Linking these two findings, the Commission ascribed JPL's superior performance to its FFRDC form of organization, and therefore recommended converting all of the NASA field centers to FFRDCs as the cure for the agency's internal ills.

The commission is arguably correct that JPL is the most productive NASA field center, but the question must be asked if the FFRDC organizational form is truly the cause. The Department of Energy's research labs are all FFRDCs as well, and their productivity today is much lower. So what other factors might account for JPL's success? How about the fact that all of its leaders are technically excellent? From Theodore von Kármán during World War II to Charles Elachi today, all of JPL's directors have been superb scientists or engineers, and the same is true of nearly all its upper managers, middle managers, and senior engineers, right down the line. That is not generally the case at other NASA centers, and it is most certainly not the case at NASA headquarters. In running a space program, it helps if you know what you are talking about.

It also helps if you know what you are trying to accomplish. JPL is mission-driven, and the missions it selects are science-driven. It develops the technologies that are necessary to enable those mission designs. The system isn't perfect; human weakness enters in, mistakes are sometimes made, and biases sometimes get into play, but overall the operation is rational and purposeful – precisely because it does not operate in the mode that the Aldridge Commission recommended for NASA. The FFRDC may be a superior organizational form to the civil service, but it isn't the decisive factor. During the Apollo period, civil service NASA centers such as Johnson Space Center and Marshall Space Flight Center had records of accomplishment at least as impressive as JPL's. But their technical leadership at that time was also superb, and they were mission-driven, too. Today, much of NASA fails to meet these two basic criteria for success.

Technical Competence and Political Convenience
The central importance of technically qualified leadership at NASA is sometimes countered by the example of James Webb, who served as the space agency's highly successful administrator during the Kennedy-Johnson years. It is true that Webb lacked a technical background, but that is only part of the story. Webb's Oklahoma country boy persona was an act used to hustle the gullible. In fact, Webb was a highly educated and incisive intellect. As one of the authors within the Kennedy administration of the Apollo program, he was passionately committed to its success, and he made it his business to learn everything necessary to understand what was going on and lead the program to victory. He could be very forceful when dealing with competing bureaucratic powers, but he never tried to dictate technical reality to engineers. Rather, he gathered together some of the top technical talent of all time, and he listened to it.

By contrast, the consequences of NASA leadership lacking in technical competence or even respect for scientific or technical considerations are amply demonstrated by the events of the O'Keefe years. In addition to the Hubble debacle, discussed above, the gross managerial failures during this period included the Orbital Space Plane program, the Jupiter Icy Moon Orbiter program, and the loss of the space shuttle *Columbia*.

First, the Orbital Space Plane. During the Clinton administration, NASA's Johnson Space Center in Houston, Texas had begun a program called X-38 to develop a crew capsule that could launch astronauts to orbit atop a medium-lift launch vehicle, thereby allowing space station crews to be rotated at much lower cost than is required for a shuttle flight. Since the Johnson Space Center is the primary NASA center with expertise in crewed flight systems, it made sense for the project to be assigned there. But apparently for political reasons, Mr. O'Keefe decided to move the program to the Marshall Space Flight Center in Huntsville, Alabama. Claiming the X-38's estimated price tag of $1.6 billion was too high, he canceled that program in midstream and set up the Orbital Space Plane program in Alabama in its place. The actual expertise of the Marshall Space Flight Center is in launch vehicles, however, and without the necessary experience, costs rapidly escalated out of control, with the estimated program budget growing to over $15 billion by the fall of 2003. Congress balked at funding this boondoggle, and the program collapsed with nothing accomplished and close to a billion dollars of the taxpayer's money down the drain.

Next, the Jupiter Icy Moon Orbiter (JIMO) intended to use advanced technology to study the frozen moons of Jupiter. This program was begun by O'Keefe himself, and could have been his greatest accomplishment – it would have been a significant scientific achievement and it would have made the essential capability of space nuclear power into a reality. The merit of this proposal lay in the fact that replacing today's radioisotope generators with nuclear power would allow a probe sent to the outer solar system to employ active sensing instruments and to transmit back vastly greater amounts of scientific data. Using nuclear power would also enable electric propulsion ("ion drive"), allowing the spacecraft to engage in extensive, highly-efficient maneuvers among Jupiter's moons.

So far, so good. However, in order to get more funding, the electric propulsion community managed to insert a requirement into the program that the flight from Earth to Jupiter be accomplished using electric propulsion, and that the trip to Jupiter not use any planetary gravity assists ("the slingshot effect"). Suddenly, under these new rules, the power needed to propel JIMO grew to 150 kilowatts in order to reach Jupiter in nine years. This is not only absurd (in the 1970s, *Voyager* made the trip in less than three years; in the 1990s, *Galileo* did it in five) but disastrous, since the nuclear reactor cannot be rated in advance for nine years of operation. In other words, JIMO would almost certainly fail before it reached the planet. Furthermore, as a result of the weight and the huge mass of the 150 kilowatt reactor and xenon propellant, the spacecraft couldn't be launched into space on any existing rocket. In contrast, had these rules not been adopted, the reactor could have been scaled down to 20 kilowatts, all the interplanetary transfer xenon propellant been eliminated, and the spacecraft thus made light enough to be put on top of an existing rocket and thrown toward Venus for the first in a series of gravity assists. These maneuvers would have allowed the spacecraft to reach Jupiter in five years on a *Galileo*-like trajectory, without needing to start burning the reactor until operations within the Jupiter system began. In other words, JIMO done the easy way could have been accomplished with one-seventh the power, one-quarter the mass, half the flight time, and a much greater success probability as JIMO done the hard way. Administrator O'Keefe apparently did not understand any of these issues. Instead, the former Secretary of the Navy wrongly equated nuclear electric propulsion for spacecraft to nuclear power for submarines, allowing them to transcend the limits of chemical propulsion and "go anywhere, anytime," without the need for such old-fashioned tricks as gravity assists. Because of his naïveté on such matters, O'Keefe failed to see this bunk for what it was, and in fact promoted it as a programmatic mantra. As a result, the program's cost ballooned to over $9 billion, and the White House declined to ask for further funding for Fiscal Year 2006. In the meantime, more money was spent studying JIMO than was spent designing, building, flying, and analyzing the data from the highly successful *Mars Global Surveyor* mission, from start to finish.

Finally, the loss of the space shuttle *Columbia* can also be traced to managerial disrespect for technical advice. No information has come to light directly linking Mr. O'Keefe to the specific decisions that led to the accident, but the accident does clearly illustrate the consequences of arrogantly insisting that technical reality conform to the management line. NASA engineers informed the agency's management that they had data showing that there could be a serious problem with *Columbia*'s thermal protection system. The managers had the means to investigate the engineers' suspicions, either by asking the Air Force to shoot high-resolution photographs of the shuttle, or by having the shuttle astronauts conduct a direct inspection themselves. Had management undertaken either course, the damage to the thermal protection tiles would have been discovered. That being the case, the crew could have attempted an ad hoc repair. It might have worked, or it might not. It is untrue that the situation was necessarily hopeless. *Columbia* actually made it most of the way back, and perhaps a crude repair might have done the trick – or if the pilot had been informed of the problem, he might have been able to fly the craft in such a way as to favor the weaker wing enough to survive. We'll never know. But certainly the managers who decided to stick with the "position" of the agency and not check the problem didn't know either. In consequence, the crew members were not even given a chance to fight for their lives.

The Aldridge Commission report did not speak to these kinds of serious shortcomings. All in all, it was a dull read, and had limited impact. Since it basically endorsed the status quo of a non-driven NASA, there was little positive damage it could do. But an opportunity to force necessary changes had clearly been lost. As a result, the key questions remain unsettled – including the need to set rational scientific goals, to ensure qualified leadership, and to decide whether program engineering will be driven by technical judgment or political convenience. The drift continued, and the Bush vision still lacked a real-life plan adequate to the boldness of its goals.

The New Space Budget

Even without a plan, the President's vision needed funding, and the members of the diverse American aerospace community lined up to show their support. This community includes a few large and many small aerospace companies; numerous government and university participants; and an array of industrial associations, technical and professional societies, and advocacy groups. These organizations differ in their prioritization of scientific, commercial, and military goals in space; in their preference for a government-led space program or a free-enterprise space industry; and in their nationalist or internationalist orientation.

Nevertheless, with virtually complete unanimity, this assemblage responded to the Vision for Space Exploration with a strong endorsement. Two organizations were formed, the industry-led Coalition for Space Exploration and the advocacy group-led Space Exploration Alliance, and nearly every outfit in the field, either through one of these leagues or on its own, commenced lobbying for the President's new policy. The unprecedented unity of the aerospace community sent a strong message to Congress that a new focus for the American space program was truly needed, and that the Moon-Mars initiative was a long-overdue step in the right direction.

While lacking in merit as a technical decision-maker, NASA Administrator O'Keefe was extremely adroit in working the congressional funding process. That fact, combined with the very clear support from the aerospace community, sufficed to reap initial funding for the Vision for Space Exploration for Fiscal Year 2005. Only about $150 million requested actually represented new funding, but preexisting programs were amalgamated to create a new Exploration Systems Mission Directorate (ESMD) with a fairly serious initial budget on the order of a billion dollars. Retired Navy Rear Admiral Craig Steidle, the former head of the Joint Strike Fighter development program, was brought in to lead the new directorate.

Moving in Spirals

Over the spring and summer of 2004, the ESMD proceeded to develop a program strategy to carry out the new space policy and created a mission architecture to implement the lunar portion of the plan. Completed in outline by the fall of 2004, this first-draft (or "Point of Departure") strategy consisted of five primary phases, or "spirals."

Spiral 1: Develop the Crew Exploration Vehicle (CEV) and its launch system and operate the CEV in low-Earth orbit.

Spiral 2: Begin short duration lunar missions. To achieve this objective, the plan proposes the following design for a transportation system. First, NASA must develop a Lunar Surface Ascent Module (LSAM) to carry astronauts to and from the Moon's surface, a medium-lift vehicle (MLV) capable of launching it, and an Earth Departure Stage (EDS) capable of delivering either the CEV or the LSAM separately from low-Earth orbit to low-lunar orbit. Carrying out a mission would require *four separate launches* – one MLV for the CEV, one for the LSAM, and one for each of two EDS vehicles. These four components would all be put into low-Earth orbit. The manned CEV would then rendezvous with one EDS, and the empty LSAM would rendezvous with the other EDS, and each would be driven separately from the Earth's orbit to lunar orbit. The CEV would then rendezvous with the LSAM in low lunar orbit, after which the crew would transfer to the LSAM for an excursion to the Lunar surface of 4 to 14 days. The crew would then ascend in the LSAM to rendezvous with the CEV in lunar orbit, transfer back to the CEV, and come back to Earth. (If this all sounds terribly complex, that's because it is. More on the implications of that complexity in a moment.)

Spiral 3: The hardware set developed for Spiral 2 is augmented by a cargo lander and a variety of surface systems, including a habitation module. Using the habitation module and associated systems, lunar surface sorties are extended to 42 days, with 90 days as a goal.

Spiral 4: A set of hardware (as yet undefined) is developed and used to perform Mars flyby missions.

Spiral 5: The Spiral 4 hardware set is expanded to enable human exploration missions to the Martian surface. The nature and duration of these missions is as yet undefined.

According to the plan, the development effort for Spiral 1 would begin immediately, with piloted CEV flight operations in low-Earth orbit commencing in 2014. Spiral 2 flight operations would begin in 2020. No dates have been set for Spirals 3, 4, or 5. At the same time, starting with Spiral 1, a set of robotic missions would be flown to the Moon and Mars to prepare or support human exploration objectives.

This ESMD plan contains many flaws that deserve severe criticism. In fairness, it should be said that most of these problems stem from weaknesses in the original presidential directive, or to arbitrary interference in the engineering design process by Mr. O'Keefe or other non-technically educated individuals. But because of these flaws, the current plan jeopardizes the success of the vision, and actually makes it possible that we will *lose* space capabilities. Put simply, the ESMD plan has too many spirals; the spirals don't logically build upon one another; the plan isn't responsive to the President's vision; and the overall mission architecture is technically unsound. Each of these four deficiencies needs to be examined in detail.

First, the point that there are *too many spirals*. As presently designed, the plan entails five spirals. There should be only three:

Spiral A: Equivalent to the present Spiral 1, but done much quicker.

Spiral B: Equivalent to the present Spirals 2 and 3.

Spiral C: Equivalent in function to the present Spirals 4 and 5.

That is, Spiral 1 should be abbreviated, while Spirals 2 and 4 should be abolished entirely as independent spirals.

Spiral 1 needs to be dramatically shortened, because the ten year timeline to develop the CEV is a dangerous stall. The decision to delay piloted CEV flights until 2014 comes directly from the original White House policy directive, which defers supplying substantial funds to the new initiative until the shuttle and space station programs can be wound down at the end of the decade. That decision was thus above the pay grade of Admiral Steidle and the ESMD mission planners to dispute. But it is a decision with unfortunate consequences. The CEV is essentially the functional equivalent of the Apollo command module which, as previously mentioned, was developed in just five years in the 1960s starting from a much lower technology base. By artificially stretching out the CEV program, the cost will be greatly increased. Furthermore, with shuttle operations scheduled to end in 2010, putting off the completion of the CEV until 2014 will leave the United States with no human space flight capability for four years. During this period, the taxpayers will be paying for a human space flight program that is not actually doing anything. This is a serious problem.

Meanwhile, Spirals 2 and 4 are unnecessary in a program seeking to achieve maximum scientific return with minimum cost and risk. Spiral 2 lunar missions accomplish much less than Spiral 3 missions, but entail comparable cost and risk. And while Spiral 4 Mars missions require less cost and risk than Spiral 5 Mars missions, the latter offer several orders of magnitude greater scientific return. Thus Spiral 2 and 4 missions are neither cost-effective nor risk-effective, and should be minimized or eliminated from the program.

This is a critical point, so let us consider it in greater detail, looking specifically at the relationship between Spirals 2 and 3. The primary distinction between these two spirals is that Spiral 3 missions have a habitation module on the lunar surface, and therefore crews can stay on the surface much longer than in Spiral 2 missions, which would offer only the limited living space of the lunar module (as in the Apollo missions). Now it is obvious that a mission that operates on the surface for forty days will accomplish much more exploration than one that stays for four days. This advantage of the longer Spiral 3 missions is amplified much further by the fact that the habitation module will have lab facilities, allowing astronauts to perform preliminary analysis of large numbers of field samples while they are on the Moon, selecting only the most interesting samples to return to Earth for further study. Thus lunar exploration during Spiral 3 will be vastly more effective than in Spiral 2.

To be sure, there are plausible objections to eliminating Spiral 2. For instance, one might argue that Spiral 3 requires a habitation module and its power supply, which is an additional development and delivery cost. But the program is committed to that cost in any case, so why not aim to use these technologies from the beginning? Another objection might be that each expedition during Spiral 2 can land at a new site on the Moon, while explorers during Spiral 3 are limited to a radius around a single lunar base. This is true, although Spiral 3 missions compensate for that loss of novelty by allowing a more thorough exploration of each site, and by being less risky because the crew will have two safe havens (the lunar module and the habitation module). And since the habitation module is also the lab module, it provides them with both the endurance and the equipment they need to do effective exploration. It makes no sense to send explorers to the Moon without the primary tool they need to do their job. As a matter of cost-effectiveness, scientific sense, and crew safety, the correct strategy is to develop and deploy a habitation module to the Moon *before* any human expeditions. The first missions don't need to be 40 days long; selecting shorter durations for initial missions is a reasonable strategy. But, for the sake of both science and safety, the habitation module should be delivered first, with crew surface duration expanding as rapidly as mission experience shows to be prudent. Deferring the deployment of the habitation module until after a series of Spiral 2 expeditions will waste money and expose astronauts to unnecessary risk.

The issue is even more clear in the case of condensing Spirals 4 and 5 into a single "Spiral C." Mars flyby missions entail significant cost and risk, but accomplish no meaningful scientific goals. Their only valid function is to test hardware. (They also test human endurance, but such tests could be accomplished much more cheaply and safely near Earth.) There is no need to develop a separate hardware set, as Spiral 4 calls for, just to conduct Mars flyby missions. It makes far more sense to just build and test the hardware for real Mars missions. This hardware can most affordably be tested by having it perform necessary work like delivering missions to the Moon or prepositioning useful infrastructure on Mars; it can even be tested, albeit at great cost, by flying an unmanned mission to the Martian surface and back. But it is irrational to send manned flyby missions to Mars. Having flown the crew all the way to Mars, they will have absorbed a large part of the risk and expense of a real Mars mission, and having done so, it makes no sense to end the mission without actually going to the surface. Flying such an abort-by-design mission before any actual missions only increases the overall program risk and cost. For this reason, Spiral 4 should be abolished.

The second major problem with the ESMD plan is that the spirals *don't sufficiently build upon one another*. The concept of "spiral development" in an engineering program involves introducing a hardware set that creates an initial capability, then improving it in subsequent phases or "spirals" by the addition of further technology. Rightly understood, therefore, spiral development involves enhancing or expanding the hardware set employed in an early phase to enable a later, more aggressive, set of objectives.

But the ESMD plan calls for designing a program that creates and then *abandons* a series of hardware sets to accomplish a progression of new goals. This is unnecessarily wasteful. Spiral 2 may be fairly said to be based on Spiral 1, since it makes full use of the CEV and its launch system. Similarly, Spiral 3 is clearly based on Spiral 2. But because the LSAM, the EDS, and the MLVs employed in the plan are all useless for Mars missions, Spirals 4 and 5 are not in any serious way based on Spirals 2 and 3. That is to say, except for the CEV developed during Spiral 1, almost none of the hardware developed during the previous spirals is appropriate for Mars missions. By contrast, with a better designed mission architecture, the Spiral 3 hardware could be directly useful for Mars missions. But that is not the case here.

The third significant flaw in the ESMD plan is that it *fails to respond to the presidential directive*. As currently constituted, the hardware used in Spirals 2 and 3 is designed to support lunar missions only, with no regard for Mars requirements. But the President's policy directive clearly specified that a central purpose of the lunar program is to enable sustained human exploration of Mars. These orders were effectively ignored by the designers of the plan.

The problem here is not merely one of formal disobedience to White House objectives. Rather, it is a matter of serious negative consequences. The ESMD plan requires a plethora of additional recurring costs and mission risks for the sole purpose of avoiding the development cost of a big new rocket – a heavy-lift vehicle (HLV). Yet, since one goal of the Vision for Space Exploration is to get humans to Mars, an HLV will need to be developed anyway. So on a cost basis, the ESMD plan will lose twice over, since it requires new hardware for Spirals 2 and 3, and then even more new hardware for Spirals 4 and 5. Furthermore, in addition to imposing maximum mission risk for lunar explorers through its own excessive complexity, the ESMD plan will also increase the risk to Mars explorers, because the ESMD lunar plan will not test the Mars mission hardware. Rather than enable human Mars exploration, the plan as presently defined would be a massive and costly detour; it would delay such missions for many decades. And since the plan would involve two different sets of hardware, it even threatens to create a situation where cost considerations will make it necessary to abandon the Moon when the time comes to proceed to Mars. By contrast, if a common transportation system were designed instead, both destinations could continue to be explored in parallel.

The plan's fourth major flaw is that it is fundamentally technically unsound. It goes to great lengths to avoid the necessity of developing a heavy-lift vehicle, employing (as described above) an astonishingly complicated mission architecture involving four rocket launches and four space rendezvous for each lunar mission – what we might call a "quadruple launch, quadruple rendezvous" (QQ) mission architecture.

Using some reasonable estimates based upon the masses of the primary components of the Apollo mission, it can be shown that it is technically possible that a QQ mission could be launched on four medium launch vehicles. But is it technically wise? Note the following factors:

  i. Each mission requires four MLV launches.
  ii. Those four launches must be done quickly, since the EDS and LSAM vehicles are carrying cryogenic liquid oxygen and hydrogen, and the manned CEV is launched last.
  iii. Each mission requires four critical rendezvous operations.
  iv. The crew flies to the Moon separate from the lunar module.

Point i speaks to the cost of the program. Using multiple MLVs to launch what could be a single HLV payload is not cost-effective. It is a basic feature of rocket economics that larger boosters are more economic than smaller boosters. The larger the launch vehicle, the less it costs to put each kilogram into orbit. So, for example, the Atlas V 500 is more than twice as economical a launch system as the Atlas IIAS, and cost projections for the next-generation HLV on the drawing boards based on the Atlas series are more than twice as economical as those for the Atlas V 500. The basic lesson here is that by adopting a strategy of multiple MLV launches, the plan will maximize rather than contain the program's launch costs.

Points ii and iii speak to feasibility. The program requires four MLV launches within just a few weeks. Three of those launches would involve cryogenic upper stages, and the fourth would involve a manned vehicle, all launched from Cape Canaveral. Such an MLV launch rate has never been accomplished with any payload and to assume that it can be done repeatedly with payloads of this complexity is wildly optimistic.

Points i, ii, and iii also speak to both complexity and mission risk. In contrast to the old Apollo mission plans, which required only one launch and a single rendezvous, the QQ plan requires four mission-critical launches and four mission-critical rendezvous. Each must be successful. That's eight big chances (in addition to lunar landing and ascent) for an operational failure that would ruin the mission.

In fact, the mission architecture is so complexly interdependent – and therefore so fragile – that a huge number of potential problems could end any given mission. The mission would fail if a mere launch delay caused *any* of the last three launches to stall so long that the propellant aboard the first payload runs out. The mission would fail if *any* of the four orbiting payloads were damaged by orbital debris while waiting in low-Earth orbit. The mission would fail if *any* of the four spacecraft should seriously malfunction. The mission would fail if *any* of the four orbital rendezvous operations failed. The mission would fail if *any* of the four engine burns needed to reach the Moon and get into lunar orbit underperformed. Just think: This mission architecture is supposed to support not just one lunar mission, but routine, repeated access to the Moon. Inserting so much complexity and vulnerability into such a transportation system is an open invitation to failure.

It is even possible to assign some rough figures to this vulnerability. Let's assume that the rockets used for this new space program will each have a 98 percent success rate. (In real life, a study of the historical reliability of the US Delta, Atlas, and Titan medium-lift vehicles shows a success rate of only about 90 percent.) And let's assume that each of the major operations in space – each rendezvous and engine burn – has a 99 percent success rate. And let's generously assume that there is a 98 percent chance that each of the last three rocket launches happens on time, and a 98 percent chance that the lunar landing is successful. Forget all the other potential failure points. Just calculating from those few assumptions, each mission would only have an expected 75 percent success rate. This means that roughly one out of every four missions could be expected to fail. If three missions are flown per year, there would, on average, be mission failure roughly every 1.3 years. Assuming a typical suspension of operations of two years after each mission failure, the program would need to be shut down for failure investigations at least 60 percent of the time.

Point iv speaks to the risks to crew. Apollo traveled to the Moon with the lunar module attached to the command module. This made the lunar module available to each crew as an emergency safe haven – which is precisely what famously saved the lives of the *Apollo 13* astronauts. Had the Apollo program used a system similar to that proposed in the QQ plan, the crew of *Apollo 13* would have died.

The central reason why the QQ mission architecture has such low reliability is because of the incredible proliferation of critical events that occurs if four launches, four rendezvous, and four spacecraft are required for each mission. Fortunately, the way to solve this problem is simple: Develop a heavy-lift vehicle (HLV) that allows the entire mission to be launched with a single booster, just as was done for the Apollo missions. This would greatly reduce program launch costs and reduce the risk of mission failure by a factor of four. It would also create a system directly useful to sending humans to Mars, which is a key requirement of the President's directive.

Regrettably, in designing this mission architecture, the ESMD planners had to act in conformity with the direction of the technically unqualified Mr. O'Keefe, who enunciated a preference that the program be conducted without heavy-lift vehicles. Such politically dictated technical decision-making is unacceptable; it is a formula for programmatic catastrophe.

Fortunately, this complicated plan is just a starting point in the design process; the ESMD is not committed to it. But it is imperative that they depart from this plan as rapidly as possible, because vacillation risks missing a tremendous technological opportunity. One of the cheapest ways to create a heavy-lift vehicle is by converting the shuttle. The shuttle launch stack has the same take-off thrust as the powerful Saturn V rocket that put American astronauts on the Moon during the Apollo era. Since the Saturn V was imprudently canceled decades ago, the United States has had no heavy-lift vehicle. But by adapting the shuttle – removing the orbiter and adding an upper stage – we can create a launch vehicle with a capability comparable to the Saturn V.

And this is precisely why delay is so dangerous. Under NASA's current plans, only about twenty-five more shuttle launches are contemplated. Absent a plan for shuttle conversion to a heavy-lift vehicle, much of the industrial infrastructure for manufacturing key shuttle components, such as external tanks, will soon be dismantled. We will be repeating the mistake of the Saturn V cancellation. Recreating such capabilities after they have been lost will cost the taxpayers billions.

Like Mr. O'Keefe's fake Hubble robotic rescue proposal, the spurious QQ mission plan merely serves to lull policy makers while critical capabilities are being lost. If such massive waste is to be avoided, NASA needs to make the case for heavy-lift vehicles immediately. But it is difficult to justify the development of a heavy-lift vehicle if flight operations for that system are not to begin until 2020. Thus we encounter again the fundamental problem with President Bush's policy. By postponing the program's goals until far in the future, important capabilities that could be used to achieve those goals will be lost before the time comes for those goals to be attempted. Under the current plan, Spiral 1 might succeed, at maximum cost, in producing a CEV in ten years. But in the meantime, the heavy-lift vehicle components embodied in the shuttle program will have been lost. As a result, in 2014, NASA will actually possess a *smaller* fraction of the hardware needed to send humans to the Moon than it does today. A decade will have gone by, along with some hundred and fifty billion dollars spent on the space program, to achieve *negative* progress overall.

Arbitrarily stretching out the program may appear to be convenient from a political point of view, as it avoids the necessity of asking for large funding increases in any particular year. But from the point of view of anyone attempting to achieve the program's mission, it is the equivalent of an order to conduct a cavalry charge in slow motion: it maximizes the losses.

The Right Way to Mars

So far we have discussed the problems that have caused NASA to drift for the past thirty years, how those problems came to the fore in the aftermath of the *Columbia* disaster, and the efforts of the administration to address those endemic problems. As we have seen, the resulting new space policy, while clearly a step in the right direction, includes so many compromises with the old way of doing business that a positive outcome remains in doubt. We must now address the question of how a rational human space exploration initiative should be done.

It is not enough that NASA's human exploration efforts "have a goal." The goal selected needs to be the *right* goal, chosen not because various people are comfortable with it, but because there is a real reason to do it. We don't need a nebulous, futuristic "vision" that can be used to justify random expenditures on various fascinating technologies that might plausibly prove of interest at some time in the future when NASA actually has a plan. Nor do we need strategic plans that are generated for the purpose of making use of such constituency-based technology programs. Rather, the program needs to be organized so that it is the goal that actually drives the efforts of the space agency. In such a destination-driven operation, NASA is forced to develop the most practical plan to reach the objective, and on that basis, select for development those technologies required to implement the plan. Reason chooses the goal. The goal compels the plan. The plan selects the technologies.

So what should the goal of human exploration be? In my view, the answer is straightforward: Humans to Mars within a decade. Why Mars? Because of all the planetary destinations currently within reach, Mars offers the most — scientifically, socially, and in terms of what it portends for the human future.

In scientific terms, Mars is critical, because it is the Rosetta Stone for helping us understand the position of life in the universe. Images of Mars taken from orbit show that the planet had liquid water flowing on its surface for a period of a billion years during its early history, a duration five times as long as it took life to appear on Earth after there was liquid water here. So if the theory is correct that life is a naturally occurring phenomenon, emergent from chemical complexification wherever there is liquid water, a temperate climate, sufficient minerals, and enough time, then life should have appeared on Mars. If we go to Mars and find fossils of past life on its surface, we will have good reason to believe that we are not alone in the universe. If we send human explorers, who can erect drilling rigs which can reach underground water where Martian life may yet persist, we will be able to examine it. By doing so, we can determine whether life on Earth is the pattern for all life everywhere, or alternatively, whether we are simply one esoteric example of a far vaster and more interesting tapestry. These things are truly worth finding out.

In terms of its social value, Mars is the bracing positive challenge that our society needs. Nations, like people, thrive on challenge and decay without it. The challenge of a humans-to-Mars program would be an invitation to adventure to every young person in the country, sending out the powerful clarion call: "Learn your science and you can become part of pioneering a new world." This effect cannot be matched by just returning to the Moon, both because a Moon program offers no comparable potential discoveries and also because today's youth cannot be inspired in anything like the same degree by the challenge to duplicate feats accomplished by their grandparents' generation.

There will be over a hundred million kids in our nation's schools over the next ten years. If a Mars program were to inspire just an extra one percent of them to pursue a scientific education, the net result would be one million more scientists, engineers, inventors, and medical researchers, making technological innovations that create new industries, find new cures, strengthen national defense, and generally increase national income to an extent that utterly dwarfs the expenditures of the Mars program.

But the most important reason to go to Mars is the doorway it opens to the future. Uniquely among the extraterrestrial bodies of the inner solar system, Mars is endowed with all the resources needed to support not only life but the development of a technological civilization. In contrast to the comparative desert of the Moon, Mars possesses oceans of water frozen into its soil as ice and permafrost, as well as vast quantities of carbon, nitrogen, hydrogen and oxygen, all in forms readily accessible to those clever enough to use them. These four elements are the basic stuff not only of food and water, but of plastics, wood, paper, clothing – and most importantly, rocket fuel.

In addition, Mars has experienced the same sorts of volcanic and hydrologic processes that produced a multitude of mineral ores on Earth. Virtually every element of significant interest to industry is known to exist on the Red Planet. While no liquid water exists on the surface, below ground is a different matter, and there is every reason to believe that underground heat sources could be maintaining hot liquid reservoirs beneath the Martian surface today. Such hydrothermal reservoirs may be refuges in which survivors of ancient Martian life continue to persist; they would also represent oases providing abundant water supplies and geothermal power to future human settlers. With its 24-hour day-night cycle and an atmosphere thick enough to shield its surface against solar flares, Mars is the only extraterrestrial planet that will readily allow large scale greenhouses lit by natural sunlight. In other words – Mars can be settled. In establishing our first foothold on Mars, we will begin humanity's career as a multi-planet species.

Mars is where the science is; Mars is where the challenge is; and Mars is where the future is. That's why Mars must be our goal.

How Do We Get There?

The faint of heart may say that human exploration of Mars is too ambitious a feat to select as our near-term goal, but that is the view of the faint of heart. From the technological point of view, we're ready. Despite the greater distance to Mars, we are *much* better prepared today to send humans to Mars than we were to launch humans to the Moon in 1961 when John F. Kennedy challenged the nation to achieve that goal – and we got there eight years later. Given the will, we could have our first teams on Mars within a decade.

The key to success is rejecting the policy of continued stagnation represented by senile Shuttle Mode thinking, and returning to the destination-driven Apollo Mode of planned operation that allowed the space agency to perform so brilliantly during its youth. In addition, we must take a lesson from our own pioneer past and adopt a "travel light and live off the land" mission strategy similar to that which has well-served terrestrial explorers for centuries. The plan to explore the Red Planet in this way is known as Mars Direct. Here's how it could be accomplished.

At an early launch opportunity – for example 2014 – a single heavy-lift booster with a capability equal to that of the Saturn V used during the Apollo program is launched off Cape Canaveral and uses its upper stage to throw a 40-tonne unmanned payload onto a trajectory to Mars. (A "tonne" is one metric ton.) Arriving at Mars eight months later, the spacecraft uses friction between its aeroshield and the Martian atmosphere to brake itself into orbit around the planet, and then lands with the help of a parachute. This is the Earth Return Vehicle (ERV). It flies out to Mars with its two methane / oxygen-driven rocket propulsion stages unfueled. It also carries six tonnes of liquid hydrogen, a 100-kilowatt nuclear reactor mounted in the back of a methane / oxygen-driven light truck, a small set of compressors and an automated chemical processing unit, and a few small scientific rovers.

As soon as the craft lands successfully, the truck is telerobotically driven a few hundred meters away from the site, and the reactor is deployed to provide power to the compressors and chemical processing unit. The ERV will then start a ten-month process of fueling itself by combining the hydrogen brought from Earth with the carbon dioxide in the Martian atmosphere. The end result is a total of 108 tonnes of methane / oxygen rocket propellant. Ninety-six tonnes of the propellant will be used to fuel the ERV, while 12 tonnes will be available to support the use of high-powered, chemically-fueled, long-range ground vehicles. Large additional stockpiles of oxygen can also be produced, both for breathing and for turning into water by combination with hydrogen brought from Earth. Since water is 89 percent oxygen (by weight), and since the larger part of most foodstuffs is water, this greatly reduces the amount of life support consumables that need to be hauled from Earth.

With the propellant production successfully completed, in 2016 two more boosters lift off from Cape Canaveral and throw their 40-tonne payloads towards Mars. One of the payloads is an unmanned fuel-factory / ERV just like the one launched in 2014; the other is a habitation module carrying a small crew, a mixture of whole food and dehydrated provisions sufficient for three years, and a pressurized methane / oxygen-powered ground rover.

Upon arrival, the manned craft lands at the 2014 landing site where a fully fueled ERV and beaconed landing site await it. With the help of such navigational aids, the crew should be able to land right on the spot; but if the landing is off course by tens or even hundreds of kilometers, the crew can still achieve the surface rendezvous by driving over in their rover. If they are off by thousands of kilometers, the second ERV provides a backup.

Assuming the crew lands and rendezvous as planned at site number one, the second ERV will land several hundred kilometers away to start making propellant for the 2018 mission, which in turn will fly out with an additional ERV to open up Mars landing site number three. Thus, every other year two heavy-lift boosters are launched, one to land a crew, and the other to prepare a site for the next mission, for an average launch rate of just one booster per year to pursue a continuing program of Mars exploration. Since in a normal year we can launch about six shuttle stacks, this would only represent about 16 percent of the US heavy-lift capability, and would clearly be affordable. In effect, this "live off the land" approach removes the manned Mars mission from the realm of mega-spacecraft fantasy and reduces it in practice to a task of comparable difficulty to that faced in launching the Apollo missions to the Moon.

The crew will stay on the surface for 1.5 years, taking advantage of the mobility afforded by the high-powered chemically-driven ground vehicles to accomplish a great deal of surface exploration. With a 12-tonne surface fuel stockpile, they have the capability for over 24,000 kilometers worth of traverse before they leave, giving them the kind of mobility necessary to conduct a serious search for evidence of past or present life on Mars. Since no one has been left in orbit, the entire crew will have available to them the natural gravity and protection against cosmic rays and solar radiation afforded by the Martian environment, and thus there will not be the strong pressure for a quick return to Earth that plagues other Mars mission plans based upon orbiting mother-ships with small landing parties. At the conclusion of their stay, the crew returns to Earth in a direct flight from the Martian surface in the ERV. As the series of missions progresses, a string of small bases is left behind on the Martian surface, opening up broad stretches of territory to human cognizance.

In essence, by taking advantage of the most obvious local resource available on Mars – its atmosphere – the plan allows us to accomplish a manned Mars mission with what amounts to a lunar-class transportation system. By eliminating any requirement to introduce a new order of technology and complexity of operations beyond those needed for lunar transportation to accomplish piloted Mars missions, the plan can reduce costs by an order of magnitude and advance the schedule for the human exploration of Mars by a generation.

The Lunar Architecture

Since a lunar-class transportation system is adequate to reach Mars using this plan, it is rational to consider a milestone mission, perhaps five years into the program, where a subset of the Mars flight hardware is exercised to send astronauts to the Moon.

This can be done as follows: First, a single booster is used to launch an unmanned habitation module which is landed on the Moon. Then, another booster is launched, sending the crew to the lunar surface in a CEV equipped with a methane / oxygen-driven ascent stage which is capable of propelling it directly back to Earth. The crew lands near the preplaced habitation module, which they then use as their house and laboratory on the Moon for an extended duration stay, after which they transfer back to the CEV and return to Earth.

This approach is much preferable to the QQ approach, because only one launch and no orbital rendezvous are required per mission, and a substantial habitat and laboratory are available to the crew starting on the very first mission. This enhances crew safety, and will make missions much more productive scientifically, as they will be able to stay longer and be much better equipped to conduct research while they are there. Furthermore, from the surface of the Moon, the launch window back to Earth is always open, as there are no orbital rendezvous phasing issues, further adding to the safety of the crew.

If the objective is to establish a permanent lunar base and not just to perform sorties to the Moon, then the production of lunar oxygen is feasible (by reducing the oxides of iron that comprise about 10 percent of Moon dirt); because of the numerous advantages it offers, this should be an early priority. If we want to visit multiple lunar sites, the most effective way is not to launch individual missions from Earth, but to employ a small rocket powered ballistic flight vehicle – a "hopper" – operating out of the lunar base camp. Using the fuel delivered from Earth by a single heavy-lift vehicle, such a hopper could make six long-range excursions if it used methane / oxygen propulsion, or ten excursions if it used hydrogen / oxygen propulsion. This compares quite handsomely to the QQ plan, which requires four major launches from Earth to visit just one site.

Thus we see that proper design of a coherent human exploration initiative allows not only Mars missions, but cost-effective lunar activities as well, using a modified subset of the Mars hardware. Approaching the design issue in this way can sharply cut overall program cost, risk, and schedule, because only one fundamental hardware set needs to be developed instead of two, and the lunar activities can be used to validate Mars mission hardware directly. This makes the rationale for the lunar missions clear, and makes it possible to continue lunar activities even after Mars missions begin, as only one transportation system will need to be supported.

The Need for Speed

Clearly, I have suggested some rather near-term dates for the human Mars mission, in significant contrast to various NASA "road-mapping" charts which situate this accomplishment sometime in the middle of the twenty-first century. Yet it should be observed that the first Americans walked on the Moon not after the hundredth anniversary of Sputnik, but before the twelfth. Indeed, it was the speed of the Apollo program that was the central factor in the program's success.

In 1961, President Kennedy committed the nation to reach the Moon before the end of the decade, and we did. But consider what would have happened if instead of choosing 1970 as his deadline, JFK had selected 1990. Had we then proceeded in such a more leisurely way, 1968 would not have seen Apollo 8 ready on the launch pad, but perhaps one of the later Mercury one-man capsule flights. But in 1968, the national mood was totally different from the Camelot era. We were in the Vietnam War, hundreds of thousands of protesters were marching in the streets, and, at the end of the year, there was a change of administration. Under those conditions, the tepid nominal Moon effort almost certainly would have been canceled – as in fact Nixon did cancel the quite successful Apollo program in real life. Clearly, if Kennedy had set his sights on the Moon in thirty years, we would not have made it there at all.

The issue, however, goes beyond the intrinsic difficulty of maintaining a political consensus in support of a program over multiple decades. There is also the matter of forcing the required technical focus for success. To use an analogy, think of two posts separated by a certain distance, say ten meters. How much rope is needed to connect them? It could take many kilometers, if the rope is allowed to be slack or tangled. Alternatively, it could be done with about ten meters, but only if the rope is pulled tight.

The Apollo era was filled with just as much human weakness as our own time. There were companies and NASA centers that were self-interested, and technologists that were obsessed with their own hobby horses. Early in the program, many fanciful and overly complex ideas were advanced on how to reach the Moon, but very rapidly, the impending deadline forced nearly all of them out. For Apollo, it was the tight schedule that tightened the rope.

It is just the same today. Mention humans-to-Mars within the NASA community, and you will be deluged with proposals for space stations and fuel depots in various intermediate locations, fantastical advanced propulsion technologies, and demands that billions upon billions of dollars be spent on an infinite array of activities which define themselves as necessary mission precursors. Representatives of such interests sit on various committees which write multi-decade planning "road maps" and exert every effort to make sure that the "roads," as it were, go through their own home towns. Under such conditions it takes not kilometers, but light years, of line to connect the posts. If we are actually to make it to Mars, however, the rope needs to be pulled tight, and only a tight schedule will suffice to do that job.

It is unreasonable today to spend ten years to develop a CEV, when in the 1960s we did it in five, or sixteen years to reach the Moon, when two generations ago we did it in eight. Embarking on the program in such a dilatory way will cost us the heavy-lift hardware of the Shuttle, which is something we can ill-afford. To believe that such slow-paced achievement is the best we can do means believing that we have become less than the people we used to be, and that is something we can afford even less.

Exploring Mars requires no miraculous new technologies, no orbiting spaceports, and no gigantic interplanetary space cruisers. We don't need to spend the next thirty years with a space program mired in impotence, spending large sums of money on random projects and taking occasional casualties while the missions to nowhere are flown over and over again, and while professional technologists dawdle endlessly in their sandboxes without producing the needed flight hardware. We simply need to choose our destination, and with the same combination of vision, practical thinking, and passionate resolve that served us so well during Apollo, do what is required to get us there.

We can establish our first small outpost on Mars within a decade. We, and not some future generation, can have the honor of being the first pioneers of this new world for humanity. All that is needed is present-day technology, some nineteenth-century industrial chemistry, a solid dose of common sense, and a little bit of moxie.

Why Now? Why Us?

So we can do it, and it should be done, but why should we be the ones to do it? Why, at a time like this, with the nation at war, with new menaces threatening to appear in various corners of the globe, and our allies drifting away, should the United States government expend serious resources on such a visionary enterprise? In my view, such considerations simply make the matter all the more urgent.

While I would not deny the necessity of military action in certain circumstances, in the long run civilizations are built by ideas, not swords. The central idea at the core of Western civilization is that there is an inherent facility in the individual human mind to recognize right from wrong and truth from untruth. This idea is the source of our notions of conscience and science, terms which, not coincidentally, share a common root.

Both our radical fundamentalist and our totalitarian enemies deny these concepts. They deny the validity of the individual conscience, and they deny the necessity of human liberty, and indeed, consider it intolerable. For them, conscience, reason, and free will must be crushed so that humans will submit to arbitrary and cruel authority.

Against this foe, science is our strongest weapon, not simply because it produces useful devices and medical cures, but because it demonstrates the value of a civilization based upon the use of reason. There was a time when we celebrated the divine nature of the human spirit by building Gothic cathedrals. Today we build space telescopes. Science is our society's sacred enterprise; through it we assert the fundamental dignity of man. And because it ventures into the cosmic realm of ultimate truth, space exploration is the very banner of science.

If the United States is to lead the West, it must not only carry its sword, but the banner of its most sacred cause. And that cause is the freedom to explore on the wings of human reason. The French may sneer, with some cause, at our fast food restaurants and TV sitcoms, but the Hubble Space Telescope can inspire nothing but admiration, or even awe, in anyone who is alive above the neck. A human Mars exploration program would be a statement about ourselves, a reaffirmation that we remain a nation of pioneers, the vanguard of humanity, devoted to the deepest values of Western civilization. But even more, it would be a declaration of the power of reason, courage, and freedom writ clear across the heavens.

Now, more than ever, we need to make those statements. Now, more than ever, we need to sign that declaration – in handwriting large enough that no one will need spectacles to read it.

Plans and Overviews of Proposed Mars Missions

Mars Society Participation at the 2003 National Space Society (NSS) Legislative Conference – Opportunities for Future Cooperation and Coalition Building, Between NSS and Mars Society in the Political Outreach Arena†

Carlos Glender
Ohio Chapter of the Mars Society

Abstract

This paper reviews the dynamics and involvement of The Mars Society delegation that attended the NSS Legislative Conference held in Washington, DC on April 6-8, 2003. This highly successful event resulted in space exploration advocacy briefings being given in the offices of more than 80 Congressional Representatives and eight Senators. These briefings requested support for specific items within NASA's FY 2004 Appropriations budget, including Project Prometheus and the Human Research Initiative, to study human factors in missions beyond Earth's orbit.

It was evident that there was considerable "common ground" on space policy issues that both National Space Society members and Mars Society members can support. To build upon the contacts developed during the NSS Legislative Conference, this paper will explore other possible opportunities for improved political outreach cooperation between NSS and the MS in the future. The possible benefits of a cooperative / unified approach would be in leveraging the strengths of both organizations – The Mars Society's strong vision and ongoing scientific research; the National Space Society's larger membership (over 22,000) and political connectedness – to achieve greater overall political support for space exploration. In addition, this paper will explore methodologies to increase cooperation between the National Space Society, The Mars Society, and possibly the Planetary Society, at the "grassroots" level, increase cooperation in the development of a mutually agreed upon space policy agenda, and increase Mars Society participation at next year's NSS Legislative Conference.

Purpose

The purpose of this paper is twofold – the first part reviews the dynamics and involvement of The Mars Society delegation who attended the National Space Society's Legislative Conference that was held in Washington, DC on April 6-8, 2003, and the second part explores methodologies to increase cooperation between the space advocacy organizations at the "grassroots" level, increase cooperation in the development of a mutually agreed upon space policy agenda, and increase Mars Society participation at next year's NSS Legislative Conference.

Legislative Conference

The 2003 Legislative Conference was set up by the National Space Society to begin grassroots advocacy of the vision to members of Congress, and to garner support for establishing a human presence in space. This two-and-a-half day event began on Sunday, April 6th, with a workshop on successful lobbying techniques and Capitol Hill logistics, and was followed during the next two days (April 7th and 8th) with an aggressive appointment schedule (which included over 80 briefings) to Congressional and Senate offices.[3] This was accomplished by splitting the combined NSS / Mars Society group into six separate teams consisting of from two to four members. Mars Society members who attended this event included Patrick Beatty, Carlos Glender, Lyle Kelly, Nick Perino and Mike Turner.

The message delivered was well received and centered on supporting the following objectives, which were outlined in the NSS briefing guide:

1. Make space access as inexpensive, robust and reliable as possible.
2. Develop a long-term space exploration architecture to provide a clear direction for the future.
3. Ensure government policy does not inhibit market forces and potential private sector opportunities.[1]

Considering the above listed objectives, support was requested for specific line items within NASA's and DOD's FY 2004 Appropriations budget. These items included:

1. Next Generation Launch Technology program – $515 million
2. Project Prometheus – $279 million
3. Orbital Space Plane program – $550 million

4. Human Research Initiative – $39 million
5. National Aerospace Initiative – $667 million[1]

The technology promoted by the above-listed line items could serves as "stepping stones" to an eventual mission to Mars, particularly Project Prometheus and the Human Research Initiative. The Mars Society delegation also promoted the vision of establishing a Human presence on Mars as a near-term goal in space at some of the legislative offices where we had made prior contact.

Increasing Cooperation

In order to develop improved cooperation in the future, it is essential to know the mission statement and composition of each of the major space advocacy societies, and build on their common ground to put together focused, well-coordinated political and educational outreach activities. The following is a brief overview of three of the major space advocacy societies.

The Mars Society, founded by Dr. Robert Zubrin, has had approximately 6,000 members since its inception, of which approximately 40% are still paying dues.[2] Mission: Promote the vision of a human mission to Mars as the next logical goal for space exploration. This is expanded upon in the Mars Society's Founding Declaration. Web site: www.marssociety.org.

The National Space Society was preceded by the National Space Institute, founded by Dr. Wernher von Braun, which was later merged with the L5 Society to form the National Space Society. It has approximately 22,000 members. Mission: "To promote social, economic, technological and political change to advance the day when humans will live and work in space."[4] Web site: www.nss.org.

The Planetary Society, founded by Dr. Carl Sagan, has approximately 100,000 members worldwide. Mission: "1. Encourages all space faring nations to explore other worlds. 2. Provides public information and supports educational activities about the exploration of the solar system and the search for extraterrestrial life. 3. Supports and funds innovative and novel research and development projects that can seed future projects of planetary exploration."[5] Web site: www.planetary.org.

Discussion

For all three of these organizations, the membership comes from people from all walks of life that are united by their interest in space exploration. This common interest can be developed, and cooperative efforts among the above-listed organizations should be encouraged. The NSS Legislative Conference was a great achievement from both the Mars Society's perspective and the National Space Society's perspective, because it furthered jointly held goals of advancing the human space program beyond low Earth orbit.

At a local level, possible venues where cooperation among the societies can take place include educational (such as tech fests, Mensa group meetings, air shows, and university presentations) and political. CY 2004 is a major election year, which provides an excellent opportunity to show grassroots support for a more aggressive human space program, with the vision of a human mission to Mars as the goal. Volunteering to help with a reelection campaign for a congressman or senator may be a beneficial way to develop valuable contacts which can be used later to promote the vision of a more aggressive human space program.

In order to develop increased cooperation among space advocacy groups, find out who the National Space Society and Planetary Society chapter heads are in your local area, and get in touch personally. By establishing communication, we may find out that we have much in common, and they may be interested in participating in Mars Society events.

Summary

In closing, please consider participating in next year's NSS Legislative Conference. It was extremely well-organized, and professionally conducted. With more volunteers, we would be able to make our presentation to even more legislative offices next year. It is a great opportunity to share our vision, establish contacts, and move humanity one step closer to landing on Mars.

References

1. *A Blueprint for Space Exploration, National Space Society Briefing Guide*, 2003.
2. Mars Society Steering Committee Meeting Minutes, August 10, 2002. www.marssociety.org/docs/2002sc_minutes.asp.
3. McMurray, Clifford R. *2003 Washington Legislative Conference Report, Ad Astra*, June / July / August 2003.
4. National Space Society Mission Statement, August 9, 2003. www.nss.org/about/index.html.
5. The Planetary Society, July 28, 2003. www.planetary.org/html/society/society.html.

Manned Mars Mission Economics – Context, Perspective and Public Perception†

Brian Enke
Southwest Research Institute,
benke@boulder.swri.edu

Abstract

The American public perceives that space exploration is an expensive endeavor. This opinion is constantly reinforced by the news media and is reflected in several recent public opinion surveys. In particular, a manned exploration mission to Mars is deemed to be vastly expensive. After a proper context and perspective are established, however, a different picture emerges.

This paper will baseline a plausible level of investment required to conduct a robust, manned Mars exploration mission. The author will then establish a proper context for this level of investment, focusing on both primary funding-source domains: public (American taxpayer dollars) and private (corporate or individual benefactors). "Take-Home Points" will briefly encapsulate key concepts.

A Message from the Author

You are about to read version 2 of this paper, dated October, 2004. I conducted the basic research for the first version in July, 2003 and presented the paper at the annual Mars Society conference in Eugene, Oregon (August, 2003). Since economic statistics are constantly changing, most of the facts and figures were updated for version 2. I have omitted a cost / benefit analysis section of the Apollo program and included a much stronger conclusion section based upon a public opinion survey in July of 2004.

Though I have compiled numerous high-profile cost / benefit analyses in my 20 years of corporate research experience, I am definitely not a professional economist or public relations expert. Some of my analytical techniques are rather simplistic and could undoubtedly be improved upon by professionals in these fields. Also, I do not attempt to quantify the highly subjective "benefit" side of the equation, as would be required in any official cost / benefit study. Rather than providing the end-all guide to manned space flight economics, this paper is intended to pose specific questions and challenge the reader's basic assumptions. I invite feedback and encourage detailed studies from greater subject matter experts.

The data within was compiled exclusively from reliable sources. Copies of the references are available to the public over the internet. I ask you, dear reader, to conduct your own research and formulate your own opinions.

To international readers, I apologize for the America-centric nature of this paper. The American government and corporate business environment are used as models. Due to the simplicity of my analysis methods, however, a similar study could easily be performed for any other space faring nation or allied group of nations.

American Public Perception (August, 2004)

The Apollo program deeply affected the psyche of the American public in the 1960's. While most of the intangible effects were highly beneficial (inspiration, motivation, etc.), Apollo also gave birth to some lingering errors in perception. Like the benefits, these errors have persisted through to present times.

The most destructive perceptional error is the high cost of space exploration relative to other public sector expenses. The roots of the problem are difficult to trace and highly speculative, but the media often reinforces this error. Sometimes, the reinforcement is deliberate. An early beneficiary of this perceptional error was the United States government.

In the early 1970's, the Nixon administration partially justified canceling the Apollo program on the grounds that space exploration was too expensive. The Apollo budget was redirected toward social programs that had suffered from perceived budget cuts precipitated by the huge, escalating cost of fighting a war in Southeast Asia. Despite the hushed fact that the entire budget for the Apollo program in 1971 was less than one billion dollars – petty cash compared to military or social program expenditures at the time – the misdirection was amazingly effective. The space program ended up bearing a shameful mantle of costliness, rather than the unpopular war effort.

Take-Home Point: As reported to the public, space investments often lack a proper context.

Though the war effort eventually collapsed anyway, the perception of costly space expenditures endures. This perception rears its ugly head whenever any ambitious space exploration project is proposed.

In a public opinion survey for the Associated Press (AP) in July, 2003, 75% of the respondents endorsed the space program as a "good investment."[1] However, only 49% supported the manned exploration of Mars (42% were opposed). A direct quote from a retired respondent in New Jersey bears witness to the current perception that Mars exploration, in particular, is too expensive:

> "We can go there after all the things wrong on Earth are fixed. I'm totally against any of it. It's a total waste of money we need for our kids, for illnesses, could put somebody's kids through college, could cure so many diseases."

Other public opinion surveys have revealed similar perceptions. In a Zogby research poll, also in July of 2003, 59% of the respondents believed that humans will set foot upon Mars within 25 years – and 18% believed it will happen within 10 years![2] However, 24% of the respondents wanted the US to end its manned space program altogether.

Superficially, the Zogby results seem quite optimistic. However, the results for the second question reveal a significant bias amongst the sampling population: Question 2: What percentage of the federal budget do you think is spent each year on the nation's manned and unmanned space programs?

Table 1 contains the results of the survey question. Cumulatively, at least 36% of the American public believes the manned and unmanned space program investments are over 5% of the federal budget, with at least 73% believing the investments are over 1% of the federal budget. Since the correct answer to the second survey question is 0.7%, i.e., less than 1%, at least 73% of the American public has a faulty perception of the relative level of space exploration funding. This perception was not corrected within the survey, so a hefty bias must be assumed within the remaining survey answers.

Less than 1%	(20%)
1% to 5%	(37%)
5% to 10%	(19%)
More than 10%	(17%)
Not Sure	(8%)

Table 1 – Results for Zogby Survey Question 2

Take-Home Point: Be wary of biased survey questions!

Other biases have plagued past public opinion surveys, sometimes blatantly. In the first question of an Associated Press survey[3] in January, 2004, half of the respondents were asked: As you may have heard, the Bush administration is considering expanding the space program by building a permanent space station on the Moon with a plan to eventually send astronauts to Mars. Considering all the potential costs and benefits, do you favor expanding the space program this way or do you oppose it? Only 43% of the respondents favored this proposal, while 52% were opposed.

The other half of the respondents were asked the same question but with "United States" substituted for "Bush administration," leading to a 48% split in the results. The attempt to tie political affiliation directly to the survey results, an interesting data-gathering approach, also significantly biased the results of this question and all later questions for half of the responding population.

However, a far worse bias permeated this same survey question. The survey was conducted a week before the official announcement of the new Vision for Space Exploration (VSE) by President Bush on January 17th, 2004. Paul Recer, an AP reporter, had previously triggered a widespread, persistent rumor in the news media that the cost of the new vision would be one trillion dollars.[4,5] Several direct references to this groundless rumor in the survey question, i.e., "As you may have heard," "expanding the space program," and "Considering all the potential costs and benefits" affected the survey results far more than the deliberately political nature of the question.

Take-Home Point: Be even warier of biased survey questions if uncorrected rumors are floating around.

Use of the term "cost" is another common bias whenever news reporters discuss space research. A more appropriate term is "investment," which implies that the outlays will be recovered over time, plus interest.

The primary route for space investment recovery is tax revenue growth coupled to increases in GDP (Gross Domestic Product). While highly subjective, various economic studies of the Apollo program have estimated the tangible ROI (return-on-investment) between 200% and 10,000%, depending upon the strictness of criteria used to evaluate the short-term and long-term benefits. Intangible ROI is even more difficult to assess but is unquestionably substantial.

While the numbers can be debated ad infinitum, the simple fact that several professional economists have attempted to assess the ROI of the Apollo program is proof of the validity of the concept. The ROI of a pure "cost" is meaningless.

Take-Home Point: Use the term "investment" rather than "cost" when discussing space research and exploration funding.

Three later questions in the AP survey exploited other common misperceptions and biases. Question 2 directly implied the inherently unproven premise that robotic exploration is more affordable than human exploration. Question 3 appealed to fear by stating bluntly that seven astronauts had been killed in a space shuttle accident. Worst of all, question 4 implied a spending level equity between space research and "domestic programs such as health care and education."

Baseline Mission Investment

Establishing a proper context for manned space exploration investments is difficult without baselining a mission first. Without some reasonable understanding of the mission goals, time frame, level of accepted risk, and investment per year, a space mission cannot be compared to other relevant governmental or private industry programs.

However, a large number of reasonable human space flight goals have been proposed by various organizations and individuals. Each requires a vastly different mission architecture, with different levels of up-front or ongoing investment. Even when the choices are limited to Mars exploration missions, at least a dozen mission profiles are currently being studied by NASA (the National Aeronautics and Space Administration) and/or various non-profit organizations.

For purposes of simplicity, we will use a well-studied Mars exploration mission proposal as a baseline mission in this paper: the Mars Direct plan, proposed by Robert Zubrin and David Baker in 1989.[6] The required level of investment for the Mars Direct plan was reanalyzed by a joint NASA / ESA (European Space Agency) task force in 2003.[7]

The Mars Direct plan is science-driven. A crew of two scientists and two engineers will reside on Mars for a period of two years before returning to the Earth. Specialized mission hardware extends the safety, flexibility, and capability of the science team. Steps are taken to minimize all known risks, so the overall risk profile of the mission is modest when compared to other potential benchmark missions. Approximately 10 years would be required to develop, integrate, and test the hardware and software (including one mission), with an average yearly investment between $2.7 billion (ESA) and $3.9 billion (NASA). The ongoing investment level to support 20 years of exploration missions, with one new mission every two Earth years, is estimated between $2.6 billion (ESA) and $3.5 billion (NASA). All estimates are in 2002 dollars.

Ongoing investment is the most useful metric for this study. Few people remember the level of investment necessary to create the first Space Shuttle (or even the name of the first Space Shuttle), yet the ongoing budget allocation for the Space Shuttle program is clearly stated within the NASA budget every year.

Take-Home Point: "Ongoing investment per year" is the best metric to use.

Take-Home Point: A robust, science-driven Mars exploration program would require a yearly investment of approximately $3.5 billion.

Context – NASA Budget

A Mars mission with human explorers may be funded privately or publicly. If public funding is used, the most likely source would be the NASA budget. To establish the investment context, we must examine the other investments in the yearly NASA budget.

2005 NASA Outlays (Proposed)	2005-Dollars ($Billion)	NASA Budget %
TOTAL NASA BUDGET	16.2	100.0
Space Science	4.1	25.3
Earth Science	1.5	9.3
Biological / Physical Research	1.0	6.2
Aeronautics	0.9	5.6
Education	0.2	1.2
Space Station, Operations	2.4	14.8
Space Shuttle	4.3	26.5
Exploration Systems	1.9	11.7

Table 2 – 2005 Proposed NASA Budget

The proposed 2005 budget for NASA is $16.2 billion.[8] Approximately half of the investment is closely related to human space flight. Major line items are listed in Table 2.

The Vision for Space Exploration (VSE) contains a specific recommendation to eliminate the Space Shuttle program after the International Space Station (ISS) has been built, around the year 2010. Barring any major shifts within the budget of NASA prior to 2010, the shuttle retirement would free a pool of $4.3 billion per year. This amount is assumed to be more than adequate for funding an ongoing program of Mars exploration, baselined in the previous section at $3.5 billion, without any budget increases, reallocation of the $1.9 billion Exploration Systems budget, or impact on ISS support. In reality, these and other areas of the NASA budget would overlap with the manned Mars exploration program, reducing its cost, while the "Moon" portion of the VSE would add to the overall cost.

The VSE road map is ambiguous, and any congressional funding increases for NASA are highly speculative. Since a robust Mars exploration program can – and should – be supported within the proposed 2005 NASA human space flight budget (without redirecting any funds from elsewhere within NASA), the rest of this document will assume the levels of investment within the following Take-Home Points:

Take-Home Point: The proposed 2005 NASA budget is $16.2 billion.

Take-Home Point: The proposed 2005 Human Space flight budget (including ISS) is $8.6 billion.

Take-Home Point (repeat): A robust, science-driven Mars exploration program would require a yearly investment of approximately $3.5 billion.

Unfortunately, $16.2 billion per year, $8.6 billion per year, and $3.5 billion per year are still just abstract numbers to the American public. These levels of investment lack a proper context in relation to other expenditures more familiar to an average American.

Context – US Federal Government Budget

NASA's budget is allocated within the United States Federal Budget. Therefore, to establish a proper context for investments into a publicly-funded Mars mission, human space flight, or space exploration in general, we must compare the NASA budget to other line items within the United States Federal Budget.

The proposed United States Federal Budget for 2005 is $2,400 billion (or $2.4 trillion).[9] The budget is complex, and the clarity of its presentation by the Office of Management and Budget (OMB) leaves a lot to be desired. However, browsing various OMB tables and sub-tables reveals numerous items of interest to most American taxpayers – items which can be used to establish a better context for space exploration investments (Table 3).

Take-Home Point: The proposed 2005 NASA budget is less than 1% of the US Federal Budget.

The New Jersey retiree quoted by the AP survey was concerned about several line items in the Federal Budget. Health care, welfare, and retirement security in the United States cumulatively cost well over a trillion taxpayer dollars each year. These important line items can all be considered *costs*, rather than investments, because they generate no direct returns. Their final "products" do not directly promote the creation of future products or directly guarantee additional government revenue. The programs preserve the health and financial opportunities of the current generation of taxpayers, retirees, and welfare recipients, or in some cases, their immediate descendants.

2005 US Outlays (Proposed)	2005-Dollars ($Billion)	US Budget %
TOTAL US BUDGET	2,400	100.0
Social Security	510	21.3
Medicare / Medicaid	478	19.9
Defense	429	17.9
Health & Human Services	68	2.8
Education	57	2.4
Homeland Security	47	2.0
Housing and Urban Development	31	1.3
NASA	16	0.7
[NASA Human Spaceflight]	[8.6]	[0.4]
[Mars Direct (ongoing)]	[3.5]	[0.1]

Table 3 – 2005 Proposed US Budget

Theoretically, if the entire NASA budget was completely eliminated, $16 billion could be reinvested into other government programs like the ones mentioned above. Due to the overwhelming percentage of Federal Budget dollars invested into these short-term programs, however, any reallocations would be practically unnoticeable.

Take-Home Point: Reallocating all $16 billion of the NASA budget into the $510 billion Social Security budget wouldn't even cover the annual adjustment for inflation. Eliminating the human space flight budget would be even less noticeable.

In fact, the cost of fraud within various short-term programs far exceeds the total level of investment into human space flight. Medicare and Medicaid fraud cost American taxpayers over $25 billion each year, three times the investment into human space flight.

Furthermore, the cold reality of the current US Federal Budget situation renders these theoretical reallocation games totally meaningless. The US Federal Budget is not balanced, nor is it required to be. The *budget deficit* is a major, perpetual component of the United States Federal Budget (Table 4).

The US Federal Budget Deficit acts as a buffer, ensuring that each line item in the budget stands or falls on its own merits. For example, the only real-world effect of completely eliminating the $16 billion/year NASA budget would be to lower the budget deficit to $429 billion, a mere 3.6% decrease. Short-term government programs like health care, welfare, and retirement security can't inherit a single dollar of the NASA budget without driving the adjusted deficit higher . . . which Congress could do anyway, independent of any decisions about the NASA budget.

But what exactly *is* the US Federal Budget Deficit? Quite simply, it is the difference between government revenues and expenditures. The US government currently borrows money to pay its expenses. Future

2005 US Outlays (Proposed)	2005-Dollars ($Billion)	US Budget %
TOTAL US BUDGET	2,400	100.0
Social Security	510	21.3
Medicare / Medicaid	478	19.9
Budget Deficit (anticipated, 07/2004)	445	18.5
Defense	429	17.9
Health & Human Services	68	2.8
Education	57	2.4
Homeland Security	47	2.0
Housing and Urban Development	31	1.3
NASA	16	0.7
[NASA Human Spaceflight]	[8.6]	[0.4]
[Mars Direct (ongoing)]	[3.5]	[0.1]

Table 4 – 2005 Proposed US Budget with Deficit

generations of American taxpayers must repay this loan (plus interest) during years of budget surpluses. In other words, the US Federal Budget Deficit is an investment into the future of the United States – just as the NASA budget, the Department of Education budget, energy or technology research, and other forward-looking line items are investments that expand the capabilities and resources of future Americans.

Long-term investments into NASA and the Department of Education are, by definition, completely compatible with the concept of deficit spending. Allocations that increase the US Federal Budget Deficit are justified if the rate of tangible and intangible return exceeds the amount of interest paid on the borrowed amount.

Take-Home Question: Since future generations will benefit from the space program, shouldn't future generations also pay for part of it?

An interest-only home mortgage is a common example of this principle in action. A consumer borrows a large amount of cash to buy a home and pays monthly interest on the amount borrowed. As long as the tangible and intangible benefits of owning the home exceed the interest paid, the new homeowner has made a good investment, even though her descendants might pay interest on the loan *ad infinitum*.

In theory, any department within the US government can take advantage of deficit spending if the rate of return justifies the expense. Deficit spending is most difficult to justify for short-term programs like Medicare and Medicaid. By contrast, the Apollo program has proven that the wealth of long-term tangible and intangible returns from the manned space program readily justifies deficit investment, even when pessimistic rates of return are assumed.

Miscellaneous US Outlays	Dollars (Billions)
TOTAL US BUDGET (2005)	2,400
Iraq war (estimated, cost in 2003)	75
Proposed highway improvement bill (2004)	53
Airline bailout (2002)	20
Farm subsidies (2004)	19
Missile defense system (2004)	9
[Mars Direct (ongoing)]	[3.5]

Table 5 – Miscellaneous US Government Outlays

Since the United States government currently does not fund a program of manned Mars exploration, our context analysis would be incomplete without mentioning various programs that *are* funded. Digging into the OMB tables reveals hundreds of context-laden government-funded programs. Some examples are listed in Table 5.

Our context analysis must also consider that the majority of funding for health, education, and welfare in the United States is by state, county and local governments. However, the budgets of all fifty states vary widely and are difficult to obtain. An in-depth study must be left to greater economic experts.

Using the 2005 state budget for Colorado (an arbitrary choice), we can easily extrapolate the approximate spending levels of all state budgets combined.[10] According to US census bureau estimates, the population of Colorado in July, 2005 will be approximately 4,400,000, which is 1.5% of the estimated 294,000,000 American citizens in August, 2004.[11]

2005 State Programs	Colorado Cost ($Billion)	Extrapolated US Cost ($Billion)
TOTAL COLORADO BUDGET	5.6	374
Department of Education	3.2	214
Department of Health Care	1.2	83
Department of Human Services	0.5	31
[Mars Direct (ongoing, reverse extrapolation)]	[0.05]	[3.5]

Table 6 – 2005 Colorado and Cumulative State Budgets – Estimated

If a fair portion of Mars Direct was funded by the state of Colorado (Table 6), the investment would be about $50 million dollars, one-tenth the outlays of the Colorado Department of Human Services.

Context – US Economic Activity

As massive as the United States federal and state budgets might be, the allocation levels are merely an anemic reflection of the United States Gross Domestic Product (GDP).[12] As the GDP grows, the Federal Budget grows. Over the past thirty years, the Federal Budget's percentage of the GDP has always remained near 20% (Table 7).

Economic Component	2005-Dollars ($Billion)	GDP %
TOTAL US GDP (8/2003-7/2004)	11,649	100.00
US Budget (proposed, 2005)	2,400	20.60
[Mars Direct (ongoing)]	[3.5]	[0.03]

Table 7 – US Gross Domestic Product

The close linkage between the US GDP and the Federal Budget allows us to expand our context of manned Mars exploration investments into a realm that is even more familiar to every American citizen. While most people don't know where the US government spends its money, they know where they spend *their* money. Statistics for "household name" corporations and products are extremely useful for comparison purposes (Table 8).

Take-Home Point: Mars Direct could be funded if one out of every eight ice cream purchases was donated to the public space program.

Take-Home Point: Mars Direct could be funded if one out of every twenty-five alcohol purchases was donated to the public space program.

Take-Home Point: Mars Direct could be funded if the impact of spam email was reduced by 5%.

Take-Home Point: More money is invested every year into new golf course construction than into manned space flight.

Miscellaneous GDP Components	Dollars ($Billions)
Microsoft Corp. revenue (2004)	37
Microsoft Corp cash reserve (mid-2004)	50
Microsoft Corp special dividend to shareholders (2004)	32
General Motors revenue (2003)	186
General Motors assets (end-2003)	288
Walmart revenue (2nd quarter, 2004)	70
State Farm insurance "administrative fees" (2003)	7
US golf course construction investments (1990's)	95
Assets of 10 random investment management firms (2003)	3,907
Europe / US ice cream purchases (yearly)	31
US alcoholic beverage sales (2002)	114
Worldwide artwork sales (yearly)	100
US cost of spam e-mail (estimated, yearly)	87
[Mars Direct (ongoing)]	[3.5]

Table 8 – Miscellaneous GDP Statistics

Facts and figures, like the ones above, permeate every aspect of the American economy. American consumers spend far more each year on any number of household luxuries (cosmetics, travel, junk food, video games, sports tickets, etc.) than their government invests into the space program.

The estimate of required investment for Mars Direct assumes no new technological innovations or improvements in economies of scale. The required investment for a privately run Mars mission using business world paradigms has never been reliably estimated, but it would undoubtedly be much cheaper than the public mission estimates. Some back-of-the-envelope calculations lead the author to believe that under optimal circumstances, a privately sponsored manned Mars exploration program could be conducted for less than one billion dollars per year.[13]

Conclusion – American Public Perception, Revisited

The perceived weakness in public support for human space exploration is largely due to the following factors:

1. Cost vs. Investment mindset.
2. Lack of proper investment context.
3. Biased survey questions

What happens when these errors are corrected? Two recent surveys have attempted to answer this question by polling the public far more objectively than in previous surveys.

In July, 2004, the Gallup organization was asked to conduct a space exploration survey by the Coalition for Space Exploration.[14] The seven questions in the survey were intentionally formulated to avoid bias and to provide true, relevant background information. For example, a question about space exploration funding mentioned that currently, less than one percent of the federal budget was invested in space exploration.

The Gallup survey found that 66% of the American public agrees it is "important" for the United States to conduct both manned and unmanned space exploration (13% disagree). 68% support (and 24% oppose) the new VSE if the total investment was capped at 1% of the federal budget. The detailed results for each question reveal strong, broadly-based public support for manned space exploration across nearly all demographic groups.

An October, 2004 marketing-oriented survey by Dittmar Associates confirms the same level of support for the VSE (69%, with 26% opposed) and similar support for NASA funding increases of up to 1% of the Federal Budget for implementing the VSE.[15] The methodology of this study was more interview oriented. Despite a potentially negative bias in the results caused by a political reference in an early question, faulty economic assumptions were corrected and the overall results of the survey seem credible.

The American taxpayers appear to be too generous, if anything. 1% of the proposed 2005 United States Federal Budget is $24 billion dollars, an amount dwarfing the $3.5 billion per year ongoing cost estimate for Mars Direct. NASA could conduct seven robust Mars exploration programs with the funds that most taxpayers are willing to invest.

References

1. Lester, Will, *Space Shuttle – AP Poll*, July 3, 2003.
2. Zogby International, *Americans Views On NASA And The Space Program*, July 20, 2003.
   www.chron.com/content/news/photos/03/07/20/nasa/finalreport.pdf
3. Ipsos, *Support For Space Exploration, But Not Necessarily Manned Space Exploration*, January 14, 2004.
   www.ipsos-na.com/news/pressrelease.cfm?id=2016
4. Recer, Paul, *NASA Starting Almost from Scratch in Moon-Mars Effort*, AP Newswire, January 17, 2004.

5. Day, Dwayne A., *Whispers in the Echo Chamber*, The Space Review, March 22, 2004.
6. Zubrin, Robert, and Baker, David, *The Case for Mars*, Simon and Schuster, Inc., 1997.
7. Hunt, Charles D., and van Pelt, Michel O., *Comparing NASA and ESA Cost Estimating Methods for Human Missions to Mars*, 2004.
8. NASA, *FY 2005 Budget*. www.nasa.gov/about/budget/index.html
9. Office of Management and Budget, *Budget of the United States Government Fiscal Year 2005*. www.whitehouse.gov/omb/budget/fy2005
10. The Official Colorado State Website. www.colorado.gov/
11. US Census Bureau, *United States Census 2000*. www.census.gov/
12. US Department of Commerce, Bureau of Economic Analysis. www.bea.gov/
13. Enke, B.L., *Shadows of Medusa*, Publish America, 2005 (publication pending).
14. The Gallup Organization: *Public Opinion Regarding America's Space Program*, July, 2004. www.spacecoalition.com/documents/gallup_coalition_report_040719.pdf
15. Dittmar Associates. www.dittmar-associates.com/

//

Economic Plan for Mars Colonization†

Brian Hanley, Konnectworld, Inc.

brian.p.hanley@att.net

Abstract

Economics is the means by which a culture is allowed to express its values. Space exploration and colonization of the planets have been assumed to be fundable only by governments, or rich individuals. However, by a judicious program of leveraging skills and human networks it is possible to develop the resources necessary over a 20- to 30-year period. This paper discusses the approach of looking at activist groups for interplanetary exploration and colonization from the perspective of economic development theory. The development of an economic infrastructure for the Mars Society group is also crucial for its long-term viability.

While tremendous technical work has been done that lays out exactly how to get to Mars, how to terraform it, and some minor economic analyses have been done of businesses that could potentially result, we need to fill in a gap — how do we get from where we are now to a colony which is self sustaining? Money. How do we create it?

Background

I have a long-standing interest in space colonization, starting from an assumption in childhood that I would be traveling to Mars and the Moon as a matter of course. But also, in my life I became affected by the huge problems of human lives in much of the world here on Earth.

Two years ago I started a private company (my third) with the mission of country development in the former USSR. For more than five years I have been active with plans for economic development in an effort to contribute in some way to alleviating the human suffering that is attendant on such economic collapse as occurred in the USSR at the time of its breakup. I have wrestled very directly with the issues regarding how to actually help groups of people effectively. This is not a simple matter, and it is one that has become visibly more complex over the two years I have spent operating a company with development goals that employs people in that region. Economics rests on a foundation of social bonds and politics — facts not as well appreciated by many Americans, living as we do in a milieu that is sound on both counts.

Initially, my efforts were focused on developing programs for government implementation. I pressured my congressperson, then senator, and after a while began to realize that it was not likely I would be successful. Even if I were to create a grassroots group, and activate it successfully, the system was in the grip of economic forces that constantly drive it to wrong-headed methods and ends, even where activated. I became aware of literature that showed that the health of nations was roughly inverse to the amount of aid which they received in the previous decade, even when corrected for initial starting conditions.[1] In this respect the breakup of the USSR was a remarkable laboratory. While I was not of a strong conservative bent, nor libertarian, I try to make my theories and actions fit the data rather than the other way around.

After this I moved on to making proposals to such groups as the Soros Foundation, Habitat for Humanity, and the like. Again, it became clear after a time that this avenue was going to have serious issues. However, I had also studied for a couple of decades, albeit at a slow pace, literature on community and ethnic group development. There is some very interesting and worthwhile work which has mostly been ignored, with the exception of books like *Making Mondragon*.[2] There are other out-of-the-mainstream works like *Heavens on Earth*[3] and a great many others, focused on how the social group, starting with next to nothing, creates wealth, often great wealth.

It seemed to me that my personal experience with development thinking, interacting with government, had strong parallels to the progression of space societies. It was, therefore, obvious to look at how the society could accomplish its goals with government playing a minor role. I remember L-5 meetings, and their focus on lobbying. NSS is similar, an important function, but not one that usually progresses toward our goal.

So it was natural for me, when looking at the Mars Society, to view the society and its goals through this bottom-up development lens. Reconstructive bottom-up bootstrap development is what I mostly believe in now. All that is required is that a group of individuals see what they can do, how to do it, and organize to create the economics which will support their goal of life. Very small outside supports, skillfully applied are all that is necessary. Everything else comes from inside.

Bottom-Up Economic Development for the Mars Society

Mars Society Assets to Economic Development
The Mars Society has very strong assets when viewed as an economic development problem.

- Social capital – This is defined as the ability of members of the group to depend on informal as well as formal agreements between each other. The Mars Society has very strong social capital among its membership. This is shown by the willingness of members to spend their time and energy as they have, above and beyond the high level of social stability which is normal to find within the USA.
- Intellectual capital – This is the collective level of education, together with the ability to apply that education in a practical manner.[4] Based on my experience with the Mars Society so far, this group has an extraordinary level of intellectual capital.
- Political stability – Operating primarily within the larger USA and European political milieu, the group is supported well by the benefits of a stable political system. Within the group, the importance of political organization and principles is well recognized. There are debates, but on all sides of that debate the base ideas are clear and strong in favor of liberty, freedom and a just society.

Mars Society Obstructions to Economic Development
- There is a lack of knowledge of how collective support can contribute to the growth of enterprises, and what the high level of return that can result can accomplish. This is compounded by a lack of personal familiarity with entrepreneurship.
- There is a common vague idea that collective economic support must have something wrong with it. Most members are probably familiar with cabals such as the Boesky clan, various Mafia families, and insider cabals like the "Skull and Bones Club" of the Bush and Rockefeller clans.
- A common perception that economics is a huge impersonal force, which cannot be affected by individuals. This is somewhat true, but for groups of remarkably small size, it is not.
- Lack of access to funding for enterprises. This problem can be solved, although it may also turn out to be phantasmal given the reach that the society has into many walks of life.
- Lack of organization and cultural orientation to creating enterprises among the members.

Discussion of Economic Development for the Mars Society
The Mars Society, to someone looking at it as an economic development problem, is extraordinary. If you have the intellectual and social capital, the effort will be successful, everything else is straightforward for those who have the will.

It will not be instant, but over the course of roughly 20-30 years, if the course I outline is followed, an economic and engineering engine will result like nothing seen before.

Social – Use the Methods of Upper Classes to Leverage into Wealth
This matter of mutual group support is key to economic success. In Europe this issue might be recognized as upper class consciousness versus working class consciousness. In the USA, this can be seen in the machinations of various wealthy familial clans, most recently those related to the Skull and Bones Club. Clearly, there are positive and negative developments possible. Equally clearly, this social strategy of mutual support makes a tremendous difference in the ability of a group to accrue wealth.

But where are the boundaries? What is the difference between a mutual support network and a criminal conspiracy? Quite simply, the boundary between a mutual support network and criminality lies in the methods and goals of the group. Gross coercion, threats, violation of securities laws or laws against drug trafficking and other criminal methods used in furthering a group's economic agenda are not acceptable. Overthrow of governments or similar acts as a goal are also not. However, everything else is.[5]

The fact is that creating a viable and sophisticated social circle, which is dedicated to producing some way of life, is completely and entirely laudable. It is the American way; it is what people came here for in the first place. From the founding of the Plymouth Colony, with its system of secular law overarching religious freedom, to the founding of Salt Lake City in Utah; from the Oneida Colony and the Seminole slave runaways in Florida to the building of St. Louis, America has been about getting together to create a different way of life. The Mars Society is right on with that tradition.

Financial – Receive Return Commensurate with Value

First, let us consider the costs of missions. In today's dollars, they are likely to cost on the order of $10 billion to $100 billion. A privately funded mission might be able to cut that cost significantly, but let us use it as a guideline. To establish a colony would require on the order of 3 to 5 missions, with proper equipment. That would tend to indicate that the float on such a venture would be on the order of $200-500 billion. Let us use that upper figure for this analysis.

Where on Earth might a group such as the Mars Society raise that much money? There is a way, though it is non-trivial. But there is nothing about going to Mars that is trivial. The answer is to grab a foothold and then leverage collective skills and cooperation. It is possible to grow one seed from next to nothing into such great wealth. Ask Bill Gates about that. If hundreds and thousands of seeds are used, the outcome becomes a virtual certainty.

The mechanism I propose is simple in concept. Create a consortium of new companies, all of which agree to allocate 5% to 15% of their stock to an entity such as the Mars Society. The Mars Society can then make money in the way that a good venture capital fund does. And if one thinks about it, it becomes clear that there are a number of ways that an organization like the Mars Society could provide value that would make the exchange economically fair.

First Rung Value Proposition – Funding

Consider, for example, the first rung of the ladder for new ventures. Once a market is identified, a product for that market and a way to reach the customer all are laid out, what is left? Funding. It is common for companies that raise private capital for new ventures to hold 5% to 10% equity afterward in addition to commissions on the fundraising. While this is not the most desirable route for new ventures, especially with the current climate for venture capital investment, it may be the only way to do it. It is also fairly common for an individual to be brought into a new venture and allocated a large equity stake simply because he has the connections to acquire venture capital from venture funds. Thus, if the Mars Society were to help with this step, serious equity stakes in new ventures would be completely appropriate and right.

Second Rung Value Proposition – Finishing the Product Successfully

Let us look at the second rung of the ladder in a new venture. The second challenge that new ventures usually have is getting their proposal to actually work. Here, again, an organization such as the Mars Society could provide economic benefit in return for equity value. There are always problems. One of the keys that distinguishes winners from losers is the ready availability of networks of experts at critical times. Venture Capital firms make such resources available. The Mars Society, being quite heavy with engineers, could excel in this area. Again, this exchange of economic value would be fair and right.

Third Run Value Proposition – Market Reach for those Key First Customers

The third rung is getting the product out to customers. But, to do that, customers need to know about the product. Often called "reach," this step of getting the message to the buyer is not as easy as it sounds. It is pretty amazing how difficult it can be to get companies to even listen to a pitch about a product that will save them significant amounts of money. Corporate deafness is a serious obstruction. An organization of supporters with links into companies all over the country, and potentially the world, could make an incredible difference for a new product. Again, such economic value provided would be worth some significant amount of equity in the venture.

Additional Economic Aid – Business Planning Help

These three things cover a lot of basic ground for new ventures. Aside from that, the society could form a group that helps visionary entrepreneurs turn their ideas into real business plans. This is something that even most venture capitalists do not do.

Goal – Unequaled Level of Success with New Ventures

By careful selection of the early ventures the society could establish an early record of accomplishment. The financial community would see it as simply good judgment, "backing the right horse." What we would understand, however, is that our good judgment was built from supporting the horse all the way to the finish line, and doing it better than the best venture capital firms can do it today. By doing this, we will make it easier for follow-on ventures to be backed. The goal would be that the Mars Society name become synonymous to the financial community with technical and financial success.

Plan in a Nutshell

In a nutshell, what I propose is that the Mars Society consider the following strategy.

1. Look for technologies that in general Mars living needs to develop that can find a market here on Earth.
2. Help those interested in such a course to validate their marketing and sales plan, technical plans, staffing and location issues.
3. Help raise money for the new ventures with connections and fundraising.
4. Be present on the board and have members available for problem solving when the new venture runs into a serious roadblock.

5. After the product hits the market, provide aid through the membership to get the message listened to. Help the company's market reach into corporations in the early stages.

Conclusion – Capital Development

How much new venture equity would need to be created to provide the Mars Society with the needed $200-500 billion?

If we assume that the society could achieve an average 10% stake in each of its early backed companies, and that each of those companies achieved a capital value of roughly $100 million, with 40% of those ventures failing completely, then we are talking about approximately $10 million in equity value per early venture. However, after 3 to 5 years it will be possible to take the income achieved by the first set of ventures and repeat it with another set, but directly funded, which will yield much higher equities in the companies in question. This will result in approximate revenue per equity sale of $50 million for each company.

Let us say that the society might be able to work with 5 enterprises in the first year. In the second year, I would expect that number could improve to another 5 for a total of 10. And again, in the third year, another 5 enterprises could be backed by the society. But watch what happens in the 4th and 5th years when returns start rolling in from the first 5. Early returns on these firms should be on the order of $10 million to $30 million. And those returns can be plowed back into funding new ventures. (Again, assuming a 40% failure rate.)

It is at this point that the equity returns become higher. Because once the society has its own money to invest, and it has the system shaken down for creating successful enterprises, the amount of equity that can be achieved is more along the lines of 30% to 60% of total shareholder value. As that value and income grows, society members can come on board to work on this concept full time.

A great deal depends on the execution in those early years. But once the system starts rolling along with appropriate guidelines and systems, the society would have a venture creation engine of great power. I believe that in the course of between 20 and 30 years, the wealth created would be enough to fund colonization of Mars privately. Not only that, the society would be partnered with a huge number of enterprises, which would have the collective expertise to build the equipment needed.

In the following spreadsheet, I have set a limit of 5,500 enterprises which an organization with a maximum overhead cost per year of approximately $1.3 billion could support active partnerships with, and a maximum of 1,500 new ventures per year. This would provide for approximately 7,100 full-time professional positions with the society at that point. Assuming 10% of that is management overhead, we end up at a bit over 1.2 line people per venture. All dollars are 2002 dollars. I have assigned an arbitrary limit of an average of $10 million per venture in seed funding for new ventures in the later years. This will allow the society to begin to accumulate money on the order of $10 billion per year in the latter third of this plan. For this model dividends are not taken into account, this is a growth model. They could add a few billion per year to the total system. Some businesses created might be very effective in creating high cash flows, but fundamentally the plan is a growth plan.

Appendix 1 – Organizational needs

Near term
- Identify the first few products / businesses for the society to back.
- Begin the process of education on mutual support economic networking – literature and workshops.
- Create the information dissemination mechanism, through the web site, or through email.
- Find out what funding networking is available now.
- Education in how to spot a confidence scam.
- Education in the difference between a con man and someone whose business plan fails. The latter will also happen, with a high level of frequency. The former, however, can be crippling. Unfortunately, questionable high tech ventures are quite common. For example, there was a company in San Diego, Silkroad, which raised millions in capital for a technology that depended on the spin of photons and grabbing them with laser tweezers, according to the White Paper on their web site.

Long term
- Creation of con-game response committee. Unfortunately, there will be instances of con men trying to take advantage of the group goodwill and social capital to raise funds for confidence games. What is important from the society's point of view is that we ensure that anyone identified as a con man be prosecuted should they be successful in taking advantage of the society. Since such frauds are a low priority for most law enforcement, it will be necessary for the society to retain a private investigator and develop a well documented case, particularly where the bag man has traveled to another country. Word gets around in that community, which will help to prevent recurrences.
- Development of the engineering / entrepreneurial / management talent for running a large venture capitalization operation.

Appendix 2 – Spreadsheet

Year	New Ventures	Per Venture Available Investment	Excess Available For Retention, Per Venture	Failed Ventures	Total Ventures Held Count	Sold Count	Return ($ Million)	Year's Retained Earnings ($ Million)	Cumulative Retained Earnings ($ Million)	Operating Cost ($ Million)	Available For Investment	Personnel Count	Mangmnt Overhead Count	Line Personnel Per Venture Ratio
2003	5	Time and attention		2	3								0	0.0
2004	5	Time and attention		2	8								0	0.0
2005	5	Time and attention		2	11								0	0.0
2006	11	2.6		2	20	3	30			1.5	28.5	8	1	0.4
2007	11	2.6		2	26	3	30			1.5	28.5	8	1	0.3
2008	11	2.6		2	32	3	30			1.5	28.5	8	1	0.2
2009	47	4.2		4	72	7	210			10.5	199.5	55	6	0.7
2010	47	4.2		4	108	7	210			10.5	199.5	55	6	0.5
2011	47	4.2		4	144	7	210			10.5	199.5	55	6	0.3
2012	173	4.6		19	291	28	840			42.0	798.0	221	22	0.7
2013	173	4.6		19	417	28	840			42.0	798.0	221	22	0.5
2014	173	4.6		19	543	28	840			42.0	798.0	221	22	0.4
2015	629	4.7		69	1,075	104	3,120			156.0	2,964.0	821	82	0.7
2017	629	4.7		69	1,531	104	3,120			156.0	2,964.0	821	82	0.5
2018	629	4.7		69	1,987	104	3,120			156.0	2,964.0	821	82	0.4
2019	1,500	7.2		252	3,131	377	11,310			565.5	10,744.5	2976	298	0.9
2020	1,500	7.2		252	4,002	377	11,310			565.5	10,744.5	2976	298	0.7
2021	1,500	7.2		252	4,873	377	11,310			565.5	10,744.5	2976	298	0.5
2022	1,500	7.1		600	5,396	900	27,000	10,650.0		1,350.0	25,650.0	7105	711	1.2
2023	1,500	17.1	7.1	600	5,396	900	27,000	10,650.0	10,650.0	1,350.0	25,650.0	7105	711	1.2
2024	1,500	17.1	7.1	600	5,396	900	27,000	10,650.0	21,832.5	1,350.0	25,650.0	7105	711	1.2
2025	1,500	17.1	7.1	600	5,396	900	27,000	10,650.0	33,015.0	1,350.0	25,650.0	7105	711	1.2
2026	1,500	17.1	7.1	600	5,396	900	27,000	10,650.0	44,197.5	1,350.0	25,650.0	7105	711	1.2
2027	1,500	17.1	7.1	600	5,396	900	27,000	10,650.0	55,380.0	1,350.0	25,650.0	7105	711	1.2
2028	1,500	17.1	7.1	600	5,396	900	27,000	10,650.0	66,562.5	1,350.0	25,650.0	7105	711	1.2
2029	1,500	17.1	7.1	600	5,396	900	27,000	10,650.0	77,745.0	1,350.0	25,650.0	7105	711	1.2
2030	1,500	17.1	7.1	600	5,396	900	27,000	10,650.0	88,927.5	1,350.0	25,650.0	7105	711	1.2
2031	1,500	17.1	7.1	600	5,396	900	27,000	10,650.0	100,110.0	1,350.0	25,650.0	7105	711	1.2
2032	1,500	17.1	7.1	600	5,396	900	27,000	10,650.0	111,292.5	1,350.0	25,650.0	7105	711	1.2
2033	1,500	17.1	7.1	600	5,396	900	27,000	10,650.0	122,475.0	1,350.0	25,650.0	7105	711	1.2
Totals	25,095			8,244			370,530		122,475.0	11,115.9				

Equity At Completion ($ Billion)	$161.9
Equity Plus Cash From Retained Earnings ($ Billion)	$284.4

References

1. *Collision and Collusion: The Strange Case of Western Aid to Eastern Europe* by Janine R. Wedel
2. *Making Mondragon: The Growth and Dynamics of the Worker Cooperative Complex* (Cornell International Industrial and Labor Relations Report, No 14) – by William Foote Whyte, Kathleen King Whyte
3. *Heavens on Earth: Utopian Communities in America 1680-1880* by Mark Holloway
4. It is possible to quibble on this definition, but let us use it here. For our purposes it is significant to emphasize practical ability to apply education. It is well worth noting that in all the community developments over 200 years in early north America, none survived except those made up primarily of craftsmen. One of those experiments ended with the death of thousands.
5. I will not open the Pandora's box of debating here what is right within an unjust society or system of law. Fertile ground, well plowed, which I will leave for others to carry the torch. There is a tangled thread in American history, reaching back into Europe as well as the Huron tribes suggesting some interesting influences on the founding of the new nation, not all of which were legal at the time.

//

The Export Economy of a Mars Settlement†

Bryan F. Erickson

Abstract

The rise of a settlement on Mars, from a nucleus of a permanent exploration outpost, will require a transition from government-backed space agency organization and funding to the creation of a true on-site economy, including income from exports to Earth, to fund the imports and continued expansion of a settlement. One ideal form of asset for interplanetary export is intellectual property, which can form the basis of a strategy for rapidly initiated return on investment between the planets.

Introduction

2004 has seen a dramatic rise in the stated goals of the United States in terms of space exploration. The US adopted as its policy the goals of sending human crews back to the Moon and thereafter to Mars. Although NASA has not yet committed itself to a timetable for putting humans on Mars, engineers have convincingly demonstrated that a dedicated effort could put people on the red planet in under ten years, with further piloted missions at each 26-month launch window, within a budget that is modestly greater than what NASA has recently enjoyed. With these plans apparently in progress, it becomes natural to compare the program of humans on Mars to the NASA missions to the Moon of 1969-1972. Despite the valuable research performed on the six piloted landings on the Moon, the program was limited in its effectiveness by the shortness of the missions; and the overall program was tragically canceled after only six landings, despite a small cost per mission compared to the sunk cost in developing the capabilities for the lunar program, due basically to the fickleness of Congressional funding. In contrast to the Moon, Mars mandates longer missions by its greater distance and the need to work around launch windows determined by its synodic positions relative to the Earth. Mars also allows longer missions by providing the resources, unique among possible target bodies in the Solar System, to allow relatively easy on-site production of fuel, breathable air, and drinkable water, while also providing moderate temperature extremes, useful radiation protection options, and even a day-night cycle close to that of Earth. It may be hoped that initial missions will spend around eighteen months on the planet at a time, taking full advantage of the period from one launch window to the next; and that after several such missions, it will become a relatively feasible jump to go from stays of eighteen months at a time to indefinite sojourns on the planet, to a permanent settlement, with people determined to make Mars their home.

Despite the popular hobby of using a speculative future on Mars as a platform for political utopianism of all stripes, the initial projection of a human presence on Mars will almost certainly require a traditional series of missions by one or several major government-run space agencies. However, it is almost as certain that such a government-backed program cannot provide the sustained level of interest and funding to set up a permanent community of settlers on Mars. Instead, private investment and innovation will probably be required, in combination with the technologies and infrastructure already developed with government backing on both planets, to allow the eventual capability of allowing private settlers on Mars. There's no doubt that the technology for settling Mars is within our grasp. Whether Mars can become home to a human settlement then comes down to how it might be financed. Robert Zubrin framed this challenge as follows:

> "Can Mars really be colonized? From the technical point of view, there is little doubt that we can eventually do just about anything we want on Mars . . . But how much can we afford? While the exploration and base-building phases can and probably must be carried out on the basis of government funding, during the settlement phase economics comes to the fore. While a Mars base of even a few hundred people can probably be supported out of pocket by governmental

expenditures, a developing Martian society, one that may come to number in the hundreds of thousands, clearly cannot. To be viable, a real Martian civilization must be either completely autarkic (very unlikely until the far future) or *be able to produce some kind of export that allows it to pay for the imports it requires.*" — *The Case for Mars*, page 218 (emphasis added).

The Economics of Settlement

A future Mars settlement may rapidly be able to supply itself with its own fuel, oxidizer, water and air; and in the medium term with its own basic construction materials and food. However, it would still need to import high-end goods, such as computer chips, from Earth for a long time. And if a Mars colony is to become a permanent and ever-growing settlement, it must make the leap from a government-financed mission of exploration, to a self-sustaining community. The wealth of a Mars settlement, and its corresponding ability to succeed over time, will depend on its ability to produce goods and services for export to Earth, to acquire currency reserves and achieve a reasonable import / export account balance. As it continues to expand and grow, it will ultimately become capable of reproducing the entire repertoire of Earthbound industrial and technological production. However, that would require a very long period of continued growth and progress, on a scale that is probably impossible without developing exports to Earth as a major component of the economic basis. Even after a Mars settlement fully matures it will still function more efficiently and be capable of more rapid growth by economic interdependence and trade with Earth.

What can the Mars settlers export? In *The Case for Mars*, Robert Zubrin lists the following options as exports from a Mars settlement to Earth, Earth orbit, the Moon, or the asteroids:

• Consumables, i.e., air, water, fuel, oxidizer.
• Food grown on Mars.
• Minerals.
• Deuterium, as raw material for nuclear fission or fusion reactors.
• Real estate and tourism, for those willing to come to Mars.
• Ideas.

Each of these investment options has a different market and a different duration of time before beginning to yield returns on the investment. A strategy for developing an export base should include export industries for the short term, medium, and long term, and in each case should focus on comparative advantages that a Mars settlement could have over Earthbound producers. Dr. Zubrin showed in *The Case for Mars* how the lower gravity to overcome at launch can provide Mars with a comparative advantage over Earth in supplying air, water, fuel, and food to other locations in the inner Solar System, including even the Moon, as opposed to lifting these materials to such locations out of the Earth's gravity well; although there will be no advantage in transporting these materials to Earth itself. This assumes the infrastructure is in place, on the Moon and asteroids, that would be able to create demand for such shipments. This could take the form of supplying telescopes, communications assets, or outposts, in the vicinity of the Earth or on the Moon, or temporary maintenance missions to any of these assets. For example, there are several expensive unmanned assets currently operating in the vicinity of Earth, but far away from low-Earth orbit, that could be reached with lower delta-V propulsion requirements and therefore more cheaply from Mars than from Earth. These include the Wilkinson Microwave Anisotropy Probe (WMAP), which operates on the far side of the Moon from Earth; and the Spitzer Space Telescope, which trails behind the Earth in its orbit about the Sun. Maintenance missions or the raw materials needed by such missions might be launched more cheaply from Mars than from Earth. Eventually, Mars could also include in such exports not only consumable raw materials, but also food and simple manufactured items, such as fuel tanks and structural components. However, developing the infrastructure to enable Mars-based provisions to supply such assets as these, and the increase in such assets to create further demand for such supplies, will both require time. Exporting fuel, water and air will therefore probably not provide significant returns on investments in Mars-based assets until the medium to long term.

High-value minerals might profitably be sent from Mars to Earth itself, although doing so in significant quantities would require a significant infrastructure for transporting such materials to the Earth's surface effectively and safely. These demands would probably push the return on investment from minerals shipments from Mars to Earth to the medium to long term. Deuterium for nuclear power has a mass-to-commercial-value ratio that should make it more easily profitable to ship back to Earth in small quantities. However, its effective collection and refinement on Mars would still require a substantial infrastructure on the red planet, pushing the return on investment on this option also out to the medium to long term.

Tourism and real estate, while not exports, still involve transfers of services and assets to people from Earth who pay Earthbound currencies into the Mars economy in return. The offerings of tourism or other services to temporary visitors and real estate and other Mars-based assets to permanent occupants, including scientific bases, companies and private individuals, will also have their growth constrained by the need to develop Mars-based infrastructure, and therefore will also not provide returns on investments until the medium to long term.

Exporting Ideas

Finally, ideas, in the form of intellectual property assets that can be licensed between the planets, have unique properties that make them a primary option for aggressive return on investment beginning in the very near term. As Dr. Zubrin said:

> "Ideas may be another possible export for Martian colonists . . . the conditions of extreme labor shortage combined with a technological culture will tend to drive Martian ingenuity to produce wave after wave of invention in energy production, automation and robotics, biotechnology, and other areas. These inventions, licensed on Earth, could finance Mars even as they revolutionize and advance terrestrial living standards . . ." — *The Case for Mars*, page 225.

Ideas that are licensable on Earth, and thereby can serve as revenue bearing exports, will take the recognized forms of intellectual property, such as patents, trademarks and copyrights. Intellectual property is unique among potential exports in that it does not suffer from the one great comparative disadvantage of economic products of Mars, that of separation from Earth, since intellectual property can be comprised in massless signals that are transmitted at the speed of light. In addition, returns on investments in copyrights and trademarks can be immediate, so that returns on investments can begin to flow right from the start. Mars also provides additional comparative advantages for exporting intellectual property, as follows.

As one example, Mars provides a unique and awesome new platform for copyrightable visual works, both of itself and as a dramatic environment for the humans living there. Mars and the people living there could be the subject of documentaries, photography, or even reality shows. Copyrighted still images and video / movie footage can be beamed to Earth cheaply. Earthbound crews can work with Mars dwellers to produce and edit final products, but the people on Mars will have rights to creative works they produce or to which they contribute, and avenues of compensation will be open from Earth to the people on Mars. Moreover, such copyrightable materials can be transmitted from day one of a voyage to Mars, and can continue earning revenue cheaply and indefinitely.

As another example, promoting trademarks from Mars through endorsements and advertisements is another cheap export that can operate from day one of a voyage to Mars. There will be a tremendous sense of wonder attached to the early human presence on Mars, which will translate into an incomparable cachet and prestige associated with such a mission. At the very least, the manufacturers of the assets used by the Martian residents will recognize the incalculable value of having footage beamed back to Earth of their products, adorned with their trademarks, being relied on by the early Mars settlers. Even simple naming rights might become a significant source of revenue. The Ansari family of Texas reportedly paid a sum in the millions of dollars just to change the name of the X Prize to the Ansari X Prize. A great potential will exist for revenue through trademark licensing and related endorsement contracts and advertising.

How much revenue is available due to copyright protected material? The top-grossing documentary has earned over $100 million domestically. The reality TV show *Survivor* is reported to have made $100 million in one season, due not only to the copyrighted show but also to several trademark licenses for product placement within the program. Even still images could contribute revenue streams; a benchmark for an analogous collection of Mars photographs might be the high-end exploration photography book *Antarctica: Explorer Series Volume I*, by photographers Pat and Rosemarie Keough, which has sold briskly for $3,000 per copy.

If documentaries, reality shows, still images, and trademark licensing and advertising and endorsement rights from Mars-based content are comparable to successful examples of these revenue sources on Earth, this could mean in the range of a quarter of a billion to a few billion dollars per year for a moderately aggressive, well-managed licensing effort. This would not likely fund a Mars settlement by itself, but neither is it anything to sneeze at; it could make a substantial contribution to the economy of a Mars settlement. As a revenue source, copyright and trademark licensing also has the great advantage that it can begin yielding revenue immediately. On the other hand, it has the disadvantage that it may actually hit its maximum value very early and decrease in income potential over time, as the novelty of people on Mars, and the planet itself, wears off. Yet, with strong creative direction and licensing management, copyright and trademark licensing have the potential to continue providing a significant component of Martian revenue indefinitely.

Which raises an important caveat on copyright and trademark licensing as a revenue source. They will never earn more than the average IMAX movie if produced by a NASA committee. We need the finest possible cameras on Mars, with transmission bandwidth to match, strong creative minds contributing to the production of content. Documentaries or reality-type programs of the Mars settlers will succeed as far as they capture the real human drama of people struggling to succeed on another planet.

Patents

Aside from licensing copyrightable works, trademarks, and endorsement and advertising contracts, intellectual property licensing from Mars could also be based on patents. Patent licensing has an even greater potential for producing revenue for the Mars settlement. Although it would not begin to yield revenue immediately, it has a greater potential to produce more and more licensing revenue from Earth to Mars as time goes on. It also requires nothing more than transmission signals to be sent between the planets, and so enjoys the same advantage over other investment options; and it also could be based on unique advantages enjoyed only by people on Mars.

Patent licensing can form a quite substantial form of income, offering excellent prospects for long-term revenue growth. The patent licensing revenues of US businesses grew from $15 billion in 1990 to $115 billion in 1999, and to a projected $180 billion in 2004. IBM alone has generated annual revenue of $1.3 billion from licensing. Rambus, a company focused on coming up with new inventions and licensing the patents to those inventions as its main revenue source, has had about a $2.5 billion market capitalization in 2004 with only about 200 employees, no inventory, and no manufacturing capability. This may be an ideal basis of comparison for a similar project at a Mars settlement, where a few hundred clever and highly trained professionals might be similarly productive in producing new inventions, and would be equally capable of securing patents on those inventions and licensing those patents, by working with Earthbound patent attorneys and managers. If such a group on Mars were able to raise a similar market capitalization in the low billions, let alone the actual revenues from the licensing deals, this could make a significant contribution to the economy of the Mars settlement.

Mars settlers would presumably have a lot to do other than work on new inventions. On the other hand, they would also enjoy some unique comparative advantages, a few examples of which follow.

First of all, as a team of very selectively chosen engineers and scientists immersed in a difficult environment, where labor is at a premium and their lives closely depend on a wide variety of ongoing engineering work, the Mars settlers will have profound new motivation – a factor not to be underestimated – to invent labor-saving or life-saving devices and innovations. Making technical innovations would be a way of life on an alien planet, rather than merely a job, in a way that must extend further than it ever could on Earth, no matter the level of devotion to one's job. On the other hand, you could always keep these inventors on Earth and offer them other incentives, perhaps not as effective, but a lot cheaper.

There is also the possibility to create an artificial comparative advantage, which could be very effective. The US government, or any other government, either together in a treaty or by themselves, could simply offer a longer patent term for inventions made on Mars.

A US patent is currently enforceable from the date it is issued by the US Patent and Trademark Office, until 20 years after the day on which the inventor(s) filed the patent application. Instead, the US could provide for, say, a 30- or 40-year term for patents based on inventions that were invented on Mars. This would offer a tremendous new advantage to any such inventors.

How much would such a longer patent term be worth to a patentee? How much revenue would it add to the Mars export account balance? For the industry most dependent on patent rights for its revenue, the pharmaceutical industry, each year under patent for a single successful drug can make a difference of hundreds of millions or billions of dollars in revenue per year. Suddenly, every prolific inventor is going to have a strong incentive to go to Mars, and their employers will have a strong incentive to send them there. The potential value of the extra patent term, and an approximate measure of the added income to a Mars settlement, might therefore be in the billions of dollars, for a single patent.

Serious proposals have been discussed, and floated in the US Congress, for offering large incentive prizes for accomplishing any of several milestones in off-Earth exploration. Such a prize has been offered and awarded in the private sector, in the form of the Ansari X Prize. While this idea remains valuable, there are decisive advantages in extending the patent term for inventions conceived on Mars as an alternative. For one thing, the system is self-guiding; whereas the prizes considered by Congress would have their amounts for various milestones fixed by a political committee, the value of a patent is determined by the market. As another virtue, an extended patent term would not cost the government any money, whereas a bounty prize commits the government or other offering agency to cough up cash. Which will be easier to convince the government to do? Which one would you rather rely on as a basis for present investment?

What about the balance of interests behind a patent? Governments grant patents to encourage the development of new inventions, and the financial security of looking forward to enforceably exclusive use of the invention for the term of the patent. Without such a policy, much of the investment that fuels technological progress simply would not be possible. On the other hand, a patent involves a compromise, as governments remove from their public the right to practice the new inventions of others until after a patent expires. This compromise is justified because the public also wouldn't be able to practice the invention if it were never there in the first place, or at least would not be there until much later, if the patent system did not exist. However, the public receives a detailed knowledge of how the invention works, from the patent document, and receives the right to practice the invention freely after the patent expires. The patentee must give up any chance for further exclusive rights after the patent expires.

Extended Term Patents

A new patent with an extended term might disrupt this balance, by removing for longer still the rights of the public to practice an invention freely. However, this would still be balanced by a dual interest in promoting the settlement of Mars as well as promoting inventions. Promoting the settlement of Mars would help accomplish a similar goal to the original goal of the patent system: to increase invention and technological progress, and by doing so to promote the economic growth of the nation. Combining the two goals into a single program with a differential patent term for inventions developed on Mars makes sense.

Does the government have the authority to make such a longer patent term? It sure does. It has changed the term in the past; it used to be 17 years from the date of issue, rather than 20 years from the date of application. Another recent innovation in the duration of patent terms provided for an extension of up to five years on the patent term, to compensate for lengthy review by the Food and Drug Administration in certain cases. Finally, in the recent case of Eldred vs. Ashcroft, 537 US 186 (2003), in which the plaintiff challenged Congress's retroactive extension of the copyright term by 20 years, the US Supreme Court said the Constitution gave unusually broad authority to Congress to legislate for copyrights as it pleases. The same constitutional mandate, in Article 1, Section 8, Clause 8, applies equally to patents as well as copyrights.

If an extended patent term is offered for inventions developed or conceived on Mars, why not broaden the range where inventors can take advantage of this change, for instance to include inventions conceived in Earth orbit or on the Moon? A crucial distinction must be applied, though, to such nearby locales. It would be too iffy about when the invention was conceived, for astronauts who might be away only for weeks or days at a time. Receiving the extra-special term should only be allowed when there is undoubted proof that the invention was conceived in the inventor's mind while she or he was away from Earth – such as on the necessarily long stay on Mars. A trip to Mars, with technology foreseeable in the near future, involves a round trip of at least a year and probably significantly more. This period of time is long compared with the time typically spent on conceiving a new invention, so it is workable for applying a separate class of patent privilege. Voyages to other bodies in the Solar System would be similarly long, and as a category, are cleanly separated in their required duration from trips in the local neighborhood of the Earth.

An extended patent term would be a simple and elegant mechanism for promoting at least long-term visits or assignments, and perhaps settlement, on Mars by well-qualified settlers. Aside from such an artificially created stimulus, there is another, potentially very great, inherent comparative advantage that Mars explorers and settlers would have in patent licensing: unique discoveries that could inspire invention, potentially including the discovery and investigation of past or present life on Mars.

The Discovery of Other Life Forms

If we discover new life forms on Mars it will be one of the handful of greatest discoveries in the history of science. It will also raise the possibility of leading to new inventions, thereby opening a dramatic new comparative advantage for the export economy of Mars. The discovery itself of something previously existing in nature can't be patented. However, the discovery of something in nature can often inspire new inventions that would not have been invented otherwise. These may include, for example:

• a purified form of the thing discovered.
• a method of making the thing discovered.
• a modified form of the thing discovered.

As a specific example, US patent number 6,448,381 is for purified forms of DNA from "bovine uterus and human placenta" that encodes heparin binding growth factor. The discovery and investigation of the gene, which exist naturally, inspired the inventors to figure out how to produce purified forms of the gene, which constitute a patentable invention.

Even a living thing can be patented, in the US at least (though not in many other nations), when it meets the other criteria for a patent, such as being produced by genetic engineering and did not previously exist in nature. This was established by the US Supreme Court in the case of Diamond vs. Chakrabarty, 447 US 303 (1980). Ananda Chakrabarty, a scientist with General Electric, applied for a patent for a new microbe he invented, using genetic engineering, that liked to eat (and thereby clean up) hydrocarbon pollution. An example of the patent claims from Chakrabarty's patent: "A Pseudomonas bacterium containing at least two plasmids, each providing a separate hydrocarbon degradative pathway." Sidney Diamond, the commissioner of the Patent Office, argued that a living thing could not be patented. The Supreme Court held that there was no such condition for patentability in the US patent law. This has remained good precedent in US patent law.

A persuasive amicus brief was filed in the Chakrabarty case by the Regents of the University of California. They argued that "Economic incentives for research conducted by Amici will be reduced by a rule excluding all living organisms from patentability. Such a rule will adversely affect commercial development of the fruits of the research, and will negatively affect the competitive stance of American industry in the competitive field of biotechnology."

How much revenue might be generated by patents inspired by Mars life forms? Perhaps a useful guide is to compare to anticipated revenue of the Earthbound biotech industry so far, as reflected in investment made corresponding to that anticipated revenue. In the past 25 years, roughly $40 billion has been invested into biotech commercialization. This is a better gauge than revenue so far from the biotech industry, both because those revenues are anticipated only after many years of research and development and have not yet matured, and because the Mars settlers also will not have to wait for investments to mature – they only need to demonstrate revenue potential sufficient to attract investment, as the Earthbound biotech industry has done.

What could Mars microbes and their DNA do that would have commercial value? It would be presumptuous to answer that. But to make the attempt anyway, we could say two things:

1. We can try to anticipate what functions would have been selected for in the Mars microbes by natural selection. Who knows what completely different physiological mechanisms Mars microbes might have developed to compensate for the extreme aridity, temperature swings, ultraviolet, and other harsh conditions of their evolutionary history – or how those might be adapted to modify Earthbound crops for better growth under tougher conditions, to clean up pollution, to produce new medicines, or for some other purpose.
2. If panspermia has functioned so that the Mars microbes share a common ancestor with Earth life forms, both the remaining similarities and the differences of its Earth-related DNA might shed a new window on and inspire new areas of understanding the genomes of Earthbound species, leading to new applications of which we might not otherwise conceive.

Conclusions

Whatever the details, there is little doubt that if present life forms, or well-preserved remnants of past life forms, are discovered on Mars, they will provide a whole new field of study, among the benefits of which will be a multitude of patents.

And even before an extended patent term is passed or Mars life is discovered, patent licensing has the potential to yield billions of dollars of revenue to Mars settlers per year, without them having to send a single rocket anywhere. Trademark and copyright licensing can form an important component of Mars settlement financing from the very beginning; patent licensing can begin to form a major component of Mars settlement financing, probably beginning in the short to medium term, potentially beginning within the first few years of a human presence on Mars.

//

Redefining the Mars Sample Return Mission†

Paul Contursi
Mars Society of New York
pcon@pipeline.com

Abstract

Shortly after the back-to-back failures of the *Mars Climate Orbiter* and *Mars Polar Lander* in 1999, the NASA Mars Exploration Program was reorganized and plans for the Mars Sample Return mission were put on hold. Now that the program seems to be back on track and attention is once again being focused on a sample return mission, an important political opportunity for the Mars Society has emerged. As it was originally conceived, the sample return mission plan was excessively costly and complex. The plan required long lead times, many launches and very high levels of technological risk. The Society's Political Task Force should take the initiative and begin a campaign for a sample return mission based on the application of In-Situ Resource Utilization (ISRU) as soon as possible. An ISRU based mission could be accomplished at a much lower cost in a shorter time with fewer technological risks. Most important, it would serve as an undeniable proof of concept for the Mars Direct mission architecture and would greatly strengthen the case for the human exploration of Mars in the near term. The presenter posits that the Society should not squander this important opportunity.

Introduction

In the summer of 1997, interest in Mars exploration reached a new zenith with the triumph of the Mars Pathfinder mission. The first successful robotic landing on Mars in a generation was the source of tremendous interest by the public and the space advocacy community alike. On the first day of the mission, the NASA / JPL web sites received 50 million hits. In the two months that followed, another three quarters of a billion hits were received. Fueled in part by the public reaction and, in part, by the success of the Pathfinder mission, NASA announced a dramatic acceleration of the robotic Mars exploration program with a commitment to a mission for each launch window over the next decade. A program that had only launched sporadic missions to Mars for a generation was being replaced by a much more robust effort that would dispatch a lander and/or orbiter to the Red Planet every 26 months. The new emphasis on Mars gave rise to more ambitious planning for future missions and the NASA / JPL team began to consider one of the most dramatic robotic challenges: a sample return mission.

Unfortunately, the enthusiasm was short-lived. In 1998, two ambitious missions were launched and both ended in failure. The *Mars Climate Orbiter* was to perform the first search for water from orbit and study the dynamics of the climate over the course of a full Martian year (687 days). Poor management oversight led to confusion between metric and English units resulting in a navigation error. The discrepancy went unnoticed until the spacecraft burned up when it entered the denser regions of the Martian atmosphere during orbital insertion. The *Mars Polar Lander* was the first attempt to explore the Martian polar region. Mission planners had high hopes that this spacecraft would return unambiguous data regarding the presence of water ice on or near the Martian surface. But the spacecraft was destroyed when a software glitch prematurely shut down the descent engine at an altitude of approximately 100 feet. A review of these failures exposed significant weaknesses in program management that needed to be addressed before the Mars Exploration Program could be resumed. As a result, the landing mission planned for 2001 was replaced with an orbiter intended to achieve objectives that were similar to the failed *Mars Climate Orbiter* (the *2001 Mars Odyssey*) and plans for the Mars Sample Return were put on hold.

Mars Opposition 2003

Mars Odyssey was a tremendous success. From its vantage point in Mars orbit, the science team could correlate the data from the new orbiter with images and altimeter readings from the *Mars Global Surveyor*. In the two years following the spacecraft's arrival, scores of papers were published in scientific journals announcing indications of substantial amounts of water ice under the Martian surface. Some researchers estimated that there was enough water in the Martian hydrosphere to fill Lake Michigan twice. While earlier theories predicted that the Martian water table would be found at a depth of a kilometer or more, the new data indicated that much of the water ice was less than a meter from the surface. If subsequent missions confirm these findings, future explorers will have easy access to plentiful supplies of water that can be converted to oxygen and hydrogen by simple chemical processes. As the summer of 2003 approached, a growing number of biologists, encouraged by their discovery of hardy life forms thriving in extreme environments here on Earth, began to express more optimism about finding evidence of existing or extinct life forms below the Martian surface.

The Mars opposition of 2003 is another factor giving Mars a higher public profile. On August 27, Mars will be closer to Earth than it has been in approximately 60,000 years. The planet will become so large and bright that it will be impossible to ignore, even in urban skies. Modest amateur telescopes will reveal a surprising level of detail. Large features, such as the light and dark areas and the south polar cap will be plainly visible in instruments with apertures as small as three or four inches.

The summer of 2003 also ushered in an era of unprecedented robotic exploration. In the space of a few short weeks during this year's launch window, a virtual fleet of spacecraft was dispatched to Mars. The Japanese *Nozomi* orbiter, arriving years late due to a propulsion problem, was finally on its way. The ESA's *Mars Express* orbiter was launched on a Russian Soyuz / Fregat rocket. The United Kingdom's first Mars mission, a microlander called *Beagle II*, rode along on *Mars Express* as a piggyback payload. The United States launched two sophisticated Mars Exploration Rovers that represented a quantum leap in capability from the little *Sojourner* rover that captivated the world during *Pathfinder* in 1997.

Mars Sample Return Missions

At the time of this writing renewed interest in a Mars Sample Return mission on the part of various officials at NASA and the ESA is stirring. Even if only some of the missions currently on the way to Mars are successful, or return more dramatic findings, it is obvious that there will be great pressure to put the sample return mission back on the front burner.

While this will be a positive development, the original mission architecture is unnecessarily complex, expensive and prone to failure. In addition, some of the programmatic difficulties that have plagued the International Space Station project have been reflected, albeit on a much smaller scale, in the current sample return plan. In light of these problems, the Mars Society should consider offering an alternative that will be faster, less risky, cheaper and more directly help pave the way for human exploration in the near term.

The original plan calls for the launch of two NASA surface packages on the same Delta II class launchers that we have used for Mars missions in the recent past. Each package consists of a lander, a rover and a solid fueled Mars Ascent Vehicle (or MAV). After landing on Mars, each rover will conduct a 90-day surface traverse that will include the collection of approximately 500 grams of soil. When the traverses have been completed, the rovers will load their samples aboard the Mars Ascent Vehicles. The MAVs will be launched from the surface into orbit around Mars to await collection.

During the following launch window, an Ariane V booster lifts off from French Guiana carrying a large orbiter (2,700 kilograms) supplied by France and Italy. Upon arrival in Mars orbit, the orbiter will rendezvous with the Mars Ascent Vehicles and transfer the samples to two Earth Entry Vehicles (EEVs). The orbiter will then fire its engines and insert itself into an Earth return trajectory. Upon close approach to Earth, the orbiter will jettison the EEVs into the Earth's atmosphere for reentry and recovery.

Even though robotic and human missions are managed by different organizations within the space agency, some of the parallels between the Mars Sample Return and the International Space Station are striking. One factor that contributed heavily to our current difficulties with the International Space Station was the political requirement to include the participation of the former Soviet Union. International cooperation is generally a positive element, but only if it adds resources or enhances scientific return. It should be noted that the United States had no trouble managing its own space station effort unilaterally when it launched *Skylab* in 1973. While there is no question that the Russians have a tremendous track record in long duration space flight aboard the Salyut and *Mir* space stations, that was not the primary motivation for the requirement of including them in the ISS project. At the time of the USSR's collapse, we became concerned that their underemployed and unemployed space engineers might succumb to the temptation of selling weapons of mass destruction to rogue states or terrorist groups. ISS was intended as a way to keep them busy in a non-threatening way. However, the necessity to include Russian participation is one of the key factors that has dramatically increased the cost and complexity of building and maintaining the ISS. Including international partners in the Mars Sample Return mission without a properly focused reason for doing so will almost certainly yield similar results.

The original mission plan requires multiple launches and the development of several different kinds of spacecraft. Most of these systems would have high levels of mission criticality. For example, a major failure aboard the orbiter will make it impossible to recover the samples waiting in Mars orbit. All told, the shopping list for the mission is long and expensive. Delta launch vehicles cost up to $75 million each and the Ariane V is about twice as expensive. The requirements also include two landers, two rovers, two MAVs, two EEVs and the orbiter. There are several variations, but no matter which one is chosen, 11-15 major hardware systems will have to be developed and/or procured.

The most imposing of the many technical challenges will be the orbiter's recovery of the samples and transfer to the Earth Entry Vehicles. Automated rendezvous and docking in low-Earth orbit has a spotty record. Performing such maneuvers under conditions in which the relative distances between Earth and Mars could delay radio signals by as much as 20 minutes represents a highly sophisticated semi-autonomous capability that will require a major hardware and software development effort.

All these factors endow the original mission architecture with a high, but as yet undetermined, price tag. Some NASA estimates are as low as $2 billion but experts outside of the space agency maintain that the price could easily exceed $5 billion.

The perception and political issues raised by such a plan couldn't come at a worse time. We are still in the throws of trying to recover from the *Columbia* accident, with all that implies in terms of critical decisions that must be made with regard to ensuring safe and reliable access to low-Earth orbit. As a result, NASA is frantically devising workarounds to maintain our space assets. At the same time, politicians and space activists alike question the value of the overpriced and poorly focused International Space Station with increasing intensity. The current incarnation of the Mars Sample Return confirms the worst fears of space critics, due to its high cost, complexity and long development lead time. It is difficult to see how the implementation of this plan, even if it is successful, will help to foster an environment in which the Mars Society can credibly argue that the human exploration of Mars is within our grasp at a price we can afford. A poorly conceived robotic mission that recovers two pounds of Martian soil at a cost of $5 Billion, offers very little hope to the man on the street that we can build a human Mars capability for $20-25 Billion.

Needless to say, this creates something of a conundrum for the Society on the political action front. Opposition to a sample return mission would be disturbingly out of character for our organization's charter. However, we might be able to offer an alternative that not only rectifies the shortcomings of the original plan but also provides a working model, on a smaller scale, of the kind of human capability that we would like to build.

In the late1990s, Dr. Robert Zubrin published an article that compared the different methods of Mars sample return. The article posits that a mission based on In-Situ Resource Utilization (ISRU) would be the fastest, most cost effective and least risky of several leading scenarios. As in the Mars Direct human mission architecture, the key to the ISRU sample return is the capability to manufacture the propellant used to launch the samples back to Earth from gases easily accessible in the Martian atmosphere. Since the Martian atmosphere is 95% carbon dioxide, simple chemical processes that have been thoroughly understood since the late 19th century can be used to produce methane propellant and oxygen on the surface. Several researchers have successfully built prototypes of this type of automated fuel factory and they have operated at high levels of efficiency.

A single Delta II class booster could launch a lander, a rover and a Mars Ascent Vehicle whose fuel tanks are nearly empty. After landing on Mars, the rover begins its traverse and sample collection as in the NASA plan. In the meantime however, an ISRU system aboard the lander uses a small amount of hydrogen from Earth to react with Martian carbon dioxide to produce methane and oxygen. By the time the launch window for the return to Earth opens, the MAV is fully fueled. No orbital rendezvous or sample transfer is needed because the locally produced propellant can be used to launch the samples directly back to Earth for reentry and recovery.

By obviating the need to include the propellants required to bring the samples home, payload mass can be vastly reduced and the mission can be dramatically simplified. Only one launcher, one lander, one rover and one Mars Ascent Vehicle are required. However, the ISRU mission is not without its own set of challenges. Considerable development will be required before today's laboratory prototypes of fuel manufacturing systems can be turned into flight ready hardware. The extreme low temperatures and the ubiquitous Martian surface dust are just two of the obstacles that mission designers will have to overcome. However, the overall level of technological risk is much less than that of the original NASA plan. One of the most important advantages is that an ISRU sample return mission is essentially a robotic proof of concept for the Mars Direct plan. ISRU systems will no longer be abstractions or laboratory prototypes; they will have a high profile as a methodology that can make Mars exploration much simpler and more affordable. Success will increase confidence that such systems could be scaled up to support human missions.

A Mars Society Role

NASA / ESA interest in a sample return mission should be encouraged. However, the Society should begin working immediately to promote the ISRU architecture as a more desirable method of obtaining pristine Martian material. We need to develop a multi-pronged plan, with a number of contingencies, to promote this alternative as soon as possible.

First and foremost, the Society should state that advocacy for an ISRU sample return mission is a cornerstone of its political action activities. Chapters and Steering Committee members alike should lobby members of the House and Senate for serious consideration of the ISRU sample return architecture. At the same time, our outreach activities should include discussion of the plan as an integral part of our remarks concerning the Mars Direct concept. In light of increased interest in space exploration on the part of the member nations of the European Space Agency and the Peoples Republic of China, our chapters around the world should coordinate their advocacy efforts along similar lines. Another important element of the Mars Society's campaign will be to join forces with other grassroots space advocacy groups in the United States and elsewhere, wherever possible.

The Mars Society membership boasts robust technical expertise that should be drawn upon to submit ISRU mission plans to the NASA Mars Discovery program. If a full-fledged sample return mission cannot be made to fit within the highly constrained Mars Discovery funding guidelines, we should consider proposing a demonstrator project in which a small vehicle can be launched from the Martian surface with locally produced propellant on a sub-orbital trajectory. This payload would also add scientific value by taking airborne images of the landing sight and surrounding areas that could provide valuable geologic context for other surface experiments. Additional opportunities to fly such systems as piggyback payloads on ESA missions should also be sought.

Conclusion

The redefinition of the Mars Sample Return mission is an important opportunity for the Mars Society. This project can represent yet another expensive example of NASA's business as usual or it can mark a turning point that will help strengthen the case for near term human missions to the Red Planet. It's up to us.

—————————————————————————————————— // ——————————————————————————————————

Applying Quality Principles in the Colonization of Mars†

Trevor Anders
Trevor.Anders@xtra.co.nz

Introduction

When people from Earth commence the exploration and colonization of Mars, they will be faced with huge challenges, both foreseen and unforeseen. They will need the very best management skills to strategize, plan, evaluate and control their activities.

Quality management has evolved during the past few decades to become a very successful management discipline, which enhances the ability of leaders and teams to optimize their intellectual resources, as well as all other resources, in setting appropriate goals and then efficiently reaching them. The means of identifying best practice, then consistently applying it, are key ingredients of quality management. Another key ingredient is a process of ensuring ongoing improvement in all areas.

Quality practice evolved over the course of the last century – especially the latter half – from fairly narrowly focused quality control checklists to broader-based quality assurance procedures, and finally to quality management. Rather than go into the history, I will cover briefly the three main regimes which embrace the universe of quality – as found on Earth today. These are:

- TQM (Total Quality Management)
- Baldrige Awards (and related)
- ISO 9000 International Standards

An integral theme for all three is continuous improvement – in value to customers, and in organizational performance (and by implication in value to employees).

Total Quality Management (TQM)

TQM is a philosophy which has grown under the nurturing of a number of "gurus" over the past fifty years or more. These include famous names such as Shewart, Deming, Juran, Ishikawa, Taguchi, Crosby and many others. While there are common themes and principles, there are no definitive criteria and no assessment system, and the term has often been used in a way that leaves its application rather nebulous. Business philosophies such as Kaizen, JIT (just-in-time) and others have linkage to TQM, and similar comments apply to them.

The generally acknowledged key concepts of TQM are also recognized within the ISO 9000 Standards and the Baldrige criteria and include:

- Leadership, and the importance of management commitment (to quality).
- Teamwork and human empowerment.
- Customer focus (expanded to all stakeholders).
- Ongoing improvement.

Baldrige Awards

The Malcolm Baldrige Award system was developed in the United States, and measures performance of an organization against a full range of business criteria. A sufficiently high score qualifies for the very prestigious "National Award."

The criteria, scoring and guidance are regularly updated and keep abreast of business trends. Organizations make submissions and highly trained evaluators score them against the criteria, and critique their business processes. Whilst significant organizational effort is required to prepare the submission, the overall process is generally seen as value adding, particularly in allowing organizations to benchmark and plan improvement. It is worth noting the following statement:

> "The Baldrige criteria and review process is not another thing to do; it is part of, and causes you to do, the only things you need to do." — *Earl A. Goode, President, GTE Directories Corp., 1994 Baldrige Award recipient.*

The following points encapsulate the core values promoted by the Baldrige criteria:

- Customer-driven quality
- Leadership
- Continuous improvement and learning
- Valuing employees
- Fast responses
- Design quality and prevention
- Long-range view of the future
- Management by fact
- Partnership development
- Public responsibility and citizenship
- Focus on results

The assessment criteria are:

1. Leadership (125 points)
2. Strategic planning (85 points)
3. Customer and market focus (85 points)
4. Information and analysis (85 points)
5. Human resource focus (85 points)
6. Process management (85 points)
7. Business (customer focus) results (450 points)

The ISO 9000 Standards

These international standards for quality management were first published in 1987 by ISO (The International Standards Organization) and updated for the second time in 2000. Derived from earlier national and military standards, they have had major international collaborative input over a couple of decades, and have been adopted very widely throughout the world.

A coordinated system operating in a large number of countries accredits qualifying bodies to carry out audits and award certification to ISO 9001 for organizations that use this standard as a model for their quality management systems. Currently, around half a million organizations worldwide are certified to ISO 9001. This demonstrates its flexibility as a base on which to build the quality management structures that are appropriate for and can be finely tuned to optimizing the performance of any human endeavor.

ISO 9001:2000 requires the organization to develop documented, controlled processes to achieve planned measurable objectives in its key business areas. Key elements are:

- Documented, controlled policies and procedures
- Strategic and detailed planning
- Quantifiable goals
- Management commitment
- Defined responsibility, authority, structure and communication
- Review of systems and outcomes by senior management
- Provision of appropriate resources and infrastructure
- Acquisition, training and development of people
- Planning, verifying and validating the design
- Design change control
- Monitoring, measuring and controlling the process
- Calibration
- Analysis of data
- Quality improvement and corrective action
- Auditing, and closing the loop

Quality in Space
Despite the best of intentions and often quite intensive quality assurance, errors occur which expose flaws in procedures or their application.

The hugely anticipated Ariane 5 rocket was ten years under development, and exploded 100 seconds after lift-off. Some fundamental quality errors were made.

The *Challenger* disaster rocked the foundations of NASA, exposing weaknesses in organizational responsibility and quality management. Hard lessons were learned from this, and today NASA applies sophisticated quality assurance processes, and has adopted the ISO 9001 approach.

To be found on NASA's web site is the following bold statement:

"On September 17, 1999, all of NASA Centers, NASA Headquarters, the Jet Propulsion Laboratory and all of NASA's Government Operated Facilities achieved ISO 9001 registration. With this achievement, NASA became the first Government Agency in the World (Federal or State) to have multiple sites under an ISO 9001 registration."

This was to the old ISO 9001:1994 Standard. The international Standard has been updated, and organizations worldwide are upgrading. Already Johnson Space Center and Marshall Space Flight Center have been certified to the new ISO 9001:2000 Standard.

With NASA committed to deploying high caliber and well-trained personnel, and to excellent quality management systems and programs, what then is the point of this paper?

It is a fact that in a great many instances, although there may be good programs, good procedures – there is not necessarily full uptake by all personnel. And, as such, opportunities for failure are created, opportunities for success are not grasped. So the point of this paper is to affirm and promote the approach that those first Martian pioneers, facing the challenges that they will, must be given the very best opportunity to work optimally and to solve problems. And hence must have sufficient training in quality management, including understanding of the principles and experience of practical application.

A hypothetical example will illustrate what I mean. I will have to treat it briefly and rather simplistically, but the principles are valid.

We will look at an early Martian Hab, with perhaps a complement of 10 people, including (for want of a better term) a Hab leader. They are six months into the current mission and it will be another six months before the next spacecraft arrives to add to or remove any crew. Some crew have started to become ill. First one, then another. Now four. It's a real worry; there seems no obvious reason.

Let us look at the different path taken by identical crews, with and without an overlay of sound quality management training and approach. (Refer Table 1).

Quality Attributes	Team has lapsed quality management	Team has good quality management
Leadership, teamwork and human empowerment	Feels commitment to solve the problem alone	Understands the importance of involving all, with lateral thinking, lateral input
Customer focus	Forgets about some stakeholders under stress	Keeps stakeholder interests and aspirations in mind
Ongoing improvement	No emphasis – locked into survival mode	Makes deliberate effort to find improvement opportunities
Good strategies, systems and procedures in place	Some gaps	No gaps, no overlaps
Strategies, systems and procedures: review and audit; equipment test	Minimum test regime followed; systems not actively reviewed for currency – accepted to be OK	Discipline and programme of review ensures it happens; audit and test schedule maintained
Good records kept, data reviewed and analysed.	Incomplete and minimal records kept.	The value of comprehensive records and data analysis understood.
Communication	Not good – deteriorates as problems grow	Good – the habit of communicating is there – people have the opportunity to contribute when they are informed.

Table 1. Team Response with and without entrenched Quality Management

So what happens?

In the first case, as people get sick, and the cause is not identified, the Leader becomes more withdrawn, has inadequate data to go on, and team dynamics for problem solving are not in play.

In the second case, the Leader is surrounded by empowered people, who contribute to the solution of the problem. One team member, contributing entirely from outside of her discipline, recalls working in a building once where people became mysteriously sick – later identified as Legionnaires disease.

This seems highly unlikely at first, but good records kept and accessed show that for a period piping was temporarily connected such that, with a failed valve a likely candidate, the hydroponics module could have contaminated piping in the main ventilation system. Bacteria resident there could be the source of the problem.

What's the point?

However simplistic the anecdote, it is quite clear that a culture and management discipline that supports the right hand column above has got a better chance of success than that reflected in the left hand column. That sort of culture and discipline does not happen of its own accord, but by design.

Thus, the people who first step out on Mars, and who have the immediate challenges of an alien environment to face with limited resources, need to have as good a grounding as possible in quality management principles, and be able to rely on well entrenched and integrated quality management systems.

Research and Education Pertaining to Mars Missions

Mars Desert Research Station Crew Rotation One – Survey Findings and Logistical Operations of an Analog Mars Exploration and Habitation Mission[†]

Troy Wegman, Mayo Clinic, Rochester, Minnesota
Jennifer Heldmann, University of Colorado, Boulder, Colorado
Heather Chluda, The Boeing Company-Rocketdyne, Canoga Park, California
and Steve McDaniel, Reactive Surfaces, Ltd., Austin, Texas

Abstract

The Mars Desert Research Station (MDRS) was occupied by its first full-simulation rotation from February 7-21, 2002. At this time, a full engineering, scientific, and public outreach mission was accomplished in spite of a continuing shakedown of the Hab and a mid-rotation crew composition / command change. The main scientific goals of this rotation included a comprehensive biology mission as well as an initial assessment of the geologic history of the terrain surrounding MDRS. "Alpha" successfully worked under engineering constraints imposed by simulated Martian conditions for the first time in MDRS history to provide both scientific and operational reports. Future Martian crews will likewise be subject to such logistical difficulties; therefore, the experiences of Alpha will be invaluable for future human missions.

Introduction

Mars Desert Research Station (MDRS) is an analog Mars habitation and exploration mission that is sponsored by the Mars Society and located in the southern Utah desert near Hanksville at 38° 24.38' N, 110° 47.51' W. The Hanksville site was chosen over other areas in the southwest US primarily due to the realistic simulation of crew isolation and the geologic similarities to the Red Planet. It is a minimally regulated, desolate area that is largely uninhabited such that there is little outside human interference to the mission. In addition to simulating the remoteness of a Mars outpost, the MDRS site is strikingly analogous to geologic conditions on Mars. Red Jurassic deposits surround the Habitat site in the form of canyons, chasms, hills, and outwash plains which simulate the terrains known to be present on Mars (Figure 1). Likewise, the desert environment of MDRS and lack of abundant macroscopic vegetation add to the Martian analog nature of this site.

During the mission, crew members reside in the "Hab," the enclosed mission living environment. The Hab's architecture resembles a cylindrical space module, similar to that of Robert Zubrin's Mars Direct mission plan, which could be flown on top of a Saturn V-style rocket[1] (Figure 2). The Hab is comprised of two levels. The upstairs quarters include private bunks, desk space, and the kitchen area, while the downstairs quarters contain the laboratory, bathroom, extravehicular activity (EVA) preparatory room, and airlocks.

MDRS is therefore a confined environment where novel stresses appear. Scientific operations would be much easier to complete without the added difficulty imposed by the simulated space mission, but a primary purpose of MDRS is to practice the method of space exploration even at the cost of final scientific return.

The priorities for Alpha and subsequent crews in order of decreasing importance were safety, simulation, science, and comfort. Crew safety is the first priority for any MDRS mission. Any necessary measures are taken to insure the safety of the entire crew. Simulation refers to the strict schedule and constraints of a space mission. The simulation of the mission was broken only if the simulation conditions compromised crew safety. For example, the removal of a space suit helmet is admissible if the helmet is obstructing one's view of dangerous terrain while riding an all-terrain vehicle (ATV). The scientific returns of the mission are the third priority. As long as the safety level of the crew is satisfactory and the experiments and/or fieldwork are being conducted in full simulation, the scientific investigations may be conducted. Crew comfort is the last priority for the MDRS simulated Mars mission. For example, meals were sometimes delayed to perform additional scientific reconnaissance, and the bulky space suit gear hindered communication, coordination, and balance during long, fatiguing EVAs. However, future Mars crews will have to work under similar conditions, and so the MDRS experience is a valuable experiment regarding crew operations and prioritization.

Figure 1. Martian-like terrain south of MDRS (Mars Desert Research Station) Habitat site.

All MDRS crews reported daily to Mission "Support" rather than Mission "Control." In contrast to the hierarchical structure of most current human space flight missions, a future Mars mission will most likely not be primarily run by a ground-based control team outlining the actions of the crew. This is not only due to the long data transmission delay between Earth and Mars, but also due to the need for adaptation to the environment, optimization of time and resources, and the arrival of unforeseen complications. MDRS crews worked through similar logistical difficulties that would have been impossible for colleagues on Earth to monitor and correct.

Figure 2. MDRS Habitat near Hanksville, Utah.

Alpha Crew

The first crew rotation of MDRS, named Alpha, commenced on February 7, 2002, and over a two-week period, selected volunteers lived and worked under simulated Martian conditions. Alpha was comprised of biologists Steve McDaniel and Troy Wegman, geologists Jennifer Heldmann and Andrew de Wet, aerospace engineers Heather Chluda and Robert Zubrin, chemist Tony Muscatello, and Frank Schubert, who led the design and construction of the project. Only six crew members participate in the mission at any given time, so for the Alpha rotation Tony Muscatello and Andrew de Wet replaced Robert Zubrin and Frank Schubert after one week. Robert Zubrin commanded the crew the first week and Tony Muscatello was commander of the second week.

The main purposes of this MDRS rotation were the following: 1) to learn what factors may be important on a Mars mission under constrained field operations; 2) to perform geological and biological field reconnaissance of the surrounding area, with subsequent laboratory analysis; 3) to jump-start the public and political outreach of a manned mission to Mars.

Landing on MDRS

The voyage to MDRS lasted about 24 hours for most of the crew, contrasted to the six-month journey to Mars. The Hab was nearly ready for occupants, as shakedown crews had prepared the living quarters. The Alpha crew began optimizing their environment immediately after entering the Hab by unpacking personal belongings, food rations, personal computers, laboratory supplies, etc. More rigorous cleaning of cabinets and floors was needed, especially in the laboratory area, where an expensive microscope on loan was to be placed. Dust can wreak havoc upon microscopic optical lenses. Some additional laboratory cabinets were installed to store laboratory reagents and supplies. The HabCom computer system was activated for communication with Mission Support, and personal computers were set up. Other crucial systems such as electricity supplied by a gasoline-powered electric generator, heat supplied by a propane-powered furnace, water supplied by a holding tank pump system, and gray water and solid waste disposal supplied by a leach field and incinerating toilet were tested for functionality. Of course, the red, green, and blue Martian flag was raised outside, and a weather monitoring station was installed.

General Daily Schedule

The commander set up a daily schedule at the first morning meeting. This schedule was to represent the simulation of an actual Mars mission. Upon waking, the crew was to eat breakfast and be ready for a 9 a.m. meeting. The goal of this meeting was to plan the EVA and other activities for the day. It was suggested that three to four people suit up for EVA each day, with more or less participating on certain days if needed. Those remaining at the Hab would serve as CapCom or perform maintenance or laboratory tasks. EVA was planned for 11 a.m. to 3 p.m. Crew members on EVA were to make sure their stomachs were full before leaving at 11 a.m. because no food was available on EVA. At 4 p.m., an EVA debriefing was scheduled to generate and send a journalistic voice report to Mission Support via email. Mission Support was Mars Society volunteers from the Denver and San Diego chapters. After the debriefing, the evening meal would be prepared and eaten. After the meal, reports would be written on laptop computers and sent via email to Mission Support. If there was extra time in the evening, an organized social activity was encouraged. The crew planned to sleep around midnight.

In actuality, the schedule varied each day. The EVAs often ran longer than four hours in the afternoon. When national media personnel visited the Hab and shot footage, this interrupted the daily schedule and EVAs were often longer to accommodate cameramen. After EVA, the crew usually replenished their energy supply with snack foods before the evening meal, which delayed the debriefing. Because of power supply limitations, cooking the evening meal often lasted twice as long because appliances couldn't be run concurrently due to power outages. Meal clean-up time varied in length and delayed the writing of the evening reports. The crew put a tremendous amount of effort writing detailed reports and often didn't go to sleep until after 1:00 a.m. Compounded by data transmission problems and limited use of computer drives, the crew often spent hours uploading and emailing files to Mission Support. These circumstances also unfortunately led to later wake-up times for some crew, which then affected upcoming daily activities. Despite these deviations from the schedule, the crew quickly adapted, primarily due to excellent command leadership. They often delayed eating, postponed sleeping, and skipped precious email time to "Earth." Alpha was extremely focused on its tasks and responsibilities.

This may be appropriate for a two-week rotation but may not be an optimum environment for long-term productivity and stability. In actuality, current space missions are planned nearly to the second. Standard operating procedures and strict schedules dictate nearly all aspects of a mission. A real mission to Mars, however, may not be run in such a manner. As mentioned previously, regardless of pre-planning and protocols developed for a Mars mission, if humans are to live on the planet for an extended time period, there will be adjustments and deviations from scheduling. On Mars, even if colleagues dictate activities, the crew will need to possess extraordinary patience waiting minutes during radio transmission delays for commands before progressing. If future missions are practiced in environments such as MDRS, these situations can be addressed before crews travel to Mars, and then limited interaction time between astronauts and Mission Support may be used most effectively.

Operations / Engineering

Significant contributing factors to the lack of a strict schedule at MDRS were operations and engineering-related factors. Not unlike current space missions, significant man-hours were devoted to maintenance, repair and improvement of Hab systems. During the Alpha rotation, water supply was an important issue. Water is a most valuable resource on a mission to Mars. Therefore, water conservation is crucial to the mission's success. Alpha's goal was to limit water consumption to 4-5 gallons/person/day. Water was not run any longer than needed. Showers were limited to Navy showers at four-day intervals with sponge baths in between. The average daily water use in the rotation was calculated at 3.8 gallons/person/day.

The goal of the MDRS water system is for gray water (drain water and urine) to be recycled and purified at 90% efficiency by barley and wheat plants grown in the greenhouse. Because the recycling system was not functional during the Alpha rotation, water was manually pumped into a leach field approximately every three days. Fresh water was supplied by an external reservoir that was pumped to a holding container near the top of the Hab. Due to water freezing in the external reservoir valve and a broken water pump, the crew hauled water through the Hab to the internal holding container, which resulted in simulation breaks. New gamma ray spectroscopy data from *Mars Odyssey* indicates high amounts of hydrogen, presumably indicative of large quantities of Martian subsurface ice.[2] Despite this discovery, data on Mars mission water conservation will be sought until its production is guaranteed in sufficient quantities or its discovery beneath the surface soon after landing on the planet.

Another crucial element in Hab operations was power supply. A 7 kW gasoline-powered electric generator supplied power, and it needed to be refilled every 12 hours. Since this was a potential hazardous spill situation, Alpha broke simulation by not wearing their suits to refill the generator. Inside the Hab, the crew was advised to conserve power and turn off all equipment when not in use, especially laptop computers. The crew soon learned the importance of this during a Mars mission when the current draw exceeded the circuit breaker rating of the generator. The crew could not use too much power at one time, which meant alternating the use of critical items such as the solid waste disposal mechanism, water heater, hot plates for cooking, and computers for data transmission. These situations altered the planned activities of the crew. Eventually the crew rewired part of the Hab and installed a second circuit line to prevent power outages. While it is unlikely that a Mars crew would have an energy supply which needs constant surveillance, it is likely they will have to respond similarly when similar energy consumption and preservation situations occur.

Human waste disposal was also a key issue in the mission. A machine similar in appearance to a toilet incinerated solid human waste. However, ash matter still remained that needed removal every couple of days. This task was the least favorite among all operations and engineering tasks and could have created an unpleasant living situation due to waste buildup. Indeed, a "space janitor" may be the most important crew member on Mars. However, everyone will share in such tasks, to maintain crew harmony.

Many hours were spent trying to establish faster data transmission and an intra-Hab computer network. Internet capability was derived from a Starband satellite system. Originally when transmitting data to Mission Support, this process was time consuming. Uploading an 800 Kb photo for emailing took about ten minutes. Many crew members sacrificed sleep to transmit data. Eventually advice was received from Mission Support on changing the polarity of the Starband and checking for unnecessary memory-hogging programs. This resulted in faster data transmission rates. Presumably, on Mars, the delay or loss of ability to communicate with fellow humans millions of miles away for unknown amounts of time will cause severe stress on the crew.

Within the two-week rotation, there were other items that needed repair, modification or installment. Some of these included stairwell support installation after a fall accident, repair of lab stools, and repair of a hinge on an airlock door. Ingenuity was required to use on-hand materials, often scraps from construction. Near the end of the rotation, a crucial improvement was made. The engineering and construction of a sound barrier and aesthetic rock wall muffled the generator noise and hid gasoline barrels. This demonstrated the engineering and teamwork skills of the crew. Again, materials found on-site, including flat boulders, were used to construct these structures. While the resources on Mars may be different from those available on Earth, the ability of a Mars crew to work together and create a functional piece of equipment from preexisting natural materials will only enhance the mission. As R. Zubrin has stated, the ability to "live off the land" will be crucial because limited materials can be transported to Mars.

Other operational factors were environmentally based. One obstacle the crew had to overcome was the strong desert winds. Winds were at times in excess of 50 mph. This usually just created loud noise in the Hab, but a situation occurred during the first scheduled EVA on February 8th. During the planning of the first EVA on that day wind strength was not considered dangerous. However, a portion of the Hab roof nearly blew off during EVA. With those in the Hab concentrating on fixing this problem, those on EVA may have been cut off from communications during the crisis. Those remaining in the Hab eventually replaced and improved the roof, and no EVA problems transpired. Also during this windstorm, damage was incurred on the greenhouse, which also then needed repair. On Mars it is unlikely that the forces of the wind would cause significant damage because of the low pressure of the atmosphere, but dust storms can create numerous problems. It is likely that there would be operating procedures for emergencies on Mars, but these can't be planned for every scenario.

Extravehicular Activities

EVAs (extravehicular activities) were performed each day in which crew exited the Hab in simulated space suits to survey the surrounding terrain and look for past or present life. Data and samples were collected and returned to the Hab for analysis and reports.

Weather conditions never caused cancellation of a scheduled EVA, and all EVAs were completed before sunset. Visual acuity decreases as darkness increases, which consequently affects travel safety.

The process of EVA began with donning of space suit gear. Although the suits were not genuine astronaut gear, they were very similar in appearance. A helmet was connected to a backpack, which provided fresh circulating air. The helmet also contained a bite valve connected to water supply source for rehydration. A communication system, comprised of a push-to-talk two-way radio with corresponding earphone and microphone, was mounted inside the helmet. US Army boots, gloves, gaiters, as well as a full-body canvas suit were worn. The gear provided a regulated environmental enclosure needed on Mars without becoming overbearingly heavy and burdensome in Earth's gravity. The time needed to put on the gear decreased as the crew became more efficient in the process, but average suit-up time was 45 minutes over the two-week period. Cross-checks were performed on all systems and personal equipment to assure functionality before exiting the Hab. The EVA crew then simulated decompression in the airlock for five minutes. The crew took careful EVA preparation and cross-checks for granted after the first few EVAs. This kind of lax behavior on Mars could lead to emergency situations.

Once outside the airlock, the team discovered performing ordinarily mundane tasks in the gear was very challenging. Dexterity was all but eliminated with large gloves. When Alpha tried to operate electronic devices such as cameras or GPS (global positioning satellite) units, they often dropped them or took many minutes to complete normally easy tasks. Even writing with pens was challenging. The crew found that kitchen sandwich bags used for sample collection could not be opened without the use of a tool. These obstacles all affected data collection during EVA, so methods were altered to attain efficiency. For example, tips of rock hammers were used to press buttons on electronic equipment. During the middle of the crew rotation, scrap pieces of cardboard were taken on EVA to record observations by writing on the cardboard with thick markers. The crew attempted various methods of EVA documentation including voice-activated recorders, but the cardboard method, surprisingly, worked best. Nonverbal communication during communication losses sufficed, and manual assistance for fellow crew members with respect to radio restoration proved beneficial.

An environmental factor which contributed to many EVA challenges was wind and subsequent blowing dust and sand. During the first walking EVA, winds were in excess of 50 mph. It was very difficult to keep documentation materials in hand. In addition, the wind reverberated inside helmets and interfered with radio communications. Dust and sand also blew into crucial components of digital cameras and eventually caused equipment failure. The crew dealt with winds by finding better ways to attach, use, and protect their documentation materials, by communicating with hand signals, and by finding wind-protected areas to communicate. Strong force winds may not a detriment on Mars, but dust will undoubtedly affect equipment. Training with optimized equipment in an environment such as MDRS will help future Mars crews prepare for similar conditions.

Navigational competence was required to successfully perform data collection and safely travel through varied terrain. Hand-held battery-operated GPS units were used to mark locations, called waypoints, for future crews. A topographical map was also used for route determination. Only a handful of times did the crew become briefly lost. Once, when barely making it back to the Hab before sunset, a crew member started to panic even though GPS was functional and the lead navigator was confident in his abilities. On Mars, an independent navigational system wouldn't include GPS technology. Such situations there could be dangerous and life threatening.

Communications between EVA crew members, and also with CapCom, were often disrupted, leading to frustration. When helmet communication would cease, crew members could still loudly speak through the helmet. However, when the crew lost communication with the Hab – due to distance, elevation or geographical blockages to the radio repeater – there could have been an emergency on EVA in which HabCom would not be able to assist. This is not an unrealistic possibility on Mars. The crew needs to know and define its limits before going on EVA and be able to deal with medical emergencies in the field, despite one's primary educational training.

Besides communication problems, other challenges included the physical exertion of EVA. The crew was quite physically fit, but the added gear and low airflow exchange through the helmet created moments when imminent rest was required. The crew members also had to maintain endurance with no sustenance besides water on EVAs that lasted up to six hours. A crew member adapted to this situation by attaching a nutritional bar inside her helmet with duct tape. This food provided additional energy during tiring EVAs. On Mars missions, the crew's diet will most likely be optimized for maximum energy, but many typical Earth foods, including animal proteins, will be limited. Adaptation to a new diet will be necessary before leaving Earth.

The crew was spared energy as all-terrain vehicles (ATVs) were used during EVA as a means for providing ease of movement and transportation (Figure 3). These small gasoline-powered vehicles can be thought of as small Mars rovers that are different in their energy source. ATVs can quickly travel across terrain that is not passable via other methods, provided that the operator is skilled in its use. ATVs were extremely helpful on EVA as the crew traveled long distances, through gullies, over hills, and allowed the crew to use time more productively. However, some crew members were not as skilled as others in their driving skills and could have become hurt. Alpha often worked in pairs to push ATVs up embankments or out of canyons to reach destinations. Future Mars crews should train with vehicles in environments similar to Mars since it takes practice to know the capabilities of a machine.

Biology Mission

A main focus on a journey to Mars will be the search for past or present evidence of life. Due to the harsh environment it is extremely unlikely that there are higher life forms on Mars. There are, however, numerous micro-organisms that flourish in harsh conditions, including the deepest, coldest, hottest, and darkest places on Earth. With the possibility of water on Mars and proposed fossilized microorganisms in the Mars meteorite ALH84001[3], evidence of past or present life on Mars may exist. In the past, landers and rovers have been sent to Mars but no definitive evidence has been found. If there is evidence of life somewhere on the planet, it is not easy to find, and without sending humans to Mars the answer may

Figure 3. Crew members on EVA (extravehicular activity) with suit gear and ATVs (all-terrain vehicles).

never be known. At MDRS, a biology mission was planned in which human labor constituted the search for life.

The biologists on the Alpha rotation were responsible for planning the mission and for the procurement of laboratory supplies needed to fulfill the mission. In a sense, this represented writing a proposal, being funded, obtaining materials, and then performing research, which are common science practices. The mission was divided into three components: primary, secondary, and tertiary. These indicate a decreasing order of importance, such that the crew accomplished the first component before starting the next. The primary mission was to locate and collect samples on EVA of biological interest, primarily rocks that harbor microorganisms. The secondary mission was the laboratory analysis of the samples, which included microscopy and Gram stains. The tertiary mission was essentially an extension of the secondary mission as it also included laboratory analysis. It consisted of screening for an enzyme called organophosphorus hydrolase (OPH). Because of limited laboratory resources during the first rotation, this simple and inexpensive enzyme test served as a means of screening for life.

Background

The biologists had little prior knowledge of microorganisms associated with rocks, but shakedown crews had scouted the area and determined their presence at the MDRS site. Rock-inhabiting microorganisms are described as lithobiontic. There are different types of lithobiontic habitats. Hypolithic or sublithic refers to the undersurfaces of rocks. A chasmolithic habitat refers to fissures or cracks in rocks. Epilithic refers to the exposed rock surface. Endolithic refers to the inside or internal spaces of rocks. These microorganisms are widespread in nature and grow in different environments.[4]

It is known that primitive microorganisms called cyanobacteria are often associated with rocks. Direct evidence of cyanobacterial activity dates back 2.5 billion years, where unique molecular fossils (biomarkers or biomarker signatures) were detected in sandstone rock from Western Australia. Indirect evidence dates back 3.5 billion years.[5] Therefore, these organisms represent some of the earliest life forms on Earth. Because of this and their ability to colonize extreme environments, cyanobacteria represent life forms that could survive on Mars given the proper conditions.

Endolithic microorganisms have been found in hot and polar deserts around the world. In hot deserts, prokaryotes are primarily found, whereas eukaryotes are primarily found in polar deserts. Bacteria and photosynthetic cyanobacteria are prokaryotes. Examples of eukaryotic organisms include lichens and algae. Certain strains of lithobiontic cyanobacteria are able to tolerate desiccation and resume photosynthesis rapidly after being exposed to water.

Cyanobacteria in hot deserts are usually members of the genus *Chroococcidiopsis*. These microorganisms typically form a green layer a few millimeters below and parallel to the rock surface.[4] *Chroococcidiopsis* also survives exposure to high levels of ionizing radiation, and it or similar organisms may have been involved in our planet's early stage photosynthetic activity.[6] Because of limited radiation protection from the Martian atmosphere, a radiation-resistant organism similar to *Chroococcidiopsis* could constitute life on the planet's surface.

Primary Mission

Collection of samples of biological interest was initially focused on desert soil, rocks, and scarce ice water found in canyons (Figure 4). One of the first encounters with lithic organisms was lichen distribution on large boulders.

Lichens, however, were not of interest as they are common in desert and arctic environments, and organisms are unlikely to be found on the surface of Mars. In the field, it was impossible to ensure sterility during collection of the samples, but the crew focused on cross-contamination prevention techniques. Samples were dislodged in the field with rock hammers and pliers and placed inside the plastic kitchen storage bags. As mentioned previously, this often required a lot of effort with bulky gloves, as did every aspect of performing biological fieldwork. Teamwork was very crucial. For example, the biologists often consulted with the geologists on rock compositions and formations at various sites.

Figure 4. Biology Primary Mission: EVA collection of samples.

The collection of samples eventually was limited to rocks possessing visible layers of discoloration which were not inherent properties of the rock. These layers, most often green in color and located a few millimeters beneath the surface, were typically found in medium-grain sedimentary rocks. Some of these rocks exhibited shiny, dark deposits on their exterior surfaces. The interior of these rocks did not contain these deposits. After the first few EVAs, most crew members were capable of locating these samples.

On return to the Hab, an Internet search was performed and a phenomenon that matched the description of the shiny coating was discovered. Desert varnish is a shiny, "painted" appearance on the surface of rocks in desert environments and composed of clays and manganese and iron oxides. These oxides are formed by varnish microorganisms during ATP production and contribute to the black, brown, and red colors of desert varnish (ATP – adenosine triphosphate – is the molecule present in living systems which stores and transforms energy into a form that the organism can use). Desert varnish, however, can form without the assistance of microorganisms and thus can be a purely chemical phenomenon.[7] Therefore, it is not an accurate marker of life. The EVA samples were divided, organized, and a portion of every one was preserved in 2% glutaraldehyde so that future crews could perform special stains as part of their individual research interests.

Secondary Mission

One piece of laboratory equipment that was crucial to the analysis of specimens was an Olympus microscope. Examination of EVA samples under magnification comprised the biological secondary mission. The samples were not cultured for microorganisms because no growth medium was available. In addition, it is difficult to keep the media and samples sterile and free from cross-contamination throughout the collection process and during manipulation in the laboratory. In addition, many environmental organisms require special growth media and incubation temperatures or just cannot be cultured. Instead, one must rely on other methods such as PCR (Polymerase Chain Reaction), enzyme analysis, or DNA sequencing to identify organisms. Often this employs 16S Ribosomal RNA gene sequencing (extracting and comparing bits of RNA), which was not a capability in the Hab laboratory. Also, PCR applications require clean environments to prevent false-positive reactions. There is undoubtedly interest among astrobiologists on the development of field detection and identification assays that could be used on Mars.

EVA samples were manipulated for microscopy with flame-sterilized tools in order to prevent cross-contamination of samples. Wet mounts were prepared by dislodging pieces and crushing them with tools into clean microcentrifuge tubes. Sterile water was added, and the samples were vortexed and then centrifuged. Fractions were removed, placed onto slides, and visualized under 400X and 1000X magnification. Samples were also prepared by scraping green areas of rock with a flame-sterilized needle onto a slide with 50% glycerol. Gram stains were also made to distinguish between bacteria and cyanobacteria in the samples.

Based on prior knowledge of lithic organisms, it was predicted that algae, cyanobacteria, and bacteria would be found upon microscopic analysis of the samples. In numerous samples, there were green- or yellow-colored cells which were highly defined in shape and structure. These were presumably cyanobacteria or algae. In particular, in numerous desert varnish samples there was a green layer a few millimeters below the rock surface that contained cells consistent with *Chroococcidiopsis*. Although culture or other methods never confirmed the identity, this organism is typically found in desert environments, as previously mentioned.

Some structures in wet mount analysis may have been bacteria, but Gram stain results revealed structures inconsistent with typical bacteria and no spores. Some of these structures did stain Gram positive, but without culture and isolation they could not be determined as living or not.

Tertiary Mission

Once microscopy was completed, some of the samples were screened for the enzyme organophosphorus hydrolase (OPH). OPH has been found in certain soil microorganisms and can be detected easily using a colorimetric substrate. It also is potentially useful in the detoxification of nerve gases. Therefore, the biology mission included not only a simple test for detecting microbiological activity, as one would on a Mars mission, but also a search for extremophilic microorganisms for industrial purposes.

The use of OPH as an environmentally sound nerve gas decontamination solution is a relatively recent discovery. Enzymes existed that were capable of detoxifying certain pesticides, but it was not known how to extract them from microorganisms. In the early 1980's, the gene encoding OPH was cloned from soil bacteria.[8] At this time, it was not known that OPH could detoxify some of the most noxious nerve agents. Study of the OPH enzyme opened the door to engineering of its activity for use as a nerve agent decontamination solution. Nerve agents, such as sarin and Russian-VX, are organophosphorus compounds. OPH hydrolyzes and subsequently inactivates these compounds.[9,10] Because the products of the OPH detoxification are water soluble, surfaces can be washed free of the compounds. OPH remains able to detoxify so long as its catalytic integrity is maintained and can be added to substances including fire-fighting foams.[11]

Paraoxon is a non-toxic and colorless oxygen analog of the organophosphorus insecticide parathion. Paraoxon is hydrolyzed by OPH into para-nitrophenol and diethylphosphate.[12] Para-nitrophenol is yellow in color. This reaction permits the use of paraoxon for OPH detection. When microorganisms containing OPH in their periplasm contact paraoxon under appropriate conditions, a yellow color is visible.

Once this enzymatic test was optimized, the screening of EVA samples began. A number of samples turned yellow compared to negative controls, indicating possible OPH activity. Many of these were desert varnish samples. However, during this rotation it was not determined if the activity that produced the yellow color was of biological or chemical origin. Later in the MDRS season, S. McDaniel and Penelope Boston on the sixth crew rotation determined by using an autoclave that positive yellow reactions were due to biological activity. An autoclave is a machine that is known to abolish all biological activity. Preliminary OPH positives were divided into two equal portions; one was autoclaved, while the other was not. After the autoclaved portions had cooled, OPH testing commenced. Yellow color was abolished in all autoclaved portions but not in non-autoclaved portions. Thus, the yellow color was indicative of OPH biological activity. Since there were also microorganisms present upon microscopy, the samples were eventually cultured for their isolation and characterization.

There is no natural-occurring substrate for OPH. Thus its function in microorganisms is unknown. There is speculation on the presence of OPH activity and conferral of a possible survival advantage to microorganisms. It may be possible that organophosphate-like xenobiotics, or chemicals that have a biological effect, are generated and degraded by OPH or similar novel enzymes as a defense mechanism between microorganisms

Alpha was surprised by the abundance of microscopic life throughout the desert. Although human labor may be crucial to the discovery of life on Mars, a different search perspective may be required. Life on Mars may not follow the principles of life on Earth, and we may be blind to its existence if we only look for what we know.

Geological Survey

Background

One of the most important contributions of a human mission to Mars will be the in situ geologic study of the Martian surface. The geological record of Mars is critical for providing clues to the history of Martian geologic evolution, as well as providing insights into geological processes of other planets via comparative planetology. Currently, our knowledge of Martian geology is derived from ground-based and orbital reconnaissance, as well as from three separate robotic landers (*Viking 1*, *Viking 2* and *Pathfinder*). Experiments in human planetary exploration such as MDRS show the immense value of sending humans to Mars due to the tremendous amounts of information that can be gathered by a field geologist compared to the quality and quantity of data collected via robotic means.

During the Alpha rotation of MDRS, a great deal of the geology surrounding the Hab was explored and documented. Since terrain at the Martian landing site was entirely new, each EVA targeted a specific type of region for intense geologic examination. Sites varied in terms of geologic setting including (but not limited to) canyons, plains, wash areas, hills, channel beds, etc. Locations studied in detail were marked as waypoints (using the GPS to get exact location coordinates), and the sites were described in terms of large-scale land form features as well as small scale rock characteristics. Each site was characterized via sample collection, digital imaging, and a narrative description, documented in individual text files sorted by waypoint. Samples are stored in the MDRS laboratory, and all digital images are stored on the HabCom computer. This initial reconnaissance of the surrounding terrain gave a fairly comprehensive representation of the various environments available for study, and future MDRS crews can use this information for planning more targeted studies in various areas depending on the scientific goals of their mission. At the end of the Alpha rotation, all data collected was combined to determine the geologic history of the landscape surrounding the MDRS Hab.

General Findings

This initial geologic survey clearly showed that the land surrounding the Hab was once a shallow marine environment. The rocks were almost exclusively sedimentary, indicating deposition in a fluvial environment. The few igneous rocks that were found originated in the nearby laccolithic Henry Mountains. These mountains were originally formed as igneous intrusions which caused the overlying sedimentary layers to rise and dome up over the intruding magma. Eventually the sedimentary layers eroded away, and now exposed igneous rock is eroded and transported to lower elevations throughout the region.

Undoubtedly, sandstones and conglomerates were the dominant rock types. Both of these rock types indicate past water activity in the region. A shallow sea must have once covered the area for the deposition of the relatively thick layers of these rock types. The waters were not always quiescent, however, as dramatically evidenced by some of the conglomerate outcrops. Some of the clasts cemented into the rock at certain waypoints are rather large (up to several inches in diameter), which indicated that the water must have been moving at significant velocities to transport such large particles.

The presence of large amounts of gypsum in the canyons also attested to the presence of a former sea. As the waters evaporated, the gypsum layers were left as deposits. Horizontal and cross-cutting gypsum layers have been discovered, and the widespread nature of such deposits attests to the extent of the past sea. Fossil findings also indicated a past marine environment. The tremendous amount of mollusk shells found in several "fossil fields," as well as an ammonite shell found at Lith Canyon, required a past marine environment.

Therefore, the overall general past geologic history of the region was determined, and future crews could examine sites of interest. These comprehensive descriptions represent more information than surface mapping or robots could provide and process in two weeks. The successful exploration and documentation of the MDRS site was extremely important in future crew's research decisions.

Summary

As the first operational crew at MDRS, Alpha successfully adapted to their new living environment and accomplished predetermined scientific goals. Information was gathered on important factors which will be important on a Mars mission such as crew priorities, scheduling, command hierarchy, and engineering logistics, among others. Geological and biological analysis in both the field and the laboratory were successfully completed. News media visited the Hab and showed millions around the world that a mission to Mars is an achievable goal if efforts like MDRS are initiated.

Alpha was comprised of a mixture of people of different sex, personality types, educational background, and age. Despite most of the crew never meeting prior to the mission, there was excellent camaraderie from the beginning. Focus, hard work, and persistence were mixed with relaxation and humor. Half of Alpha was composed of three people in their 20's with the others in the 30-50 age bracket. This was beneficial as the younger crew's sense of adventurousness was mixed with the older crew's caution. There were four men and two women on the crew, and this ratio worked well. Despite a change in crew leadership and composition mid-way through the rotation, the crew switched roles if needed. For example, a biologist focused on engineering and operations tasks after the crew changeover while still contributing to the biology mission. An engineer was also involved in the geology and biology fieldwork. Others assisted in the engineering of the barrier walls. Multitasking led to the success of the mission when complications arose.

Despite specific engineering and scientific goals, there was an appropriate amount of flexibility throughout the mission. This was required, as mission simulation breaks and unforeseen circumstances often arise when conducting a mission such as MDRS. Despite these additional complications and having no prior MDRS experience, a complete biological and geological survey mission was accomplished while performing operational and engineering tasks. Alpha's scientific mission may have been the most comprehensive of the four month season because more microscopy and geological surveying was performed than during other, subsequent rotations.

On this MDRS rotation, there were many deviations from how an actual Mars mission would be run, especially since the Hab was still in partial shakedown mode. Most Hab systems needed optimization. However, the Hab or its equivalent will always be a work in progress until humans are launched to Mars. MDRS is a first step in learning about the human exploration and habitation of Mars. Practicing in an environment similar to Mars can only increase the likelihood of success during a Mars mission. Lessons learned on the Alpha rotation have been communicated to Mars Society members supervising the Euro-MARS project. Euro-MARS is the European equivalent of MDRS and will become operational in Iceland in 2003. Another Mars Society Hab is being planned in Australia. Eventually, it is likely that future Mars crews will train in environments similar to what the Mars Society has initiated.

References

1. Zubrin, R., with R. Wagner. 1996. *The Case for Mars: The Plan to Settle the Red Planet and Why We Must.* Simon and Schuster, New York.
2. Boynton, W. V., Feldman, W.C., Squyres, S. W., et al. 2002. *Distribution of Hydrogen in the Near Surface of Mars: Evidence for Subsurface Ice Deposits. Science.* 297:81-5.
3. McKay, D. S., Gibson Jr., E. K., Thomas-Keprta, K. L., et al. 1996. *Search for Past Life on Mars: Possible Relic Biogenic Activity in Martian Meteorite ALH84001. Science.* 273:924-930.
4. Friedmann, E.I., and R.O. Friedmann. 1984. *Endolithic Microorganisms in Extreme Dry Environments: Analysis of a Lithobiontic Microbial Habitat,* p.177-185. *In* M. J. Klug and C.A. Reddy (eds.), Current Perspectives in Microbial Ecology, Proceedings of the Third International Symposium on Microbial Ecology, Michigan State University, 7-12 August 1983. American Society for Microbiology, Washington, D.C.
5. Summons, R.E., Jahnke, L.L., Hope, J.M. and G.A. Logan. 1999. *2-Methylhopanoids as Biomarkers for Cyanobacterial Oxygenic Photosynthesis. Nature.* 400:554-557.
6. Friedmann, E. I., Hofer, K.G., Billi, D., Grilli-Caiola, M., and R.O. Friedmann. *Ionizing-Radiation Resistance in the Desiccation-Tolerant Cyanobacterium Chroococcidiopsis.* Abstract of the First Astrobiology Science Conference. 3-5 April 2000. NASA Ames Research Center, Moffett Field, California.
7. Dorn, R.I., and T.M. Oberlander. 1981. *Microbial Origin of Desert Varnish. Science.* 213:1245-1247.
8. McDaniel, C.S., Harper, L.L., and J.R. Wild. 1988. *Cloning and Sequencing of a Plasmid-Borne Gene* (opd) *Encoding a Phosphotriesterase.* J. Bacteriol. 170:2306-2311.
9. Dumas, D.P., Durst, H.D., Landis, W.G., Raushel, F.M., and J.R. Wild. 1990. *Inactivation of Organophosphorus Nerve Agents by the Phosphotriesterase from Pseudomonas diminuta.* Arch. Bioch. Biophys. 277:155-159.
10. Rastogi, V.K., DeFrank, J.J., Cheng, T.C., and J.R. Wild. 1997. *Enzymatic Hydrolysis of Russian-VX by Organophosphorus Hydrolase.* Biochem. Bioph. Res. Co. 241:294-296.
11. LeJeune, K.E., Wild, J.R., and A.J. Russell. 1998. *Nerve Agents Degraded by Enzymatic Foams. Nature.* 395:27-28.
12. Dumas, D.P., Caldwell, S.R., Wild, J.R., and F.M. Raushel. 1989. *Purification and Properties of the Phosphotriesterase from Pseudomonas diminuta.* J. Biol. Chem. 264: 19659-19665.

//

A Closed Mars Analog Simulation – The Approach of Crew 5 at the Mars Desert Research Station†

William J. Clancey
NASA-Ames Research Center
and University of West Florida
bclancey@arc.nasa.gov

Abstract

For twelve days in April 2002 we performed a closed simulation in the Mars Desert Research Station, isolated from other people, as on Mars, while performing systematic surface exploration and life support chores. Email provided our only means of contact; no phone or radio conversations were possible. All mission-related messages were mediated by a remote mission support team. This protocol enabled a systematic and controlled study of crew activities, scheduling and use of space. The analysis presented here focuses on two questions: Where did the time go – why did people feel rushed and unable to complete their work? How can we measure and model productivity, to compare habitat designs, schedules, roles, and tools? Analysis suggests that a simple scheduling change – having lunch and dinner earlier, plus eliminating afternoon meetings – increased the available productive time by 41%.

Highlights: Protocol And Activities

The fifth crew occupied the Mars Desert Research Station (MDRS) near Hanksville, Utah for two weeks in April 2002. With the author serving as the commander, the MDRS5 crew included Andrea Fori, Jan Osburg, Vladimir Pletser, David Real, and Nancy Wood. This group was selected from over 200 applicants to provide a balance of science, operations, and reporting expertise. The MDRS facility is very similar to the Flashline Mars Arctic Research Station (Clancey, 2000b, 2001b) designed as an analog of a Mars surface habitat. MDRS includes six private staterooms, galley, workstations, and meeting / eating area on the upper deck, with a laboratory, toilet, shower and EVA preparation rooms on the lower deck.

After a few hours of overlap with the previous crew for orientation, we devoted the first day to setting up equipment and working with Mars Society volunteers to improve the station. The second day, Monday, April 8, began a twelve day period of the formal simulation. During this period, we were alone, except for two short visits by a contractor resupplying fuel and water. All other conversations with humanity were restricted to email, with all mission-oriented (non-personal) communications restricted further to a single point of contact (a member of mission support staffed by the Northern California Mars Society) serving as CapCom ("capsule communicator," a NASA term stemming from the Mercury program of the 1960s). All reports, requests, and assistance was first directed through CapCom. By protocol, after a secondary contact was established (e.g., someone to advise our work on the greenhouse), further conversations on the same topic were not mediated by CapCom, but were always copied to him.

We organized two special activities:

- A simulated multiple-failure EVA situation (Appendix 2).
- A full-day open house for the international press. (German: ARD TV, RTL TV, and Der Spiegel; American: TechTV, Phoenix Fox-10 TV; Swiss: FACTS; Norwegian: Dagbladet Daily; and British: Sunday Telegraph of London).

Consequently the two-week simulation period had three special days before the Sunday of departure: a day of rest (when we conducted the simulated EVA failures), the open house on Saturday before departure, and the clean-up day before the open house. In sum, there were ten actual simulation days, Monday-Friday of the first week and Saturday, plus Monday-Thursday of the second week. This distinction turns out to be very useful for analyzing the group's behavior.

The crew investigated the following issues during their stay:

- If there is life on Mars, how do you take a soil or rock sample that includes it?
- "Expedition memory": Can a geologist understand the work performed by previous rotations to develop a geology primer of the region?
- What is the effect of chores (life support maintenance) on science productivity?
- What are the psychological effects of growing plants in the Hab?
- How do plans develop and change during the mission? How do individual and group activities interact?
- How can Earth's mission support understand and assist Mars surface exploration? Can possible EVA targets and routes be suggested using reports from previous crews?
- How is public and private space used? How can the Habitat's layout be improved?

Methods And Data Analysis

To address questions about productivity, planning, and layout, the following data were collected:

- Time lapse video of upper deck throughout the rotation (every 3 seconds, 320 x 240 pixels) from 730 a.m. until midnight (or on several occasions until everyone is asleep).
- Video recording with sound of all planning meetings.
- Log of crew location and activities every 15 minutes on two consecutive days ("snaplist").
- Personal crew logs of awake and sleep times, plus time devoted to being the cook for the day ("director of galley operations" or DGO).
- Written plan of daily proposed and deferred tasks in a table by date and person.
- Written crew ("post-occupation") surveys.
- 97 reports posted on the web, with completion dates, including commander's daily log and health and safety officer's (HSO) daily reports.
- Approximately 1,000 time-stamped digital photographs.

Effect Of Scheduling On Productivity

To address the question of "where does the time go," we need to consider what time is available, subtracting sleep time, group activities (especially meals and EVA operations), and unscheduled interruptions (especially power failures).

To begin, chores were formally assigned to strictly share the work on a rotating basis (Table 1). The actual time devoted to cooking and cleaning up varied between approximately 200 minutes and 350 minutes per day, with an average of 4 hours 23 minutes (Figure 1). The two people assigned to refueling the generator spent about 45 minutes a day (EOP and EOA, usually 3 times/day). Sleep time varied between 8:09 and 6:21 (h:mm, Figure 2). Duration is most strongly affected by the late evening movies and required attendance at the 9 a.m. planning meeting. Note how sleep duration changes together for the group (4/14 was the Sunday without a meeting).

The extent of group activities, including meals, meetings, movies, etc., can be determined within a few seconds accuracy from the time lapse record. The camera was placed on a tripod (Figure 3), using a wide-angle lens. As previously reported (Clancey 1999, 2000a, 2001a), the frames are examined manually, creating a spreadsheet of events and start / stop times, which is then processed by a computer program to produce tables for graphing. Table 2 summarizes the total time available to each person to work inside the habitat, taking into account their sleep, time devoted to chores, EVAs (and assisting others), lost time from power failures, movies, lunch, dinner, and meetings.

The big surprise here is the range of two hours between A and V because of V's shorter sleep and DGO time. These figures do not include irregular chores (helping the fill the water tank every few days), personal hygiene, or breakfast. The striking and important conclusion is that crew members do not have the same time available to do individual work. Note that N has the most time because she participated in fewer and shorter EVAs. A has almost fewer 1.5 hours available per day, which must be considered in evaluating her sense of frustration at lacking enough time to do her work.

	A	B	D	J	N	V
4/08/02		EOA		EOP		DGO
4/09/02			DGO	EOP		EOA
4/10/02		DGO		EOP		EOA
4/11/02	EOA			DGO		EOP
4/12/02	EOA				DGO	EOP
4/13/02	DGO & EOP		EOA			
4/14/02	EOP		EOA			DGO
4/15/02			DGO & EOP		EOA	
4/16/02		DGO	EOP		EOA	
4/17/02		EOA		DGO	EOP	
4/18/02		EOA			DGO & EOP	
4/19/02	DGO	EOP		EOA		
4/20/02		EOP		EOA		DGO
4/21/02				EOP		EOA

Table 1. MDRS5 Chore Assignments – (columns correspond to crew members; DGO = cook; EOP and EOA = engineering officer primary and secondary, for refueling generators)

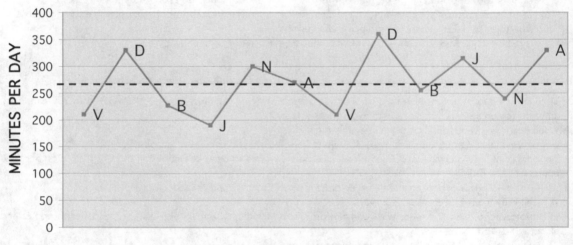

Responsible Person for Day

Figure 1. Total time devoted to cooking and cleaning by the assigned crew member, April 8-20; average 4 hours 23 minutes.

To this point, we can relate individual differences and understand roughly how much time is available for laboratory analysis, writing reports, and so on. The initial question of "where does the time go" is now all the more mysterious – Why did the group report at the planning meeting on the fourth day (4/11) not having enough time, when everyone has at least nine unscheduled hours per day, and the average is 10.5? What the averages disguise is the change that occurred after this meeting (see Daily Schedule, Figure 4).

Creating the daily schedule chart was the most pivotal part of this analysis. It shows trends that were not visible or even known to members of the crew, such as the how having dinner earlier leaves open time for additional work before starting the movie. This graphic helps us understand at a glance how scheduling changes how much time is available when it is most needed, especially before and after dinner. The chart also makes us more aware that people do not operate on a 24-hour day (except for perhaps the journalist), and that every hour is not equally available for doing productive work. In practice, only the time between 9 a.m. and 10 p.m. is universally available. Some people may rise early and work on email before 9 a.m. and others may work well past midnight. But if we are to understand how group scheduling affects productivity, we need to shift from studying individual averages to considering what time is *practically available* for everyone. Also, we need to examine the data in terms of variables that changed, looking for effects on productivity.

Figure 5 reorganizes the daily schedule data to show what time is equally available to everyone, comparing the two five-day periods previously described, omitting the rest, cleaning, and open house days. Productivity is increased by shortening the morning planning meeting and eliminating the afternoon tutorial activity. Again, we shift here from

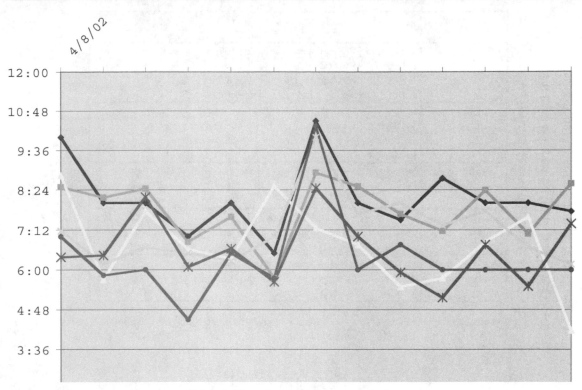

4/8/02

Figure 2. Sleep duration by crew members.

Figure 3. Video camera set up for time lapse recording (right), with an example frame (lower right). (Pletser is about to surprise the group with a box of candy.)

studying individuals to studying the resource available. Furthermore, productivity metrics can't be absolute; we must compare something. Given that the group discussed its productivity problems on the fourth day, a comparison of the first five days to the second five should reveal a significant cumulative change. In summary, our objective here is to *measure a valuable resource* that schedule changes (for example) might have affected.

What other changes might be useful? Lunch is already short (about 40 minutes); more might be lost in group cohesion by eliminating it as an organized activity. By making dinner earlier, more productive time is available in the evening. In the first week, the group worked as hard after dinner, but had to do this by making the movie later or skipping it. This effort probably cannot be sustained; one might further argue that 13 hour days are too long.

Notice that the shorter meeting time allowed an earlier lunch, plus it provided more work time (tempered somewhat by starting the meeting later). In turn, the EVA could be scheduled earlier (helped by improved weather that made waiting for the cooler part of the afternoon unnecessary). Furthermore, the increase in productive time after EVAs is not at the expense of EVAs, which increased by 53% total person hours.

Productivity Metrics

To this point we have considered only how much time the crew worked per day and how the schedule affected available individual time. Can we measure directly what the crew accomplished? I consider here the crew's reporting activity and to what extent daily plans were completed.

ACTIVITY PLAN	GEOLOGISTS' SCRIPT
\<EVA PREP\> (-,-,-)	1. Drive in EVEREST with backpacks, helmets, suits, all equipment 2. Start Minibooks & GPS 3. Don suit with boots, gaiters, radios & headsets 4. Put on backpack & helmet & connect cables
Checking equipment (20, 5, 20)	1. "Start CHECKING EQUIPMENT activity" 2. "Start tracking my location every 60 seconds" 3. "Start tracking my biosensors every 5 minutes" 4. "Start WALKING TO TOP OF CANYON activity"
Walking to top of canyon (10, 0, 10)	{Astronaut 2 improvised a voice note during the walk}
Sample fossils (10, 5, 0)	1. "Start SAMPLE FOSSILS activity" 2. Sample bag, voice annotation, association, photo 3. "Start WALK TO HEAD OF CANYON activity"
Walk to head of canyon (10, 0, 10)	\< Walk carefully down the hill and proceed to the head of the canyon to the south (your left) \>

Table 2. Time available to each person per day, after subtracting sleep time, group activities, and chores. Average is 10 hours 37 minutes.

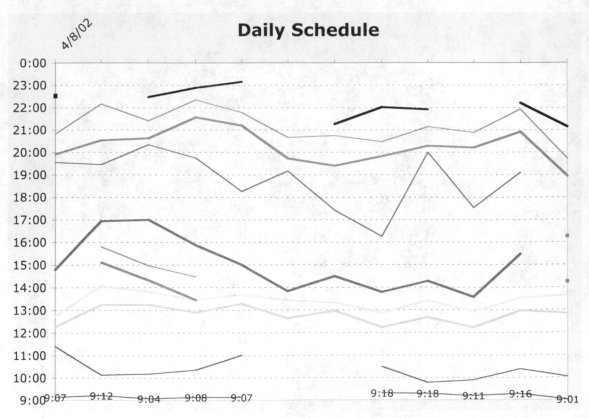

Figure 4. Extent of regular group activities (colors correspond to start and stop times of each activity, e.g., on 4/8 dinner began about 20:00 and ended about an hour later).

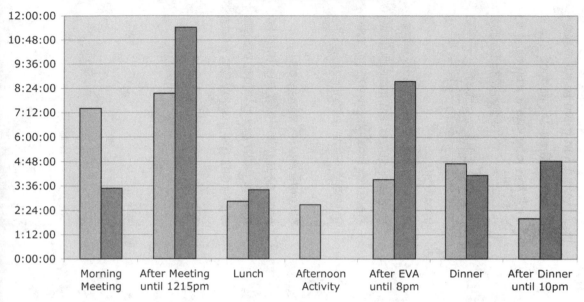

Rescheduling Effect on Productive Time
(shorter & fewer meetings, earlier lunch & dinner, increased available individual work time from 23.3 hrs to 32.8 hrs = 41% and total EVA person-hours increased by 53%)

Figure 5. Affect of rescheduling on individual's available productive time (9am – 10pm).

Report Writing
The crew wrote 97 reports over 12 days, totaling 57K words (Figure 6; for all MDRS reports and photographs, see www.marssociety.org/MDRS/2002Dispatches/). The commander, HSO and ESA scientist wrote extensive daily logs (including French translations, not included in the total); the biologist and geologist wrote daily activity notes and weekly reports; and the journalist wrote five crew bios and two daily life stories. The number of words increased by 26% in the second week, with the most on the last Friday, suggesting an end-of-mission "completion effect" of submissions by the scientists and journalist (Figure 7).

The length of reports is perhaps the crudest metric of productivity. Nevertheless, reporting is an important part of surface operations work and the word count is objective. To be fair to the geologist and biologist in particular, other metrics must be considered: Number of samples taken, cultures grown, work in the greenhouse, the EVA area explored, number of waypoints found or logged, repairs, etc.

Task Productivity
Experience in FMARS simulations (Clancey, 2000b, 2001b) suggested further study of how the crew plans daily activities. Each day the commander edited a table in a word processor to indicate what each person planned to do. Tasks were copied over from the previous day as necessary, with text changed to a strike-out font, to indicate lack of completion the previous day. Different formats were tried; the use of a simple table with one column per person and extra columns for group

Communication	Astro1	Astro2	HabCom
broadcast_alert	2	0	4
create_location	1	1	0
download_images	3	1	0
GPS	3	2	0
GPS_start_logging	1	2	0
initialization	2	4	4
location_with_samplebag	1	1	0
newData	0	0	9
new_activity_started	2	1	7
SampleBag	3	1	0
start_specified_activity	2	0	2
storeData	27	13	0
TakePicture	0	1	0
voice_annotation_with_location	1	1	0
voice_annotation_with_samplebag	1	1	0
VoiceAnnotation	7	1	0

Table 3. Plan for Saturday, April 13 (strikeouts for previous day indicate tasks that were not started or abandoned; repetitions such as A's "EVA64 report" indicate continued work)

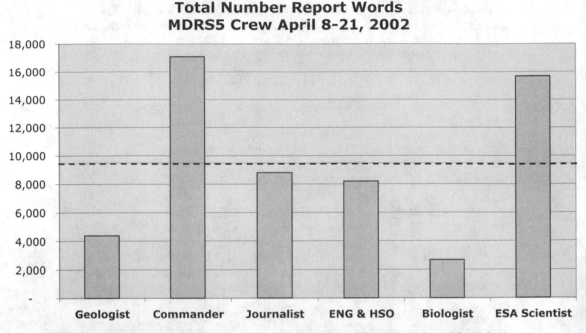

Total Number Report Words MDRS5 Crew April 8-21, 2002

Figures 6 (above) and 7 (below). Total number of words in reports released for web publication by the MDRS5 crew and output per day compared over the two weeks.

Daily Total Crew Report Words Completed
(26% increase in second week)

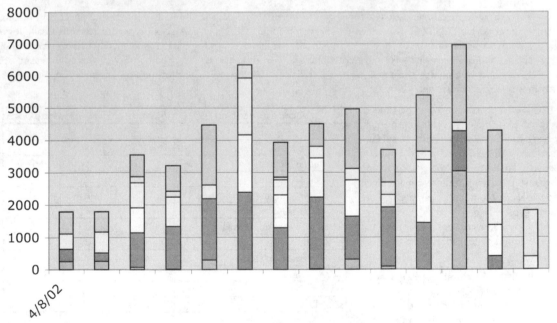

activities worked best (Table 3). These plans were analyzed by tallying the number of tasks proposed and deferred each day (Figure 8). On average two tasks were proposed per person/day; 60 were completed in the first six days, 72 in the second.

Note that April 14th is Sunday, the rest day. A strikingly even number of tasks are proposed (15/day) with many additional tasks in the last week for wrapping up projects. Group tasks include EVA objectives, tutorials and drills. Like counting report words, this measure is very crude – some tasks take a few minutes, others take days. Nevertheless, the comparisons between days should have some validity. Excluding the day of rest, whenever the group as a whole slept more than the previous night, the number of completed tasks increased compared to the day

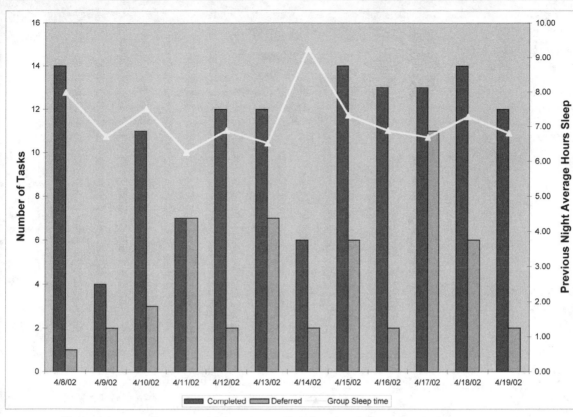

Figure 8. Tasks completed and deferred each day, plotted against average sleep time per person.

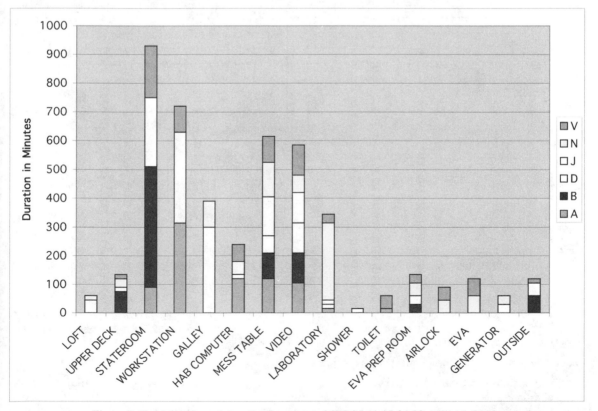

Figure 9. Total time in each location by person, MDRS5 11:15-24:00 April 15, 2002.

before. Otherwise, when the crew slept less, the number of completed tasks decreased or stayed the same. Although intuitive, the pattern is almost too good to be true and merits additional study.

Layout and Use of Space

Because no visitors were allowed inside the habitat during the 10-day closed simulation, we can completely characterize how space was used. The time lapse allows in principle determining, for example, when and for how long people used their staterooms. However, analyzing this data frame-by-frame is very time consuming and has not yet been attempted. Realizing this difficulty in advance, I instead recorded on paper where people were and what they were doing, every 15 minutes over a period of one and a half days (Figure 9). After this time, the patterns were obvious and it was clear that additional recording (at the 15 minute grain size) would provide little or no further information.

Direct observation showed that each person tended to spend most of their time in only a few (two to four) places. The three people with internet connections in their stateroom (B, D, V) spend most of the day there (A uses the stateroom to read before dinner; on other days J uses his upper bunk as a work table). Of the other three crew members, two (A and J) are working at the workstation bench of the upper deck (where they can connect their computers to the internet); on this day N is more often found in the laboratory of the lower deck. On this day D was the DGO, hence he is in the galley. The mess table, used for meetings, meals and video, is used evenly by everyone. ("Outside" refers to required work without suits; J worked with the water pump; B photographed an EVA activity.)

	11:15-24:00 April 15, 2002	7:00-24:00 April 16, 2002
Loft	1%	0%
Upper Deck	3%	3%
Stateroom	20%	20%
Workstation	16%	12%
Galley	8%	5%
Hab Computer	5%	7%
Mess Table	13%	16%
Video	13%	9%
Laboratory	7%	1%
Shower	0%	1%
Toilet	1%	1%
EVA Prep Room	3%	5%
Airlock	2%	0%
EVA	3%	14%
Generator	1%	2%
Outside or Greenhouse	3%	2%

Table 4. Comparison of crew use of different habitat areas over two day. Occurrence >= 5% shown in bold. See text for explanation of how percentages are calculated.

Comparison across two days shows fairly consistent use of space (Table 4). The second day EVA was longer and N worked at her workstation instead of the laboratory. Percentages are calculated by treating a person in a location as a "visit"; thus 6 visits are observed every 15 minutes; percentages given are based on number of visits in that location over the period indicated. Staterooms are treated as one place rather than six. Thus if everyone stayed in their staterooms for the entire observed period, it would be recorded as 100%. Similarly, 16% at the workstation on the first day could be one person for the entire period or some combination (see Figure 9). Notice that about 25% of the day is spent sitting at the mess table. The lower deck is obviously underutilized; it is occupied only 8% of the time (though many visits to the toilet were not observed at the 15 minute grain size).

Data about use of space can also be used to measure how often people move around. A "move" is defined as a person changing location from one 15 minute observation to the next. The first day has 52 observations; D moved the least, 17 times; V moved the most, 28 times; the average is 21. The second day has 69 observations; A moved the least (17); D moved the most (30); the average is 23. Further data are required to establish individual differences, if any. A person may also be very reactive – prone to get or check something when the idea occurs to him or her. Such short and frequent movements would not be caught by the 15 minute grain size.

Various commercial devices are now available for logging movements (e.g., used by retail stores for investigating where people look, what they touch, etc.). Although our time lapse video has this information, automated tracking is necessary for practically analyzing detailed movements over more than a few hours.

Crew Post-Occupation Survey

Crew members completed an individual written survey after the closed simulation ended (Appendix I). For the items ranked by importance, the most important (average rank >=4) were clean drinking water and sufficient power, closely followed by diet, adequate EVA suits, and toilet. Ranked next were free time, and showers.

From the open questions, the following patterns are apparent:

1. Everyone observed that the crew interacted harmoniously; our reactions to each other were without exception upbeat and cheerful.
2. Everyone reported either insufficient time to accomplish objectives and/or inadequate leisure time. Computer network problems and/or interruptions are cited by everyone, but only two people mention unnecessary report writing (in such a simulation, the audience can be ill-defined).

3. Four out of six (4/6) said the most important problem is the toilet facility (followed by power).
4. Providing stable, sufficient power, without refueling, would have the greatest effect on morale and productivity. Power outages halted all activity (loss of about an hour per day) and exacerbated the computer network difficulties (only three internet lines, tenuous connections, and low bandwidth). (A new diesel generator was later installed with greater power output and less maintenance required.)
5. Tool usage posed problems cited by everyone, either from lack of access to a tool (computer or camera), lack of knowledge to use a tool, or being interrupted by requests for assistance to use a tool.
6. The best moments were outdoors (5/6), with three people mentioning a particular EVA in which they participated together (this is remarkable given that surveys were privately prepared).
7. The worst moments were all different and had no resemblance to each other (e.g., as commander, my worst moment was when a crew member felt ill during dinner; nobody else mentioned this event, not even this crew member).
8. For those having a computer connection, the stateroom was an important place, for others it was just a place to sleep; half mentioned the importance of privacy.
9. Everyone wanted a better EVA suit and more than one shower/week.
10. Everyone would have continued a third week (though one person couldn't get time off from work); everyone would stay for a month if family and work allowed.
11. Food and habitat temperature were non-issues for this rotation.
12. The group provided varied and imaginative ideas for habitat improvement, with a surprising number of suggestions focusing on kitchen equipment.

In response to the question of "where did the time go," every crew member would try to change how they used their time on subsequent simulations:

• Geologist: wanted to find a way to work uninterrupted for longer periods of time; couldn't concentrate on reading and writing.
• Commander: never read any of the crew's reports (!), too much time writing logs.
• Journalist: didn't write enough "daily life" stories; on reflection would have done this instead of writing biographies.
• Health and Safety Officer: too much time being "Mr. Fix-it"; didn't have time to create a systems model of the habitat as planned.
• Biologist: insufficient lab time, too much group time (ironically, this person had the most amount of available time per day, considering sleep, EVAs, and chores).
• EVA Scientist: insufficient EVA time, too short and lacking adventure.

Individual differences in what individuals didn't say are also intriguing:

• Only one person (the commander) didn't complain about lack of time (rather I wanted entire days off, to do something entirely different).
• Only one person (the journalist) didn't mention outdoor activities as a highlight.
• Only one person (the biologist) didn't complain about interruptions or network problems, perhaps because she spent so much time working alone in the laboratory, away from her computer.
• Only one person (the EVA scientist) didn't complain about the toilet and mentioned the importance of skill training in advance (reflecting his professional experience).

These differences may reflect mood or experience on the day of the survey. In any event, we are reminded that individuals will react differently to identical circumstances, sometimes because of different roles, but also because of temperament and past experience. The comments highlight that differences must always be expected, even in a crew that experiences itself as being especially harmonious. Further, over a long mission, the crew might enjoy varying its routine, so for example, one person could be left alone to work and eat independently for a day or a week, according to his or her personal need for concentrated effort and variety.

Discussion: Systematic Work System Design

In this section I summarize the framework of analysis and modeling that orients the empirical study carried out in MDRS5 and comment on how the analysis is informing ongoing research on work systems design (Clancey, et al. 1998; Clancey, 2001b, in press; Sierhuis, 2001). In this context, the work system (Figure 10) refers to the design of the MDRS habitat and its tools, the crew's roles and assignments, the daily schedules, protocols for interacting with mission support, and practices for using the habitat, tools, and interacting with each other.

The framework suggests that, in order to understand how a given work system design causally influences productivity (the quality and quantity of work products), work system studies should focus on changes in resources brought about by schedules, facilities, roles, processes, etc. For example, this paper has shown how the crew's change in the daily schedule, prompted by a desire to be more productive, resulted in a significant increase in the amount of time available for individual work, such as report writing. The next step is to show how a change in resources actually affected work products (e.g., number of samples processed).

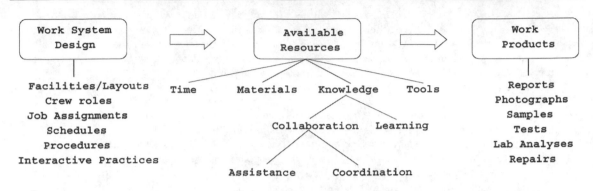

Figure 10. Different work system designs (facilities, tools, crew roles, etc.) affect the resources available, which affects the quality and quantity of work products.

In previous reports about the Haughton-Mars Project and FMARS (Clancey 2001b, in press), I have mentioned a simulation method for representing work practices, using a tool called Brahms (Clancey, *et al.* 1998; Sierhuis, 2001). The question arises whether a Brahms simulation of MDRS, which is so tedious and expensive to construct, would add anything to a spreadsheet analysis. On the one hand, the analysis given here reveals what a generic habitat model must contain, and how it must be parameterized (the work system design elements). For example, the DGO schedule would be represented as a document, posted on a whiteboard near the mess table. The schedule is uniformly visible where it can be referenced by individuals and during meetings.

The most obvious use of a Brahms simulation would be for what-if analyses, in which we would experimentally vary the work system design. For example, we could have the simulated crew follow different daily schedules, to show how this influences productivity. Or we could have the crew adapt different protocols for (not) interrupting each other. A simulation is useful when many variables interact and have longer-term consequences. For example, suppose that the crew were not allowed to interrupt each other during the morning work session, between the planning meeting and lunch. Individuals needing assistance may have their work delayed, affecting the entire crew later in the day. Or on the contrary, the crew may learn to depend more on mission support, which even with the time delay may be a better resource than other busy crew members.

What if one person were always the DGO or the generator were always handled by one person? One scientist thought such changes would improve her productivity. However, ethnographic studies of scientific expeditions organized in this manner (Bernard and Killworth, 1974) show that severe interpersonal difficulties may occur, as all must live together in an hazardous, isolated environment, but some are reduced in status to being servants to the others. But other changes are possible: What if the DGO assignment could be freely traded by crew members, so they could work on different days or do something else in exchange or be freed of the task, in recognition of another contribution to the mission?

Based on the analysis of MDRS5, the most pressing use of a Brahms simulation model would be to help us formulate and understand how interactions in crew activities and their individual problems cause interruptions for others and thus reduce productive time (including time to get back to work). However, creating such a model would require observations that were not made during MDRS5. Thus, we see the familiar pattern that a study reveals a topic of interest, and we must return to the field with different questions and additional observational methods. In particular, individuals must be tracked more closely during the day, with a better understanding of how they organize time, in order to understand the cause of interruptions and their effect.

Lessons Learned About Analog Research
Of paramount importance, given the effort and expense to build a research station like MDRS, is determining what can be learned from an analog activity and how the activity should be managed and controlled as a scientific investigation. If we are learn to live and work on Mars by doing authentic work in these habitats, we must at least be learning from each occupation how to configure and use these Earthbound research stations. The key findings from MDRS5 are:

1. Adjusting the group activity schedule creates more *useful time* for working (before lunch, before and after dinner).
2. Interruptions significantly affect productivity: power failure, group activities, assigned chores, requests for assistance, computer network problems, incoming email.
3. Timing and counting activities (systematic recording) is essential for detecting patterns and making work system design recommendations. The process is iterative; a successful outcome is determining what observations need to be made on subsequent controlled simulations.

From a work systems design perspective, the chief finding is that rescheduling group activities can increase the amount of available, useful time for individual work. Specifically, we shortened the morning meeting (most likely because less planning was necessary), we had earlier (but not shorter) meals, and we eliminated the after lunch "tutorial." In this analysis, time after 10 p.m. is not deemed as available (not useful for getting work done). An important consequence of shifting the day downwards was that movies could begin earlier (by 10 p.m.) and people could get to bed sooner (by midnight). This lowered fatigue on the subsequent day and increased the number of proposed tasks at the planning meeting.

Secondarily, observations (supported by the post-rotation crew survey) showed that interruptions negatively influence productivity. These include: power outages, group activities, attending to assigned chores, requests for assistance, computer network problems, correspondence by email, lack of access or knowledge how to use tools (such as digital cameras and GPS).

Methodologically, the analysis demonstrates that systematic recording enabling timing and counting activities (e.g., time lapse) is essential. In particular, the above findings are based on comparative analysis that measures total group activities, resources, and products in different time periods, before and after a change in the work system. The analysis requires identifying resources that affect work product quality and quantity and that are affected by the work system design (WSD). Whereas measures of productivity may be always incomplete and problematic (e.g., number of words written, number of photographs, duration and extent of EVAs), one can often measure intermediate factors that are universally available for different purposes (e.g., useful time available, internet bandwidth, number of chairs) and then show how these influence productivity. Nevertheless, although the crew reported lack of time as the chief frustration, one solution is to redesign the reporting products required and hence the schedule. Thus it is essential to keep in mind the relationship:

WSD (e.g., schedule) —> RESOURCES (e.g., time) —> PRODUCTS (e.g., reports)

Changing the WSD affects resources, but one may also redefine the outputs required.

Finally, the study showed the importance of knowing what you want to measure in advance, so recordings are taken systematically. Nevertheless, this understanding develops and changes as the study proceeds. For example, the experience in FMARS 2001 suggested that full time lapse over the entire rotation was desirable. Then during MDRS5 I discovered that following individuals to understand how they were interrupted was also necessary.

Lessons Learned About Metrics

This study has considered different metrics for describing the work system and productivity. Given the well-known importance of metrics for engineering, it is useful to summarize how the concept carries over to work system design:

• Metrics, as considered here, require *counting*, which requires *systematic observation over some domain of analysis*.

• A *domain of analysis* consists of a time period and some combination of one or more of the following: places, kinds of events, individuals (or a group), artifacts (e.g., a tool), and information (e.g., a task in a plan). For example, I measured how often people moved during 15 minute time periods; I measured how many tasks were proposed by the group each day; I measured the duration of power outages. I could use the time lapse to count other things, such as how long people used the habitat computer and how many times people visited their stateroom during the day.

• An observer must avoid two extremes: 1) observation focused only on predefined metrics (in order to be precise) which produces no surprising kinds of observations (e.g., the journalist noticed how the chairs were repeatedly moved to five different areas), and 2) observation so general and unsystematic (in order to be unbiased) it produces no measurements at all (e.g., my first-timer's experience during HMP-98).

• Metrics of value will not all be anticipated; you will discover what can be counted in the data if you have cast a broad enough net (e.g., the EVA reports enabled observing that every possible group of two people shared at least one EVA). You cannot know for sure in advance what counts will produce meaningful metrics.

• Work system design metrics are *multidimensional*: they always include a time dimension, and they are comparative (planned vs. actual, one time period vs. another, one individual vs. another).

• Cumulative group productivity metrics may be essential where different roles would make averages misleading or irrelevant (e.g., number of person/hours devoted to EVAs in a given time period).

• Data often needs to partitioned to reveal key variables (e.g., the logic of viewing the rotation as two 5-day periods rather than 12 uniform days or two weeks). An overview trend graph (e.g., Figure 4) may show a gradual change as a new policy or strategy is adopted, thus suggesting a "before and after" partition. Preferably a comparison should be based on a change in the work system design (e.g., scheduling change).

• Resources should be identified and measured, especially when products are difficult to quantify.

Lessons Learned About Photographic Observation

The following are some lessons learned about using photography to record human behavior:

• Regular observations are more amenable to analysis than raw video or unsystematic photographs. Decide in advance some category you want to photo-document (e.g., the phases in putting on a suit, all uses of duct tape, use of staterooms). Create a list and refer to it during the simulation.

- Common photographic mistakes:
  - Focusing on individuals and not photographing the context (better to have a wide angle camera fixed on an entire meeting table, than to keep focusing on individuals).
  - Clear, audible sound is more important than close-up images.
  - Forgetting to set and synchronize clocks in recording devices.
  - Not bringing enough film, tape, or storage (in the guise of reducing costs).
- Observational heuristic: When two or more people are together, photograph the setting, the details of tools and documents, people pointing, and people using tools. Ask yourself what they are doing and document how the interaction develops.
- Time lapse is especially good for capturing the start and end (and hence duration) of group events (especially meetings and eating). When placed appropriately, it can show when people enter and leave different spaces (e.g., staterooms).
- Don't worry about watching, let alone analyzing all your video – the trick is to capture the moments that need to be studied. For example, by recording every planning meeting, I captured the discussion of the fourth day when productivity problems led to schedule changes, and this was one hour well worth studying.

Lessons Learned About Analyzing Behavioral Data
The following are lessons learned using spreadsheets and charts to analyze behavior data:

- Prepare spreadsheets to organize the data and then charts.
- Use macros (e.g., programmed using the Visual Basic language built into Microsoft Excel) to count occurrences, producing additional tables and charts (e.g., counting how many times people moved from the original spreadsheet that recorded where people were every 15 minutes).
- Create many kinds of charts to discover meaningful trends or comparisons. For example, using Microsoft Excel's Chart Gallery preview option, explore what kinds of graphics produce readable (and hence potentially useful) displays.
- Find ways to meaningfully partition data by grouping people, places and days in different ways.
- Categorize periods by abstracting how time is used: "morning individual work time," "pre- and post-sleep," "productive work time" (9 a.m.-10 p.m.).
- Compare periods to understand schedule changes, periodicity and carryover effects.
- Identify recurrent events not recorded systematically (e.g., interruptions).
- Look for invisible patterns and practices (e.g., frequency of movement).

Conclusions and Next Steps

The fifth rotation of MDRS met all of its primary objectives – in field science, in reporting, and in the study of human exploration. This rotation introduced a number of creative innovations to the Mars Society's analog program: the closed simulation protocol, systematic recording of the entire rotation, a multiple failure IVA / EVA simulation, a plant growing psychosocial experiment, a resident journalist, an international press open house day, and many improvements to the habitat (bread maker, video / slide screen, medical inventory, lab tools inventory, computer connections document). The group generated 97 reports, including a geology primer for the region and comprehensive reports on the greenhouse data logger and soil cultures of the surrounding environment. The data recording methods were entirely successful, allowing discovery of unexpected trends in group behavior, showing how schedules influence productivity. The analysis was related to ongoing work in work practice simulation to show how statistical data informs modeling, while revealing questions for further observation.

The following are ideas for future work:

- Use of staterooms: Use the available time lapse to determine the number of visits and duration per person.
- Experiment with an internet connection in all staterooms: Does everyone choose to work inside and ignore the outside workstation area? Or do one or two people adopt the external area as a personal desk?
- Develop tools for automating EVA navigation and science data logging (Clancey, et al., in press).
- Study interruptions by following ("shadowing") individuals for at least three hour periods. Record the causes, durations, and any apparent disruption in other activity.
- Ask individuals to track time spent on activities of importance to them (i.e., let them decide what to record).
- Provide verbal EVA reports and let mission support write the formal reports
- Compare use of public and private space in MDRS and the EuroMars habitat.

Although the closed simulation was beneficial for carrying out a systematic study of use of time and space in the habitat, this is by no means the only scientific way to use MDRS or other analog research stations. In particular, one can carry out controlled protocols for shorter periods of time, say a few hours or a day, in which experimental equipment and procedures are used. MDRS's setting makes it especially attractive for research on EVAs, including different configurations of suits, rovers, robotic assistants, agent-based software, local CapCom monitoring, and remote mission support.

Finally, it is worth reviewing the scope and dimensions of the observations and analysis I have carried out in the HMP, FMARS, and MDRS settings since 1998:

- The organization and activities of a field science expedition:
  - Multiple phases / rotations with handovers
  - Multiple field seasons, showing development of exploration
- Metrics of work practice based on the rotation, subgroups, individuals
- Individual scientific discovery and reporting genre
- Ensemble behaviors (micro or interaction analysis; e.g., process of starting and ending meetings)

My impression is that carrying this work forward will benefit from focusing more on individual's experiences as they plan, learn and cope with the broader expedition setting.

Acknowledgments

This work was made possible through the vision, volunteer work, and financial contributions of the Mars Society and its supporters, especially Robert and Maggie Zubrin, Frank Schubert, and Tony Muscatello. Mission support was provided by Mark Klosowski, Frank Crossman, and other members of the Northern California Mars Society. Discussions with Boris Brodsky, the Brahms modeler for FMARS and MDRS, have been very helpful. Related work pursued by Maarten Sierhuis and Alessandro Acquisiti, in simulating how ISS crew members follow schedules, has inspired several aspects of MDRS5 observation and analysis. The crew members provided essential information for carrying out this study, as well as helpful comments for improving the presentation and this paper. For more information about Brahms see www.agentisolutions.com and papers at http://WJClancey.home.att.net. Funding for this research has been provided by the NASA's Intelligent Systems Program and the University of West Florida.

References

- Bernard, H.R. and Killworth, P.D. 1974. *Scientists and Crew: A Case Study in Communications at Sea. Maritime Studies and Management*, 2: 112-125.
- Clancey, W.J., Sachs, P., Sierhuis, M., and van Hoof, R. 1998. *Brahms: Simulating Practice for Work Systems Design. International Journal of Human-Computer Studies*, 49: 831-865.
- Clancey, W.J. 1999. *Human Exploration Ethnography of the Haughton-Mars Project 1989-1999*. Presented at the Mars Society Annual Meeting. Boulder, Colorado. In R. Zubrin and F. Crossman, *On to Mars: Colonizing a New World,* Burlington, Ontario, Canada: Apogee Books, published 2002, CD-ROM.
- Clancey, W.J. 2000a. *Visualizing Practical Knowledge: The Haughton-Mars Project.* (Das Haughton-Mars-Projekt der NASA – Ein Beispiel fur die Visualiserung Praktischen Wissens). In Christa Maar, Ernst Pöppel and Hans Ulrich Obrist (Eds.), *Weltwissen – Wissenswelt. Das globale Netz von Text und Bild*, pp. 325-341. Cologne: Dumont Verlag.
- Clancey, W.J. 2000b. *A Framework for Analog Studies of Mars Surface Operations: Using the Flashline Mars Arctic Research Station.* Presented at the Mars Society Annual Meeting. Toronto. In R. Zubrin and F. Crossman, *On to Mars: Colonizing a New World,* Burlington, Ontario, Canada: Apogee Books, published 2002, CD-ROM.
- Clancey, W.J. 2001a. *Field Science Ethnography: Methods for Systematic Observation on an Arctic Expedition. Field Methods*, 13(3), 223-243, August.
- Clancey, W.J. 2001b. *Simulating "Mars on Earth": A Report From FMARS Phase 2*. Presented at the Mars Society Annual Meeting. Stanford, California. In R. Zubrin and F. Crossman, *On to Mars: Colonizing a New World,* Burlington, Ontario, Canada: Apogee Books, published 2002, CD-ROM.
- Clancey, W.J. (in press). *Simulating Activities: Relating Motives, Deliberation, and Attentive Coordination. Cognitive Systems Research*, special issue on situated cognition.
- Clancey, W.J., Sierhuis, M., Kaskiris, C., van Hoof, R. (in press). *Brahms Mobile Agents: Architecture and Field Tests. Proceedings of the AAAI Symposium on Human-Robot Interaction.* North Falmouth, MA. November 2002.
- Perrow, C. 1999. *Normal Accidents: Living with High-Risk Technologies*. Princeton: Princeton University Press.
- Sierhuis, M. 2001. *Modeling and Simulating Work Practice*. Ph.D. thesis, Social Science and Informatics (SWI), University of Amsterdam, SIKS Dissertation Series No. 2001-10, Amsterdam, The Netherlands, ISBN 90-6464-849-2.
- Stuster, J. 1996. *Bold Endeavors: Lessons from Polar and Space Exploration*. Annapolis: Naval Institute Press.

Appendix 1 – MDRS5 Crew Survey

1. Did you go anywhere to be alone or to seek quiet?
2. What is your overall impression of your stateroom in a word or phrase?
3. Did you use your stateroom during the day? For what purpose?
4. Was there sufficient work space?
5. Did you ever work on the lower deck? For what purpose?
6. Were you productive? In what way or why not?
7. What was your most memorable positive moment (or moments)?
8. What was your most memorable negative moment (or moments)?
9. What is your assessment of the crew selection in terms of quality of knowledge, skills, and temperament?
10. Did you feel you had enough time for your research, i.e., sufficient field and lab time?
11. What reports did you write? Was this too much or less than you desired?
12. Would you be happy if you were staying another week?
13. Would you participate in a longer simulation of 4 weeks? 8 weeks? 12 weeks?
14. Did you have sufficient electricity? Water? Food?
15. Was the Hab temperature comfortable? Was the ventilation adequate?

16. Was the amount of unscheduled time appropriate given the group's goals? Sufficient from a personal perspective?
17. Any suggestions for improving the EVA suits and equipment?
18. Did you like the evening group activities, or would you prefer to use the time for personal reading, working, etc.?
19. Did wearing the suits hinder exploring this area? In what way?
20. What is the one thing you would most like to see improved in the Hab?

Please rate the following for their importance to you on a scale of 1 to 5, 1 being Not Important and 5 being Very Important:

1. Clean drinking water –
2. Sufficient and reliable electrical power –
3. Unscheduled time for sleep and unscheduled activities –
4. Reliable and effective toilet facilities –
5. Capable and simulation realistic EVA suits and equipment –
6. Wholesome and varied diet –
7. Work areas with adjacent space for personal items (e.g., notes, drinks) –
8. Entertainment (e.g., DVD movies) –
9. Quiet and comfortable private space –
10. Cleanliness (showers, hot water) –
11. Personal storage areas –

Appendix 2 – Simulation of a Normal Accident

At the midpoint of the closed simulation, the crew had one relatively unstructured day. During the afternoon, a previously agreed upon script was enacted as an experiment on time-delayed communications and emergency management. The scenario was deliberately designed to avoid serious concern by mission support, meaning that none of the events would suggest that escalation was required (e.g., notifying Mars Society headquarters). Nevertheless, the events were designed to be realistic and dramatic.

The scenario was inspired by a framework provided in *Normal Accidents* (Perrow, 1999), such that multiple problems occurred with different kinds of causes: becoming lost during an EVA (*human error*), a stuck zipper (*mechanical failure*), wind and heat (*environment condition*), and radio problems (*system design*). According to Perrow, accidents in complex systems cannot be completely eliminated. Rather they must be viewed as part of normal operation, however infrequently they may occur. The design of a work system must accordingly take into account tools, procedures, training for handling emergencies, and critical situations. The purpose of our experiment was, first, to develop a method for simulating multiple failures in the MDRS setting and, second, to understand how multiple failures would be coordinated by the commander and mission support.

Time, (min. since start)	Actor	Action	Actual time
0	N	Reports she is alone	15:45
5	B	Reports no comms with J &A	15:50
10	D&V	Report zipper on Greenhouse door is stuck, so they are trapped inside	15:55
15	N	Reports she is very hot and walking back to the hab (but goes via waypoint 102 to fetch camera)	16:00
20	B	Reports via email that UPS is beeping, and he must refuel in 30 minutes	16:05
25	D&V	Ask for permission to cut or rip open the door	16:10
30	J&A	Report they are back at potholes, but N is not in sight (N has returned by way of the wind catcher hill so she can follow the main road)	16:16
35	D&V	Are arguing about what to do	16:20-16:24

Table 5. Script excerpt for multiple failure accident.

The method for this simulation included a detailed script, indicating what individuals would do and what they would report at five minute intervals (Table 5). Special for this experiment, audio recordings (using the Radio Shack DSSPlayer) of communications between remote teams and the habitat CapCom were transmitted to mission support by email with an average five-minute delay (roughly the time required to download the file from the recorder to the computer, to compose an email, and for the message to be received). Additionally, mission support responded with a simulated five-minute time delay. Coordination and communication occurred using hand-held walkie talkies. The CapCom would prompt crew members according to the script to remind them of their lines. For each recording, CapCom would then say, "Okay, record now," and the crew member would call in using the usual protocol (e.g., "CapCom, this is Nancy. It's really windy out here . . ."). The script took two hours to enact, as designed. Mission support suspected something unusual was occurring, but for the most part viewed the scenario as a real series of events.

The plot: CapCom is on the MDRS upper deck monitoring a simultaneous IVA (D and V working inside the greenhouse) and an EVA (N, J, A walking to a site about 100 meters south of the habitat). The beginning appears normal. After a radio check, N, J, and A suit up, egress, and proceed to a previously visited location to deploy and retrieve sample containers. Meanwhile, D and V have entered the greenhouse to do something with the data logger. The problems begin as indicated in the script (Table 5), with new important information being received by CapCom every five minutes. The script is such that with the time delay, any sense of priority or problem resolution plan must be reconsidered just as it would be transmitted by mission support. Even with the minimal time delay (five minutes), mission support is unable to stay on top of a fast-changing situation, one that requires revising plans as new information comes in. At best they can follow along. CapCom, too, is overloaded. At best he can put the crew on "safe hold" and focus on the most important problems. For mission support, the situation changes too quickly to provide any useful advice. In a real Mars emergency, the time delay is likely to be longer and the information available perhaps less complete, in particular, automated methods would be required to keep mission support aware of what is occurring, as CapCom will be too busy interacting with the crew and local systems to be sending and receiving email. Even within this scenario, CapCom often forgot to check whether mission support had sent a message.

Appendix 3 – Suggested Research Station Improvements

The following are some of the crew's suggestions for improving MDRS or similar habitats:

* Recognize that the EVA prep room can serve as a private meeting area; provide chairs that stay there.
* List available tools and movies in the manual (so future crews know what to bring)
* Show new crew members how an upper bunk can be used as a workbench; provide a stool (perhaps opening from the wall) to enable using that area instead of the built-in desk.
* Provide the following in every stateroom: a mattress, a porthole, an internet connection, book shelves, more clothes hooks.
* Lower deck is only used for work by two people, an inefficient use of open space. Consider how it might be outfitted to draw the group downstairs for variety (e.g., MDRS7 used it as a music studio).
* Provide better kitchen appliances for more efficient use of crew time. The crew made extensive use of the crock pot (slow cooker) and bread maker.

On the Road to Mars as a Member of the Mars Desert Research Station Crew 8†

Phil Turek, Cerritos High School
paturek@earthlink.net

Abstract
A Mars colonization concept named Earth Mars Ambassadors (EMA) calls for building a new culture on Mars, a culture focused on the application of lessons learned from the exploration and colonization of Mars to the problems facing life on Earth. As a member of MDRS crew 8, I created a photo story showing the greening of Mars – one step to the successful selling of the EMA concept. Images from the photo story and an update on the EMA concept are presented.

Introduction
Most people consider discussions of manned missions to Mars to be premature here in the year 2003. However, just as it typically takes a full decade to develop new space flight hardware, so too it may very well take a full decade to properly prepare the crews who will fly in such space ships. To support meaningful manned missions to Mars early in the next decade, it is necessary to put into place a plan for training future astronauts and conservators now.

The Earth Mars Ambassadors (EMA) project contains such a plan. The vision embodied in EMA calls for educators to dispatch teams of students now to explore ever larger regions of Earth. Students in the lowest grade levels may explore no farther than their own school building. Students in middle school may explore a county-sized area, while high school students may visit sites spread across the nation. Graduate students may cycle through facilities such as the simulated Mars bases maintained and operated by The Mars Society. During this first phase of EMA, students practice working together as a team solving unexpected challenges encountered in the field. In addition, they will gain first hand familiarity with problems facing humanity on Earth. Finally, this first phase will help create a societal expectation that we're going to Mars specifically to help life on Earth – an expectation crucial for securing public support and funding for eventual manned missions to Mars.

During the second phase of EMA the best teams from phase one (now grown up and fully trained astronauts) go to Mars. While all of the first humans to Mars are expected to be NASA selected and trained astronauts, the EMA crew members are intended to serve as more than just Mars surface exploration astronauts. They are to generally stay and build a new culture on Mars, one that highly values all forms of life. In addition to studying Mars, the EMA crew members will be working to establish life on Mars transported from Earth, most likely initially in relatively small biologically supportive enclosures. It is at this time that the EMA crew members learn to successfully nurture selected species of plants and animals brought from Earth.

The third phase of EMA is the most important. During this phase conservators from the new colony on Mars come to Earth as ambassadors from Mars, ambassadors with a mission to help solve problems facing life on Earth. These former students will know well various aspects of our planet, for they are the same former students who will have studied Earth so diligently during phase 1. After living on Mars for a time span of at least four years, these conservators will have earned the title of Martian Ambassadors, a title meant to help confer influence onto these individuals, allowing them to be more effective in their efforts to change things here for the better.

The Greening of Mars Photo Story

When you look at the flag used by The Mars Society you'll notice it has the three colors Red, Green and Blue. Red represents the original state of Mars – a red planet. Green represents the greening of Mars – bringing life from Earth, adapting it and establishing it on Mars. Blue represents a possible future Mars that has been terraformed to be as much like Earth as possible.

Nearly all simulation work to date at the Mars Society's research stations has involved just a Red Mars – carrying out basic scientific investigations of the conditions on Mars and the processes behind them. During the 4 December 2002 EVA we wanted to tell the story – with pictures – of the greening of Mars. This is a story of astronauts living on Mars, using our best technology to adapt Earth life to the harsher conditions found on Mars, and planting that life outside existing protective structures. This is a story not of humans planting flags or footprints on the Red Planet, but of humans planting life on the new world.

Two members of crew 8 (Sini Merikallio from Finland and Derek Smith from Canada) went on an extensive excursion in full simulation mode. I went along out of simulation as a photographer. This allowed me the freedom to arrange some shots to better tell the story of the EVA.

The 4 December EVA called for a sequence of arranged shots telling the story of the greening of Mars. In this story space suited astronauts are to be shown planting seedlings on the barren Martian surface. As time goes by the seedlings grow into adult plants. Additional images were taken from other excursions to round out the photo story.

Figure 1. A wizened Mars Conservator emerges from a greenhab with a precious cargo of seedlings.

In Figure 1 a space suited Mars conservator (portrayed by Dusty Samouce, MDRS crew 8 Commander) carefully brings out a box of genetically hardened plants suitable for growing in the relatively harsh Martian environment.

In Figure 2 the Conservator makes his way from the greenhab past the base support strut and various items of field equipment.

Figure 2. The seedlings are no longer protected within the greenhab.

The cargo is carefully loaded onto the back of the ATV as shown in Figure 3. It seems appropriate that the oldest Martian is conveying to the youngest Martian the task that lies ahead for her for the day. The enthusiasm the MDRS crew 8 members had for doing this photo shoot is evident on Sini's face and in Dusty's pose.

Turek-F03

The Martians head off on an excursion as shown in Figure 4. The seedlings are not the primary mission for the day. Rather, they represent a Martian cultural activity. Much as Johnny Appleseed made a point of planting apple seeds wherever he went, so too the Martians make a point of looking for appropriate locations to plant life wherever they go.

Spotting a place to plant a seedling, the young Martian digs a hole and plants one from the box. The white splotches on the ground in Figure 5 are due to natural chemical processes, and contribute to the impression this is an alien environment.

Figure 3. The cargo is lovingly secured to the ATV.

Figure 4. The Martians head off on the day's excursion.

Figure 5. A seedling is planted on Mars.

The sequence of images designated Figures 6, 7, and 8 shows that the plant has taken to the Martian environment.

Against a scenic backdrop Derek hands Sini the seedling for planting in Figures 9 and 10. It was Sini who had picked this site during our excursion.

Figures 6, 7 and 8. A time-lapse sequence showing the growth of the seedling on Mars.

Figure 9. At another site another seedling is selected for planting.

Figure 10. A scenic spot is picked for this seedling.

It only takes a few minutes to do the job (see Figure 11.) In a short time the Martians will be back to going about their business for the day. The plant, however, will change the Martian landscape for years to come.

Figure 11. The Martians plant the seedling.

In Figure 12 a potent nutrient broth is given to the freshly planted seedling. This concoction will give the plant a fighting chance to adapt to its new home.

Figure 12. A squirt of a nutrient broth welcomes the seedling to its new home.

Over time, the seedling grows (see Figure 13.)

Figure 13. Perhaps two years later the plant has visibly grown.

Long after the Martians did their work, the seedling has grown and spread, colonizing the spot with its offspring (see Figure 14).

Figure 14. After several years, a number of plants now occupy the site.

The Mars Desert Research Station, the Utah desert, and the MDRS volunteer crews are a unique combination of resources that allowed this exciting story to be shown with images.

A Progress Report on the Earth Mars Ambassadors Program

I'm convinced American society will broadly embrace manned missions to Mars when a detailed plan is put forward that is perceived to be rational, reasonable, and credible. The Mars Society seems the natural choice to serve as a home base for the EMA concept, a place where the concept can flourish and be popularized. However, the organization currently seems to have some drawbacks that make it less than ideal as a home for the EMA concept.

As an example of the difficulty with The Mars Society, a NASA life sciences researcher put forth a suggestion to The Mars Society at this convention. Referring to the possibility of life on Mars, he categorized three possibilities:

1. There's no life on Mars. In that case, there is no problem regarding possible contamination of Martian life by Earth life.
2. Life on Mars is found to be biologically similar to life on Earth. In this case, too, contamination of Martian life is not a serious issue.
3. Life is found on Mars, and it appears to be from a "second genesis" – it began and evolved completely independently of life on Earth. In this third case, it might be desirable to preserve the possibility that we someday decontaminate locations on Mars where we may have introduced life from Earth.

The surface of Mars is thought to be destructive to unprotected life from Earth. If true, then decontaminating the surface of Mars from Earth life is a relatively simple task, consisting of little more than exposing such a site to the natural Martian environment (particularly sunlight, which is UV rich at the surface of Mars relative to sunlight at the surface of the Earth).

The NASA scientist proposed it might be advantageous to design Mars missions for such an eventual possibility – the possibility of eventually wanting to decontaminate places where life brought from Earth might reasonably still be found.

This suggestion seemed reasonable to me. A number of similar analogies come to mind of what the scientist was talking about. Most cars, for example, are equipped with a spare tire – as a reasonable precaution in the event one of the four tires in use becomes flat. When the US army invaded Iraq this past spring, one of its Abrams tanks became disabled in Baghdad. The US army brought in a tank mover – a vehicle designed to transport disabled tanks – to remove it from the city. Since the 1970s US culture has been promoting the idea of "pack it in, pack it out" when hiking – that you remove your trash when you go someplace, such as visiting a park. Importantly, standard engineering practice for spacecraft mission design recommends roughly 10% of resources should be allotted to removing a spacecraft from service at the end of its operational life.

Yet, the NASA researcher's suggestion was aggressively attacked at this Mars Society convention. It was condemned as being a stifling constraint on Mars exploration. A distinguished and highly respected leader of The Mars Society stood before the audience and asserted that the adoption of the NASA researcher's proposal would lead to nothing less than the end of life as we know it on Earth.

I feel such rhetoric is harmful to the cause of Mars exploration and development, and I do not want the EMA concept linked to an organization that casually slips into such rhetoric. The Mars Society is a wonderful organization; it has done much to stimulate grassroots support for manned missions to Mars. It has been successful precisely because it has chosen to oppose preexisting conventions regarding the best way to conduct manned missions to Mars. Unfortunately, the organization at times seems to choose to also oppose good ideas with equal vigor. I do not want EMA associated with an organization that is perceived by the general public to be hostile toward environmental concerns, that exudes disdain toward constructive suggestions submitted in good faith by members of its own ranks, and which seems willing to casually alienate mainstream American society without good cause.

EMA is a good concept. When students learn of the three phase plan for bringing Martian Ambassadors to Earth, they enthusiastically endorse EMA. The best solution for EMA regarding The Mars Society would be for me to publish a book about EMA. Whether I have the stamina to publish such a book remains to be seen, but it would best serve the needs of the program.

On a separate front, my congressman is well placed (as Chairman of the House Science Committee's Space and Aeronautics Subcommittee) to guide congressional deliberations on the EMA concept. The stumbling block here is his opposition to any consideration of manned missions to Mars until such time as the Moon is commercially developed. Specifically, he is currently advocating the commercial development of a new propulsion system, one that will dramatically reduce the cost of space travel and open it to large numbers of space travelers in the near future.

Here, at least, I have a solution that may open the door to my congressman's endorsement of EMA. I have invented a new form of space travel that is predicated on recycling orbital energy and momentum. Because the system involves structures that function as both a catapult (for launching spacecraft) and as a catcher's mitt (for landing spacecraft) I name my invention a CataMitt Recycler. Shown in Figure 15 is a current model of a technology demonstrator structure intended for use in low-Earth orbit. This invention is the subject of my thesis research I'm conducting toward a Master's degree in aerospace engineering. After the thesis is submitted (in March 2004) for publication I will be ready to share it with my congressman. Hopefully at that point he'd become receptive to learning about EMA.

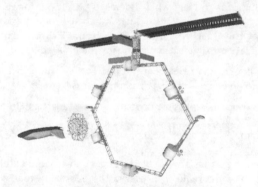

Figure 15. CataMitt Recycler about to receive an approaching spacecraft.

One last recent observation regarding EMA is worth mentioning. I attended a space conference in Los Angeles at which a number of space advocates who had been exposed to the EMA concept over the years were in attendance. Despite their current complaints about the space program, not one of them recalled EMA or the solutions it offered to any of the specific issues they raised. Observe the cover art used at this Mars Society convention shown as Figure 16 and reproduced with permission. It shows only the first two phases of EMA – an Earth exploration phase on the left, followed by a Mars exploration phase on the right. If EMA had taken root here, then the missing third phase – showing Ambassadors from Mars assisting us here on Earth – would have been included in the cover art.

Left from "Lewis and Clark at Three Forks" by Edgar S. Paxson, 1912, courtesy of the Montana Historical Society - Right ©2002 by R.D."Gus" Frederick

Figure 16. Cover art from this convention is missing the third phase (Martian Ambassadors come to Earth) of EMA.

The solution here is to introduce EMA as a new idea. I am going to promote EMA as a three step program to bring Martian Ambassadors to Earth. As a second advertising slogan, I'll pose the question, "If we have the foresight to decide to spend $87 billion (this year) rebuilding just two countries (Afghanistan and Iraq), then why don't we go ahead and spend only $50 billion (spread over 10 years) to build a whole world? (Mars)."

Visions from the Past

When I first created EMA I employed several artists to generate sketches showing various aspects of the concept. During my time at MDRS a number of those early images were borne out. Here is a collection of predictive images made in 1990 and images taken at MDRS in 2002.

Figure 17. 1990 sketch of colonist on Mars contemplating Earthrise.

In 1990 I wanted to show a phase two astronaut at home on Mars, gazing out the window of the habitat and watching the Earth rise. The astronaut's thoughts were being recorded in a journal that was key to tying together the three phases of EMA. The sketch shown as Figure 17 was created by Chris Butler.

One early morning during my stay at MDRS I caught this image (Figure 18) of Derek working at his laptop, gazing and contemplating life on distant worlds.

In 1990 I asked space artist Chris Butler to create a sketch for EMA showing a researcher on Mars using the same basic skills that I teach my science students in my own classroom here on Earth. Chris created a sketch (Figure 19) of a Mars colonist looking through a microscope.

Figure 18. Derek gazing out the main Hab window at MDRS in 2003.

Figure 19. Sketch from 1990 of Mars astronaut using basic science skills.

In 2002 I caught an image (Figure 20) of Dusty showing me his computer controlled telescope at MDRS.

Figure 20. Dusty showing off his telescope at MDRS in 2003.

In 1990 Chris created this sketch (Figure 21) of a space cadet – one whose foundation was based on getting a good education.

Figure 21. Space Cadet in 1990.

Artwork courtesy of Chris Butler

During one of my turns at taking an excursion Sini snapped this shot (Figure 22) of a somewhat older space cadet.

In another early EMA sketch (Figure 23) I wanted to show that people on Mars would do very human things that we on Earth could easily relate to – things such as watching a sunset together.

During one MDRS excursion I took a shot (Figure 24) of our space suited shadows falling onto desert rocks.

What of other images from the early days of EMA – how have they played out? One piece of EMA calls for students to be directly involved with the search for life on Mars. First, the youngest students are approached and asked to carry out an experiment that would test for life. The astronauts on Mars would then perform the test and report the results back to the students. Next, the collective students at the next higher grade level would be asked to put forward their own experiment to test for life on Mars, and again, the astronauts on the surface of Mars would carry out the experiment and report the results back to the students. In this manner students would be leading the discoveries – not just following the mission as if it were already in some history book. In addition, students at higher grade levels would strive to put together more sophisticated experiments than had been created by the grade level below them – pushing students at ever higher grade levels to excel in their studies. At the same time, students at lower grade levels would be pulled into following the unfolding story, leading them to also deepen their own understanding of science so they might continue to follow the more complex experiments that follow their own. Figure 25 shows an eighth grade biology classroom, with the students actively directing the Mars surface astronauts as the teacher and NASA liaison mentor passively watch.

This year NASA is creating a new program specifically aimed at directly involving students in the exploration of Mars. I have successfully enrolled my high school to be one of 54 schools in the United States that are participating in this pilot program. My students at Cerritos High School have been assigned to Rover Watch. Using current Thermal Emission Imaging System (THEMIS) data from the *Mars Odyssey* spacecraft, as well as current Thermal Emission Spectrometer data from the *Mars Global Surveyor* spacecraft, we are being tasked with monitoring temperature conditions on the surface of Mars at the Mars Exploration Rover landing sites for a one week period sometime after the rovers land in January of 2004. Our responsibilities include informing NASA if temperature conditions threaten the health of the rovers. This program is named Mars Exploration Student Data

Figure 22. Older space cadet at MDRS in 2003.

Figure 23. 1990 sketch of astronauts watching a sunset on Mars.

Figure 24. Space suited shadows at MDRS in 2003.

Teams (MESDT). A second program involves teams of students directly with the Mars Excursion Rovers after they begin operations in January. Shown as Figure 26 is another early EMA image. The setting is the Jet Propulsion Lab. I had asked the artist to show journalists directing a rover on Mars in doing a photo shoot for a publication similar to *Life* magazine. Perhaps this, too, will soon come to pass.

Figure 25. Students lead the search for life on Mars in this 1990 EMA sketch.

Figure 26. The public directly participates in Mars rover operations in this 1990 EMA sketch.

The MDRS Remote Science Team – Earth / Mars Communication and Collaboration†

Shannon Rupert Robles and Stacy T. Sklar

Abstract

In August 2003, a remote science team was established for the Mars Desert Research Station (MDRS) that included the fields of biology, geology and multidisciplinary human factors. The need for a rigorous, coordinated science program at MDRS dictated a new, more active approach to the science being conducted there. The MDRS Remote Science Team (RST) completed its first successful field season beginning operations with Crew 20 and concluding with Crew 29. The group of twenty-one scientists was selected from past MDRS and/or FMARS crews, and so was able to share with incoming crews their knowledge of MDRS and the surrounding area, the expertise of their diverse disciplines and the challenges of doing science during a simulation. In addition, if a crew did not have any direct science goals, the RST developed experiments for these crews that either added to, or advanced, the science being conducted at the station. This first formal experience at running a remote science team looked at ways to accomplish science, via email and other forms of communication, between the crews and scientists back on Earth. The biggest challenge was finding adequate ways to communicate with the crews in order to facilitate a truly collaborative effort between the Earth-based scientists of the RST and the Mars-based scientists at MDRS, both prior to and during a rotation. Despite these challenges, we were able to complete a number of successful experiments, including continuation of a long-term microbial ecology study and a rotation in which all science was directed remotely by the RST. Other RST accomplishments during its first field season included setting the lab up for microbiology by soliciting several thousand dollars in donated supplies and equipment and resurfacing the lab countertops.

Introduction

In the NASA approach to telescience, an automated or human controller directs a robotic instrument deployed at a remote location or in space. In effect, these ground controllers direct the payload, whatever it may be. There are no human-human interactions, because that which is being controlled is not human. Early in the manned space flight program, the newly formed Mission Control Center needed to develop protocols for directing astronauts while they were in space. This was the first instance of ground controllers directing people. As the space program advanced, more control was given over to the astronauts; however, most of the flight was still controlled by people on the ground, including those in "back rooms" who assisted and advised the flight control team. In effect, for Mission Control to work, ground controllers needed to direct the humans in their endeavors.

Any mission to Mars will require much less direction from ground controllers here on Earth. Missions will be considerably longer and the time delay in communications will make it critical for astronauts to train and develop the ability and expertise necessary to respond quickly and correctly to any situation, without the benefit of help from a ground team. Hence, the Mission Support structure put into operation at the Mars Society's first two analog research stations was not one in which the ground controllers directed the humans, but rather assisted them as needed. The structure was a simple one; a Mission Support Director led a CapCom, who in turn led a group of discipline experts in science and engineering in their communications with the crew. This approach has stood the test of time and works well. Yet even as the position of Science Officer was expanded to include a Biology Back Room, interaction with the crew continued to be more one of support than direction. The need for a vigorous, coordinated science program that allowed for both effective communication and collaboration between the on-site scientists and those who were remote dictated a new approach to how science was done. For remote scientists to contribute to the long-term success of science being done at Mars analog sites, first we needed to develop procedures that allowed the Mars / Earth scientists to effectively work together. In its first year of operation, the Mars Desert Research Station's Remote Science Team (RST) addressed some of these challenges.

Background

The Mars Desert Research Station (MDRS) began operations in the spring of 2002. During the first field season, six crews completed rotations of two weeks duration. While Mission Support was operational during this time, the science support for these early crews consisted of the presence of a Science Officer on the Mission Support team. Assistance to the crew was limited to procurement of some supplies, information on operating the lab's equipment and the editing of and commenting for clarification on the daily science reports. There was no active participation in the scientific research by the Science Officer, who rotated not only between but also within crews. In addition, there was limited communication between Science Officers, and there were no procedures in place to pass down information. This limited the ability of the Science Officer to actively collaborate in the science being conducted by the scientists in the field. During the second field season, ten additional crews undertook rotations of between two and four weeks and the Mission Support team began taking a more active approach to science support. While some of the crews did not require or want outside assistance, several crews, most notably crews 14 and 18, were assisted in mapping, experimental design and selection of sites of scientific interest around MDRS. During the 2003 FMARS field season, some of the ideas generated at MDRS during the 2002-2003 field season were put into practice with the implementation of a Biology Back Room, made up of several experienced MDRS and FMARS crew members. This first formal experience at running a science back room looked at ways to accomplish science, via email, between the crew and scientists back on Earth. The experiment was a success and in August 2003, the decision was made to establish a Remote Science Team for MDRS that would include the fields of biology, geology, psychology and human factors. Twenty-one scientists, all former crew members at either MDRS or FMARS, volunteered to pioneer this effort. It became the responsibility of this newly created RST to decide the direction the RST would take in its approach to operating a remote science team, and the goals and actions required to be successful in that approach.

There has been a need for coordinated science programs at analog stations and for ways for scientists in the field to communicate, if not collaborate, with the field crews since the beginning of the Mars Analog Research Station

Program. Even before FMARS and MDRS were constructed, people were thinking about how a Mars crew would communicate with people on Earth. There are two ways for a crew to do effective science. The first is to advance the science – to contribute to the overall body of science already being done at the station in a new and innovative way. The second is to add to the body of science either by repeating something that has already been done, or by continuing an experiment started by another crew. Both of these approaches to science are valid and vital to the success of the science program at Mars analog research stations. The Remote Science Team (RST) needed to combine the best aspects of telescience, mission control and mission support in their contribution to these science efforts. The RST members shared their knowledge of the station, the surrounding area, and the challenges of doing science during a simulation. They also shared the expertise of their diverse disciplines in support of any science being conducted. In addition, if a crew did not have any direct science goals for their rotation, the RST designed experiments for these crews that added to, if not advanced, the science at the station. The approach to any given crew was flexible, and was determined prior to each rotation by the crew commander and the RST Coordinator. Communication of goals and needs by both of these key personnel were paramount to a successful rotation. Finally, the RST would become a storehouse for the valuable science being accomplished at MDRS.

Initial Approach to the RST

Prior to the beginning of the 2003-2004 field season, a set of protocols for operation of the RST was defined. Very early in the season it became obvious that several of the original operational goals could not be implemented, mainly due to uncertainty in crew assignments. This was unexpected, as crew selection and assignment had not been as difficult in previous field seasons. As a result, crew / RST communication prior to a given rotation was not optimal in most instances and lead to a less formal method of communication than originally planned. Overall, contact with crew scientists and commanders was haphazard, rather than organized. The shortest time the RST had to communicate with a crew was one week prior to the rotation; while in another instance there were six weeks between initial contact and the beginning of the rotation. While time was not the only, nor the major, factor affecting the working relationship between the RST and a given crew, generally more time between initial contact and the rotation created a better working relationship between the crew and the RST.

We have included here the original protocols as envisioned by the RST prior to the field season. Much of this was first suggested in a meeting at the Mars Society Convention in August 2003, when a Working Group for MDRS was formed. Following each numbered section is a paragraph addressing how successful each goal was, and how things were changed in the event the proposed goal could not be met.

Original Goals of the RST and Their Implementation

1. The RST Coordinator (RSTC) will coordinate with each crew commander (CDR) to plan protocols for interaction with the crew. The CDR and RSTC will contact each other at least one month prior to the rotation. The responsibility for this contact will be with the CDR and should be done as early as possible. The CDR will provide the RSTC with a "Summary of Science Objectives," which the RSTC must promptly review and return with questions and comments. In addition, the CDR and RSTC will create a "Plan for Interaction" which will define the parameters for the crew's interactions with the RST. These documents will define for each crew the parameters first discussed by the MDRS Working Group in August 2003. The RSTC will contact the team scientists that best fit with each crew and create the schedule of participation based on availability.

This goal was unworkable from the start. As mentioned earlier, this was partly because crew selection was often finalized without much time before the rotation actually began. In some instances crews only had time to work out the logistics of their rotation before arriving at MDRS. But more importantly, as time went on, it became clear that a more uniform way of dealing with initial contact and defining science objectives was needed. This original plan was created out of respect for the commander's right to define how his / her rotation was run, but it had the effect of creating more confusion that flexibility. In an effort to correct for this, during the 2004-2005 field season, all crew members will receive an email from the RST Coordinator detailing the role of the RST as soon as crew selection is finalized. The email will also include a request for a short summary of science being proposed by each science crew member. One of the original goals of the RST was to advance from requesting a short summary of science objectives to making it a requirement for each scientist to prepare a proposal for peer review by the RST prior to going to MDRS. The RST would be responsible for evaluating each proposal and providing feedback to the scientists. The role of the RST in this peer review process would be one of supporting the scientist. The RST would not be able to accept or reject a crew member based on this peer review process. The benefits to the process are twofold: one, it would strengthen the scientists work, as he / she would benefit from the experience of the RST; and two, all science being done as part of regular crew rotations at MDRS would be known to the RST, an important aspect of a coordinated program.

2. The RST will read all reports and provide feedback and guidance to the crew / CDR through the Point of Contact (POC) (currently the RSTC) on a daily basis. The Mission Support Director (Tony Muscatello) will be copied on all correspondence between the RST and crew. It will be the responsibility of the POC to distribute

the reports to RST members who are scheduled for duty. It will be the responsibility of the crew or CapCom to get the reports to POC. All feedback and guidance will be relayed to the crew via the POC.

This goal was successful for the most part. The RST Coordinator acts much like the Mission Support Science Officer in terms of reviewing all reports and providing feedback on a daily basis. The role of Science Officer in Mission Support was amended to Science Editor. The Science Editor edits and comments on the crew reports for posting on the MDRS web site, benefiting not only the crew and Mission Support, but the RST as well. For the entire 2003-2004 field season, with the exception of the two weeks the RST Coordinator was at MDRS, the RST Coordinator, and not a POC, was on duty. The original goal of having RST members rotate as POC for different crews did not happen. As the season progressed, the RST Coordinator began posting all science reports and some of the requests from the crew to the RST email list, and RST members provided expert advise as needed. The crew sent the reports through the Mission Support email system, which was very efficient. All feedback from the RST was relayed to the crew via the RST Coordinator. The Mission Support Director was not copied on all correspondence, as we soon discovered not all communication was mission critical. For example, if a crew member was having some difficulties running the autoclave, that series of emails between the crew member and the RST Coordinator were not copied to the Director or the RST. The one exception to this was Crew 25, when all correspondence between the RST and the crew went through Mission Support.

3. The RST will provide support and assistance to the crews as requested through interaction with the Point of Contact (POC). The Mission Support Director must first approve any support requiring purchase of materials and supplies. Support and assistance will be governed by the "Plan for Interaction."

This worked well. As noted above, the RST Coordinator was the POC for most of the field season. Most science related supplies and equipment were provided to the crews by the RST through donations and loans of major equipment. Some crew scientists brought their own supplies and equipment. The "Plan for Interaction" idea was never formalized, and was never used by any crew.

4. The RST will develop protocols and provide direction for crews that do not have a defined science program.

Aside from supporting the crew scientists, in several instances the RST did develop protocols and experiments for the crews who did not have a defined science program. The most successful of these was Crew 25, which will be discussed in greater detail elsewhere in this paper. In most cases, this idea proved hard to do in practice. One crew, for example, had a good start on an experiment they were doing for the RST, but the crew member in charge had to leave unexpectedly in the middle of the rotation. The work was picked up and completed by another crew member, but it was difficult for both the crew member and the RST, because everything had to be explained and learned all over again. Another crew, who asked for the RST to design an experiment for them, changed the long-term, multi-crew experiment in such a way that it had to be abandoned because the crew didn't tell the RST about the changes until the end of the rotation and they were not conducive to the long-term goals of the experiment. Crews would also decide not to finish an experiment they had requested from the RST, simply abandoning it before the end of their rotation.

5. The RST will provide documentation for the MDRS web site that contains a summary of each crew's science goals and objectives, a summary of the science completed during their rotation and a feedback report of the RST's interactions during each rotation. The RSTC will provide these documents to the MDRS Webmaster / MDRS Librarian (Gus Frederick). A standard system will be established early on and kept consistent. There will be a standard file-naming scheme and standard formats, (i.e., Excel for data tables, Word for shared reports, PDF for final reports, etc.) Only by standardizing reports will we eventually have data readily accessible by some search / database mechanism. An on-line searchable database is the one of the main goals of the RST. The key to this working smoothly is early adoption of standard formats and names. Protocols for naming can be found at: www.marssociety.org/mdrs/filename.asp. The web site will also contain information and biographies about the RST.

Since the RST was unable to get each crew to provide their science goals, the first part was never implemented. This meant that we did not create the summary database of all science being done at MDRS. We had a group of software engineers willing to help with this, and it would still be a good goal. The web site was operational at the beginning of the season, thanks to Gus Frederick, the MDRS Webmaster, although we are still in the early stages of developing it as focal point for all of the science at MDRS. Crew reports are becoming increasingly standardized, allowing for easier access and the ability to do searches in the future.

6. The RST will develop a summary list of all science done at MDRS and reference location of papers written. The RSTC will be responsible for procuring and posting a Summary of Science Accomplishments report from the CDR within two weeks of a crew's rotation. It is the responsibility of the CDR to make sure the report is written. The RSTC will coordinate with Publications (Frank Crossman).

This has not been done, mainly due to manpower and time constraints. It is something that should be done. There is a trend toward crews posting mission goals at the beginning of their rotation and a summary report of their accomplishments at the end. If this trend continues, it will be possible to easily keep an organized summary of science being done at MDRS. No follow-up to crew publications has been done.

7. The RST will assist in continued maintenance and upgrading of the lab at MDRS. Prior to the field season, the RST will clean and inventory the lab. This inventory will be updated and left at MDRS, along with instructions on how to maintain and operate the scientific instruments in the lab. These instruments are both donated and on loan, and so care should be taken not to damage them. The lab will be stocked with donated supplies (if possible). Each crew will be required to report which and how many of the lab's consumable supplies they have used via an email to the RST POC at the end of their rotation. They should also note any broken or malfunctioning equipment. This is to insure that future crews will know the exact state of the lab before they arrive at MDRS. The Musk Observatory will not be considered part of the MDRS science labs.

This was done for the 2003-2004 field season. The RST collected thousands of dollars of donated equipment and supplies, much of which was used or is still in use at MDRS. In addition we were given loans of several expensive instruments, including a bench top autoclave, lab oven, digital thermometer and pH meter. This equipment allowed scientists at MDRS to do microbiology for the first time. Unfortunately, this equipment was little used and was returned to the lenders at the end of the field season, although several other pieces have since been donated. The lab was cleaned and inventoried prior to the field season. It was also cleaned at the end of the field season, and, with the help of the San Diego Chapter of the Mars Society, new Formica countertops were installed. Unfortunately, a scheduled trip to MDRS in October to update the inventory was canceled. In the future, it should be a priority to keep the inventory updated, as it has proven to be the key to many crews' successes or failures.

8. The RST will document the effectiveness of the RST by proven methods and provide a self-study and review at the end of each field season. Lessons learned will be implemented as needed. The assessment technique to be used is TBD. RST POC will keep a Daily Communications Log and Pass Down, which will be distributed to all RST members via a communications platform still TBD. A meeting of the RST will be held each year at the Mars Society's annual convention to review the past field season's effectiveness and suggest improvements. Notes from this meeting will be sent to all RST and MDRS Working Group members for review and comment, before a final summary document will be developed. It will be the responsibility of the RSTC to schedule the meeting, request input, disseminate the information and implement the recommended changes.

This is being done. As we have learned what works, we have changed to become more effective. The Pass Down Log will only be needed if the RST uses a rotating POC, which has not been done yet. The yearly meeting was not held at the convention. The RST email system has proven to be an effective means of communication between RST members.

Accomplishments in the First Field Season (2003-2004)
The RST interacted with crews 20-29 to varying degrees. What follows is a brief summary of the interactions between crews and the RST and what was accomplished as a result.

Crew 20 – Microbial Respiration Experiment
This crew worked on a microbial respiration experiment that has been ongoing at MDRS since the first field season. Mid-rotation, the crew member doing this work had to leave due to an emergency and another crew member completed the work. Although this situation gave the RST an opportunity to train another crew member on an experiment in the middle of a rotation, this was very challenging and, unfortunately, the resulting data could not be used due to anomalies in the measurements between the two crew members.

Crew 21 – Remote Science Team Study and Collaboration
One of the questions the RST wanted to address was how to determine what was required for a crew to continue work started by a previous crew, because we feel very strongly that the ability to reproduce science is imperative for a strong science program at MDRS. We determined that a good start was to have a crew scientist replicate the results of another crew scientist's work. This study, the first formal project of the RST, looked at collaboration between different crews and between crews and the RST. Both authors were directly involved in this project. Rupert Robles was Commander of Crew 21 and Sklar was the RST Geologist. Shannon Hinsa was the RST Coordinator for the rotation and Tiffany Vora was the RST Biologist. Using a data set created in Season Two by Vora of Crew 11, we tested how well another crew could relocate her samples sites, and replicate her data. We had detailed data for five sample sites. Could Crew 21, using her database, find their way back to her sites and repeat the sampling? We came up with a set of protocols that we would use to test this, which also allowed for communication with the RST for clarification. The field crew had a printed copy of the database information for the five sites that included a written description of the site, GPS coordinates, and a photograph, taken using the dissection scope, of the collected samples. There were in situ photographs for only two of the five sites. During several EVAs, the crew attempted to locate the sites. In the end, only one sampling site was positively identified, and that was due to it being a very distinct location containing a unique sample site. For two other sites, the sample type and general vicinity were located, but we had no way to be sure we were collecting the identical sample as in Vora's database. For the

remaining two sites, we knew we were very close to the location, but could not confirm this, nor locate the sample. Two additional descriptors would have helped: an in situ photo for all samples and scale indicators, for example, how big or small the site was that held the sample. But we felt we needed even more information. We determined that what would have helped the most were better photographs of the site which put the sample and the sample site in the context of the general area it came from. In other words, if the sample was located under a boulder at the bottom of a rock outcrop, then a photo showing not only the boulder, but another showing the rock outcrop and then another showing the rock outcrop with it's outcrop neighbors would be a bigger help that more accurate GPS coordinates or a database with more written site descriptors. For our purposes, we felt a picture would be worth a thousand words and a series of photos, taken from agreed upon perspectives, would be an even greater help. Good as Vora's database was from an Earth scientist's perspective, we found that we were still extremely challenged by it's limitations. Clearly, a more thorough methodology was going to be needed for one crew to replicate an earlier crew's experiments. This was a simple project, but the outcome was the knowledge that things were going to be harder than we thought to replicate someone else's science.

Another activity completed by Crew 21 was scouting and selection of sites for an experiment to be conducted by Crew 22. The crew was instructed to find suitable sites in both the Morrison Formation and in Mancos Shale for the experiment. They were given a description of the site requirements and possible locations. They went out and found sites, carefully taking site photos and GPS coordinates. Unfortunately, these sites were never considered by the new crew, who selected other sites for the experiment.

Crew 22 – Surface Erosion Processes
The intent of the RST-designed Surface Erosion Processes Experiment was to have crews monitor a set of precisely excavated holes placed at cardinal directions on two hills over the remainder of the field season. The commander of Crew 22 had asked for the RST to provide an experiment early enough that the RST was able to develop this experiment, our first attempt at multi-crew collaboration. Unfortunately, this collaboration was plagued by communication problems. While the crew had enough time to review the experiment and ask for and receive answers to any questions they may have had, this creative crew altered the methods for the experiment and did not ask permission for, nor notify the RST of, these changes. These changes included increasing the number of holes the crew was required to dig. The original experimental design called for a total of 32 holes (16 each in a single hill in the Morrison and in Mancos Shale), but the crew had doubled that number. Since the holes needed to be 30 cm in depth and were being dug in sim, it was quite a bit of extra work for the crew. Midway through the rotation, the crew decided not to complete the experiment. Unaware at the time that the procedure had been altered, the RST requested that they complete the experiment and the commander did so. Only at the end of the rotation did the crew inform the RST that they had changed both the experiment and the pre-selected sites scouted by Crew 21. The changes were not wrong, just incompatible and impractical for the purpose of the experiment. In this case, we had a completely competent crew who didn't understand what we were trying to accomplish in terms of crew collaboration.

Crew 23 – Good Communication and Our Fifteen Minutes of Fame
The RST had little contact with Crew 23 prior to their rotation, so communication during their rotation was limited to some exchanges between the crew geologist and the RST for clarification on reports and findings. This communication, however, was all very positive. In addition, one of the crew was a reporter from England who corresponded with the RST Coordinator about the purpose of the RST and included a paragraph on our work in an article he wrote about his experience at MDRS.

Crew 24 – Microbiology and Support
This crew gave us the opportunity to look at how we support crews. We exchanged over one hundred emails with this crew, mostly on support issues such as how to run the autoclave and how to do microbiological procedures. While very little in the way of collaboration occurred, our communications with this crew's scientists were always professional and friendly.

Crew 25 – Directed Science
Crew 25 gave us the opportunity to take all that we had learned up to that point and put it into practice. When the crew's original commander decided not to participate less than a week before the rotation, the RST was asked to direct the crew's entire rotation. The crew consisted of five students and the last minute addition of a long-time Mission Support Science Officer and CapCom. What was most impressive was the way this crew, at the last possible moment, pulled together and, even with a complete change in their rotation's direction, did remarkable well. The work done by the RST and Crew 25 is quite possibly the most important work we have done for crew / RST interactions to date. This was due, in part, to what we had learned along the way, but it was mostly due to the crew's willingness to embrace the experiment and put 100% of their energies into making it a success.

We started by giving the crew a daily schedule. But our goal was to direct their research, and not their rotation, so it was up to them to decide if that schedule worked best for them. They did alter the schedule but those changes, such as staying up later than scheduled, never affected their work.

The crew was given the task of characterizing the sites where they collected their samples through photo documentation. We had taken a lesson learned from Crew 21, that even something as simple as returning to a sampling site could be a challenge, and has developed a methodology for documenting a site that we felt would be complete enough for another crew to return to the sample site. The approach we used was similar to landing on the surface of another planetary body in a robotic craft such as the MER rovers. The crew was instructed how, when exploring an area, to approach their investigation using perspectives going from global to regional to local perspective to microscopic. They would be required to document each sample site using the following five perspectives (the global perspective was omitted from the photo documentation and regional was taken from local USGS topographic maps):

1. Pan perspective (Similar to MER rover pan cam): Once at location, first locate it on the map, document GPS coordinates, take a panoramic (Pan) image with scale, using another persons' height as the scale including direction N, S, E, W.
2. Outcrop and/or Single Image perspective: Once at location, document GPS coordinates, take image with scale, again using another person or Jacobs' Staff if available, indicate where the samples are being taken (photo image with annotations including scale and direction).
3. Rock perspective: Document GPS coordinates, take image with scale of individual rock type and direction (in situ).
4. Image of rock back at Hab with scale (ruler): Label sample in photo so correlation of sample numbers and/or tags with Outcrop and slides for Microscopic viewing can occur.
5. Microscopic perspective: Take image using 3X magnification on the dissecting microscope.

This photo documentation methodology, created by Sklar and further refined by both authors, would lay the foundation for a more detailed exploration methodology later developed by Sklar and Rocky Persaud.

The RST created a list of twelve geology EVAs for the crew. Each EVA had a destination with coordinates, a priority of its importance to the RST, an estimated time to complete the EVA objectives, and a detailed list of EVA objectives, including notes of the significance of the selected location. Sklar also created a Geology Primer so that they would have a basic idea of what they were investigating on each EVA. It was the responsibility of the crew to determine which EVAs to complete each day. Once they had decided which EVAs to do, they sent their EVA Plan to the RST Coordinator, who approved it prior to the EVA. Although they did not get to plan their EVA locations, they did plan which EVA to do each day and who should be on the EVA team. They were extremely good at this and all EVA Plans were approved by the RST as submitted.

In addition to the twelve geology EVAs, there were five biology EVAs planned for the crew. They had been given the protocols for the Microbial Respiration Experiment and a list of sites to collect samples. As with the geology EVAs, each of the biology EVAs had a pre-planned destination, GPS coordinates to the site, a site description, the EVA's priority and estimated time for completion. These EVAs had to be completed early in the rotation in order for the samples to be processed and the data collected. The crew exceeded expectations. They completed the EVAs, conducted the experiment, and sent updated Excel sheets daily with their data. This was a valuable lesson for the RST. By getting the data updated daily, the RST was able to point to possible problems and correct for them. Had this been done with Crew 20, we would have been able to use those data. As an added benefit, at the end of the rotation, all data were in Excel format, and it was a simple matter of exporting them into the master file to include them in the Microbial Respiration study. This was the best work done by a crew for the RST. What was interesting, in retrospect, was that since the work itself was going so well, the crew and RST started to collaborate on the experiment. For example, one set of samples had initial masses that deviated from the norm. The RST noticed this and asked the crew member what she thought. Although the crew member was not a scientist, this prompted a series of email exchanges between the RST and the crew member that benefited both parties and led to another lesson: when communication is working well, collaboration can occur.

The volume of email between the RST and the crew during this rotation was staggering. Several hundred emails were exchanged over the course of the rotation and during the week prior to the rotation. Some of this could have been avoided, because there were numerous occasions when the RST responded to the same question from several different crew members, but most of the email was necessary in order for the RST to direct the crew's activities and for the crew to understand what the RST was asking them to do. The RST Coordinator personally took forty-two vacation hours from work in order to keep up with the demands of the communication required to remotely direct the rotation. And this was with a willing and able crew. Had their been personality conflicts, or had the crew's abilities been challenged by the workload, there would have been even more hours spent communicating.

Crews 26, 27 and 28
The time between crew selection and their rotations were short for crews 26, 27, and 28. Crew 27 worked very hard on GreenHab and Observatory projects, which are not managed by the RST. Crew 28 was lead by an RST member, and one lesson learned from this is that communication is much more relaxed when the crew member and the RST know each other.

Crew 29 – Mobile Agents

The NASA Mobile Agents project has its own RST, of which both authors are members. However, the Mobile Agents RST will be addressed in another paper and so will not be discussed here.

More information on Crews 20 – 29 can be found on the MDRS web site at www.marssociety.org/mdrs/fs03/.

The Future of Remote Science

Remote science isn't new, but it hasn't been practiced much lately, at least in terms of human-to-human interaction. While it is true that there are several remote science teams currently operating robotic craft on this and other planets, the only reference we have for communication between scientists on a planetary body and scientists on Earth are the Apollo missions, and they occurred more than three decades ago. Many of the lessons learned by the Apollo astronauts and Mission Control, with its science back rooms, are still true today.

The biggest challenge was and is communication and finding ways to improve it. This isn't a challenge for remote scientists directing robotic craft because communication is limited to the remote scientists commanding the craft and the craft doing what it is told to do. But while good communication is critical for a working relationship between field crews and the RST, it is very hard to accomplish due to technical issues such as what gets lost in translation when you put it in an email, personality conflicts, the "us versus them" mentality that can develop between the on-site crew and the remote scientists, and even the lack of time to communicate effectively because writing an email takes time and effort. Possible ways to eliminate these challenges are for remote scientists and the crew to communicate and even collaborate prior to the rotation, like the Apollo astronauts and their back rooms; to develop documentation methodologies and systematic procedures that aid in communication; or even to find different way to communicate other than email. In the future, the challenge of communication between the field crew and the remote scientists will need to be continually addressed and new ways of exploring how to communicate should be developed. It should be the role of the RST to facilitate this.

In order to do good science, however, we also need more rigorous science crew selection criteria, including a standard of education and a better way to assess crew member's capabilities. The MDRS RST currently spends most of its time supporting crew efforts because the crew members themselves are not capable and/or experienced enough to do what they are trying to do. The burden of crew selection should not be put on a single person, either; it should be the responsibility of a committee that includes members of the RST. They are a valuable resource because they know what it takes to do analog science. The committee should also be responsible for recruitment of scientists. In the future, the RST's role should evolve from one of support to one of guidance and direction.

There should be deadlines for application to a crew and for submission of science goals. Every prospective science crew member should be required to submit a proposal for a project to complete at MDRS. It may be that the scientist, but not his / her project, is selected for a rotation, but submission of the proposal will at the very least indicate some of prospective crew member's scientific thought processes. Those whose proposals are not accepted can apply to work on a project for the RST. Our experience has been that crew members who have a well thought out proposal do better and accomplish more than crew members without a plan. With only two weeks to complete their work, deciding what to do once they arrive at MDRS is counterproductive and a waste of valuable research time. We acknowledge that crew members are volunteers, but they still need to be held to a certain standard and be held accountable as productive members of a research team.

In addition, science crew members need to be forthcoming about their research. In a number of instances, crew members would not share their science goals and/or results with the RST. Peer review of proposals needs to be implemented; at the very least there needs to be a required summary of results. While it is true that station personnel do not manage science being done at research stations all over the world, all stations are aware and have a record of what science is being done at their site. That needs to be the case for MDRS as well.

Only once we have reached a high level of quality and consistency in the scientific research being done at MDRS will the crews and remote scientists be able to effectively collaborate on projects. In order to make that happen we need better communication methods, a rigorous crew selection process and accountability in the form of peer reviewed proposals and expectations of a crew member's ability to do research. This is not to say that anything done to date is bad; because we are so new at Mars analog research we are still on a steep learning curve. But the future of remote science is dependent on the ability of the field crew to effectively complete research in the field. Only when the RST advances its role from supporting the crew to collaborating with the crew will we have returned to the level of Apollo era remote science. We need to reach that level and beyond before we send a human mission to Mars.

Acknowledgments

The authors wish to thank the founding members of the RST: Melissa Battler, Dr. Brent Bos, Dr. Penelope Boston, Dr. William Clancey, Jonathan Butler, Dr. Jonathan Clarke, Dr. Steve Dawson, Julie Edwards, Brent Garry, Jennifer Glidewell, Dr. Shannon Hinsa, Elia Husiatynski, Dr. Steve McDaniel, Vuong Nguyen, Rocky Persaud, Dr. Maarten Sierhuis, Jody Tinsley, Dr. Tiffany Vora, and Dr. Nancy Wood. We would also like to thank all the members of

Crews 20-29, especially MDRS Crew 21: Melinda Capes, Nick Hall-Patch, Ashraf Hegazy, Mike Kretsch, Gus Scheerbaum and Crew 25: Kevin Sloan, Amy Blank, Dennis F. Creamer, Daniel Hegeman, Ryan Kobrick and Jason Schwier. Special thanks to Tony Muscatello, Robert Zubrin, Gus Frederick, Paul Graham, Gary Fisher and the San Diego Chapter of the Mars Society, especially Tim Sommer, for all their support, and Edward Martinez for his insight on the direction of the manuscript.

//

Mars Base Zero – Summer 2002 Productivity†

Ray R. Collins, Frances J. Collins,
Ruth Freeburg and Debi-Lee Wilkinson, ISECCo
http://isecco.org; Ray@isecco.org

Abstract

The International Space Exploration and Colonization Company (ISECCo) has been committed to the advancement of space colonization since 1988. Critical to the human habitation of space is life support technology which includes sustainable food production. ISECCo has built a semi-Closed Ecological Life Support System (Semi-CELSS) facility in preparation for designing a fully closed system. We are currently operating the semi-CELSS, named Mars Base Zero, to achieve a balanced diet that minimizes crop area and maximizes yield. Mars Base Zero uses strictly soil-based techniques such as one would use on Mars. The 2002 operation of Mars Base Zero and crop yield are discussed at length.

Introduction

Long-duration space missions and future colonization will rely on closed ecological life support systems (CELSS) to provide most of the needed food, air and water. With this in mind, the International Space Exploration and Colonization Company (ISECCo) began its experiments into developing closed ecosystems in 1988. Our current test platform, named Mars Base Zero, is a 24 x 36 foot (about 7.3 meter by 11 meter) greenhouse with an 8 foot (2.4 meter) wide apartment attached to one end.

Results

Mars Base Zero is a semi-closed ecosystem designed to provide food for one person. While the structure is not yet complete, we have been doing some preliminary experiments with productivity. In the spring of 2002 we planted about 485 square feet (45 square meters), which is a little more than half of the eventual maximum crop area of 864 square feet (80 square meters). The total available crop area is broken into six equal squares, each square being 12 feet (3.6 meters) on a side; currently only three of these are filled with dirt. We also had potted plants and planting trays in a square that was not filled with dirt.

During the 2002 season we did not close the structure, which is to say that the south half was completely open to the weather. This means that, for the most part, the climate inside Mars Base Zero was the same as the local weather, with the possible exception of the back corners where some heat would be retained by the sheltering structure overnight. Unfortunately for us, 2002 turned out to be an unusually cool summer. June and July were 2° F (1° C) colder than average, and August was 2.5° F (2° C) colder. A couple degrees may not seem like much, but it really clobbered the wheat and some of our other warmer-weather crops. The open structure also caused some problems with the soil getting too wet in the potato square. It started wet from snow melt, and never really dried out because it rained enough to keep it soggy.

On September 28, 2002 Ray Collins moved into Mars Base Zero. The next day he closed himself inside and for the following week he lived exclusively off Mars Base Zero produce. The trial run ended on October 6 in mid-afternoon. For the nine days Mars Base Zero was occupied, and the 166 hours Ray was closed inside the only food available in Mars Base Zero was what was harvested (on the days Ray was not closed inside the entire day – i.e., the first and last day – he did not eat exclusively from Mars Base Zero). The diet was primarily potato, though there was a practically unlimited supply of carrots, cabbage and turnips. Other foods that were in high demand, like onions and peas, were strictly rationed to assure they would last. Although never really hungry (he always ate whenever hungry), Ray reported a low average calorie consumption (averaging 1,675 calories a day) with some minor symptoms of calorie restriction: weight loss (2 pounds over the 9 days) and difficulty keeping warm when inside temperatures ranged down to 25° F (-4° C) – Ray normally does not get cold at these temperatures. (Note: the furnace was not set up for this trial, so the inside temperature was basically the same as outside, in spite of the fact that plastic was put up closing the south side of the structure.)

During the 2002 season we had some notable successes, such as a broccoli plant with a 4.5 pound (2 kilogram) head and potato plants producing 5.5 pounds (2.5 kg); and some notable failures: the beans didn't produce at all; the wheat did very poorly. Overall, the crops we would consider to be completely successful were the potatoes, carrots, cabbage, cauliflower, broccoli, lettuce and the herbs. The crops that did the worst were beans, peas and wheat. Intermediate crops were onions, spinach and turnips. Potatoes have almost always proven to be our most productive crop so we planted an entire square – the single largest area for any crop – in potatoes. Wheat is probably the most useful crop one can grow, so we devoted quite a bit of area to it (83 square feet – 7.7 square meters) but it failed. In order of increasing area we also planted Early Scarlet Globe radishes, self-blanching cauliflower, Salad Bowl leaf lettuce, spinach, Bloomsdale long standing spinach, Miracle Sweet tomato, soy beans, red beans, Royal Chantenay carrots, black beans, pinto beans, navy beans, dry onion bulbs, Purple top turnips, Green Arrow tall peas, and Provider bush green beans. We planted tomatoes, Lemon Balm spice, Candy mint, Siam Queen Thai Basil, and celeriac in pots. Shogun broccoli, Mid-season cabbage, and self-blanching cauliflower were planted in both the planting squares and pots. (See also Table 1.) The following is a description of each plant, with comments and a description of next year's plans.

Crop	Area planted, m^2	Growing period, plant-harvest, days	Total yield, grams	Total yield/m^2	Grams yield/m^2-day, (yield/m^2 divided by days)	Calories per m^2 per day
Broccoli – in pots	0.56	5/25-9/15 : 119	1020	3642	30.6	9
Cabbage in Pots	0.28	5/25-9/30 : 128	6519	8809	34	16
Cabbage-Mid Season	0.74	5/25-9/29 : 128	3261	4407	34	9
Carrots – Royal Chantenay	0.93	5/26-10/1 : 129	3371	3664	28.4	12
Cauliflower in Pots	0.28	5/26-8/19 : 91	2762	9864	108	27
Cauliflower-Self Blanching	0.12	5/26-9/2 : 105	1156	9632	92	23
Celeric	0.09	6/2-8/26 : 91	280	311.6	31.8	12
Lettuce-Salad Bowl	0.14	5/28-8/26 : 98	1297	9267	94.6	17
Lettuce – 2nd crop	0.14	7/13-9/6 : 56	133.1	951	17	3
Onions-Dry Bulbs	1.53	5/26-9/2 : 105	2125	1389	13.2	5
Potatoes – All	11.14	6/2-10/1 : 121	31.03	2,785	23	24
– Purple	4	121	4.33	1,080	8.92	9
– Red	1.14	121	2.4	2,087	17.25	18
– White	6	121	24.3	4,050	33.47	35
Radish – Early Scarlet Globe	0.56	5/26-7/13 : 49	189.8	340.7	6.95	1
Spinach-Bloomsdale	0.28	5/28-7/13 : 56	357	1275	22.76	5
Spinach (2nd crop)	0.14	7/13-9/6 : 56	65.15	465.4	8.31	2
Turnips-Purple Top	1.86	5/26-9/28 : 128	3279	1763	13.8	4

Table 1: Crop yield data

Green Beans

We planted the green beans in what turned out to be the worst spot in the garden. During the winter of 1999-2000 Ray and Frances spent the winter living in Mars Base Zero, and we had an emergency space heater set up in the area where we planted the beans. In the process of fueling the heater small amounts of diesel fuel got spilled. This may be why the beans (and also peas) did poorly. The location is also shaded in the early morning. Since beans grow quite well here in Alaska their failure was probably a combination of the soil contamination and the lack of sun in the morning. Though we did get a few pods, the yield was basically zero.

Next summer we'll put the beans in a different location that has more sun. We may also dig out some of the contaminated soil so what we do plant will grow better! Green beans should be planted by June 15 to harvest in the middle of September.

Peas

Although we did get a few peas, we were overall very disappointed in their productivity. The plants got about 3 feet tall (1 meter), which is a great deal shorter than they normally get to in Alaska (6 feet – 2 meters – is common). They were planted against the back wall so we could use the wall to attach the fence for them to climb. Possible causes for their slow growth: lack of *Rhizobia* bacteria (a bacteria which fixes nitrogen for plant use); shading (like the beans, the peas were planted in the very back of Mars Base Zero); lack of water (planted along the very back edge, which may not have gotten properly watered on a regular basis) and possibly some of the soil was contaminated, since some of the peas were planted near the area where the fuel spills occurred.

Next summer we will probably plant the peas in the same location. To insure they grow better we may pre-germinate them; dust them with *Rhizobia* bacteria; and make a better effort to keep them well watered. We may also remove some of the possibly contaminated soil. Peas should be planted by June 15 to harvest in the middle of September.

Onions

Onions are one of the most productive crops that can be grown (in terms of calories produced per square foot per day). Unfortunately that isn't so in Alaska. We've grown onions many times, but in general we've had very low productivity. This summer Mars Base Zero did not prove any different. Though they did grow, the largest wasn't much more than 2 inches (5 cm) in diameter, and the total yield was fairly low. Onions are sensitive to low fertility, and it is possible we didn't have fertile enough soil. Alternatively, the soil may not have been warm enough for optimum growth. By late August temperatures were cool enough so the onions started to senesce for the winter, so we harvested them.

Next summer we'll try growing onions in pots and trays, to try to get the soil warmer to encourage faster growth. We can also plant them earlier in the spring, to extend the season. It may be worthwhile to start a few in April, to see if we can get full-sized onions before it gets too cold for them.

Tomatoes

Tomatoes do not grow very well outdoors in Alaska. We planted them, but they frosted before we got any harvest. Next year we'll start the tomatoes a little earlier, probably the middle of March, and attempt to keep them a little warmer.

Turnips

Turnips usually grow very well in Alaska. Unfortunately they are sensitive to bolting (flowering before growing into harvestable food). We planted our turnips in the middle of a back square, in what may be one of the warmest areas. This turned out to be a mistake because the turnips tended towards bolting rather than growing large tubers. Though we did have a fair harvest, it is nothing like we were expecting.

Next summer we will plant the turnips closer to the outer edge of the greenhouse, and try to keep them a little cooler. Late may planting should yield a good harvest by the middle of September.

Lettuce

The lettuce grew nicely at the south end of one of the troughs. From a little more than 1 square foot (0.3 square meters) we harvested lettuce leaves for 6 weeks totaling 50 ounces After the 4th week we started a second planting of lettuce and spinach in the open space that had been growing spinach (see spinach). The second planting had a good yield, but by the time the lettuce really began to mature the cooler temperatures had set in and continuing harvest slowed down. The first harvest was at 7 weeks, with a more robust harvest the following weeks for the first planting. The second planting (planted at the 8th week) yielded the first crop at week 14.

Next year we want to plant about the same area. To have a harvest in the middle of September we will want to have a planting in the middle of July. Naturally we'll plant some earlier so we can have fresh lettuce for our volunteers during the summer.

Spinach

We planted equal amounts of spinach and lettuce in the front of the greenhouse. Unfortunately, spinach tends to bolt due to the long daylight hours here in Alaska. We first got spinach about week 7. By week 8 it had started to bolt, and by week 9 it had nearly all bolted. Even so, we did harvest 12.6 ounces (340 grams) from 56 plants from our first planting. The second planting also bolted but we managed to get 2.3 ounces (64 grams) over 3 weeks of harvesting. Timing of the plantings was the same as for the lettuce.

Next summer we'll plant our spinach much later in the summer, and we'll plant it in the back of the greenhouse to reduce light and hopefully inhibit bolting. We'll also try other varieties and we hope to find a bolt-resistant one. Another variety of spinach (mustard spinach) grows more like lettuce and Debi-Lee has grown it in an Alaskan garden more successfully than the taller varieties. Seven weeks before the middle of September is late July, so we should plant our spinach in late July.

Cauliflower

We harvested four lovely and tasty heads of cauliflower yielding about 12 cups (about 3 liters) per head. Interestingly, the three plants in pots did better (they could have been harvested 2 weeks sooner) than the one plant in the regular square. It was situated three quarters of the way to the back (to the north, which is shaded in the morning and evening) and may have done better farther south, where there was better sun. Also the pots probably had better soil temperatures. All the cauliflower heads yielded very close to 2.5 pounds (about 1 kg) each. The first three heads were harvested at week 13 – although the other two of them could have been harvested at week 12, but we didn't have anyone available to do harvesting then. This number of weeks refers to the length of time since they were transplanted into Mars Base Zero; the plants were about a month old when transplanted. Total growth time was 16 weeks.

Next year we will plant all our cauliflower in pots. To have a harvest during the middle of September, we will need to plant our cauliflower (from seed) in the middle of May. Four to six plants will probably be adequate again.

Broccoli

Conversely, the broccoli plants grew better in their corner of the 12 foot by 12 foot (3.6 meter) squares than in the 1+ square foot ($1/3$ meter) pot. However, only one of two plants in the square survived. These were started four weeks early (in pots) and one did not survive the transplant, possibly because of not enough water. The first broccoli (all 4 pounds – 1.8 kilograms – of it!) was harvested at week 11. Since the broccoli was from pre-plants (like the cauliflower, 4 weeks old), total growth time was 15 weeks.

Next year we will plant all our broccoli in the squares. We'll want to plant four to six plants from seed in the 3rd week of May to harvest in the middle of September.

Dry Beans

We experimented with black beans, navy beans, pinto beans, red beans and soy beans. We planted them in the middle back crop square. Most of the beans were pre-sprouted in rag dolls (a rolled paper towel, kept moist). They all grew very well, reaching 3 to 5 feet (1-1.5 meters) in height and climbed nicely around the posts. They did not have enough warm weather time to produce beans but did have a few flowers. We had a number of freezing nights that slowly killed them back in the last month (September) of operation. The soy beans did well compared to earlier experiments which indicated that they need a hot climate. With better summer weather and starting them early enough in the spring we may coax a crop from them.

Next summer we'll plant beans again, but we intend to start them very early (April?) in the hopes that we'll get a long enough season to actually have something to harvest in September. Keeping the greenhouse heated in the fall will help.

Cabbage

We planted two cabbage pre-plants (four weeks old) in the 12 foot (3.6 meter) squares and two in pots. Like the broccoli, only one grew in the squares but it outgrew the potted ones. Cabbage is a good Alaska crop since it both grows well and survives a hard frost. Although not as large as we've seen it grow in gardens, our cabbage was a good size, weighing 6 to 8 pounds (3+ kilograms) each.

Next summer we'll plant the cabbage from seed in early May to have a harvest in the middle of September.

Carrots

Actually, the carrots did better in total yield per day at one ounce (28 grams) per square meter per day over the full 19 weeks of operation than the potatoes at 0.8 ounces (23 grams) per square meter per day: calories per square meter per day is the most important factor for Mars Base Zero. Since carrots have fewer calories per ounce (or gram) than potatoes; their productivity is actually a lot less in terms of calories produced per square meter per day. See Table 1.

This carrot variety yielded small – 1 to 6 inch (2.5-15 cm) in length – but normal diameter, that were very sweet. Early harvest is determined by crowding. We actually had a terrible germination rate; only half the carrots planted came up (67 of 145) though most of the carrot patch looked full. Very likely germination failure was due to insufficient water. Next year we will put a layer of plastic over them until they sprout, which will not only keep the soil moist, but it will heat the soil as well (encouraging faster germination).

There should be no change in how we plant carrots next summer; we'll plant them as soon as the greenhouse is warm enough. We do want to see how well they grow in trays, however. So we'll plant them both in squares and in trays to compare growth rates and yield.

Celeriac

We planted one stalk in a pot and harvested a root ball almost the size of a softball. Celeriac can be used somewhat like celery, and is a good source of selenium – a mineral that is very hard to get enough of on the Mars Base Zero diet.

Next summer we'll plant another pot of celeriac, probably using a pre-plant again.

Wheat

Wheat is a very marginal crop in Alaska, and any summer that is even slightly cooler than normal will cause the wheat to fail. We had a cooler than normal summer this year, and with the open structure our wheat did very poorly. The poor overall health of the wheat led to serious aphid infestation, which only aggravated the situation. If we'd been relying on wheat for food we'd have gone pretty hungry!

Although wheat is such a marginal crop for us, it is so useful as a food that we'll try again next year. We expect to increase our total area planted by 30%, so we'll have the area to try the wheat again. We'll also try a few other grains like oats and barley for variety.

Potatoes

In the past, potatoes have been our best crop. They did not disappoint us this year either. We planted 138 square feet (11.14 square meters) in potatoes. The most productive potato plant yielded nearly 6 pounds (2.6 kilograms) of potatoes. We were a little worried that the soil was too wet (it started wet, and never really dried), but they grew well anyhow. We will probably plant the same amount of area next year, with no change in fertilization or watering schedule. Total yield in potatoes was 68 pounds (31 kilograms). If the entire planting area had been planted in potatoes we would have been producing an average of 2,600 calories a day, which is marginally enough to support a person.

This year we planted three different varieties of potato. The white potatoes (Yukon Gold) out-produced the others with a total yield for the season of 0.8 pounds per square foot (4 kg/m^2); red potatoes produced second best with a yield of 0.4 pounds per square foot (2 kg/m^2), and the purple did worst with only 0.2 pounds per square foot (1 kg/m^2). The Yukon Gold variety produced 3.2 calories per square foot per day (35 calories per square meter per day). If all 864 square feet (80 m^2) were planted in these potatoes the yield would be 2,800 calories per day – continuously.

We were careful to keep the potatoes in the front half of the greenhouse to prevent the soil from getting too warm since potatoes fail to set if the soil temperature is more than 68° F (20° C). The soil temperature in the back parts of the greenhouse never exceeded this, so it may be an option to plant them further back.

Next year we'll plant about 50% more potatoes that we did this year, and we will use only white potatoes, though we may try a few different varieties of them, such as Yukon Gold and Bake King. We'll also try to keep the soil somewhat dryer. We want to try some experiments with crowding, for some of our data indicated that the potatoes were getting too crowded because the margins did far better than the centers. However this could also be due to warmer soil temperatures, so we will also experiment with planting potatoes in trays and farther back in the greenhouse, to see how well they grow in warmer soils. Some of the data indicated the boardwalk (a board which runs over the center of the potato patch to assist in weeding and watering – the entire 12 foot by 12 foot (3.6 meter) square was a solid bank of potatoes) may have reduced productivity of the plants immediately beside it (due to shading). We need to investigate this possible effect too. (The plants seemed to be growing unimpeded by the plank, but we need to justify this observation quantitatively).

Conclusion

Mars Base Zero was not fully operational in the summer of 2002. In spite of that, we had a crop sufficient to support a person for more than the week we actually had someone closed up inside. The crop was lacking in certain dietary aspects, partially due to the failure of certain crops. (For example vitamin E consumption was less than 20% of RDA; calcium was around 35% and riboflavin was 40%.) However we feel the trial was a complete success since we managed to close the door this early in our experiments. Further trials will stretch our operating capabilities, and as we work toward getting the structure finished with all the available crop areas planted we will be able to have longer and longer closed periods.

For Mars Base Zero to support a person it needs to continuously produce food. Our current method of operation does not allow continuous production, but it will allow us to build the operational skills needed to run Mars Base Zero continuously when the structure is complete. Although our planting techniques are different now than they will be when under continuous operation (for example, with continuous operation new plants can be planted before the old ones are harvested) we can get a general idea of total productivity. If our goal is to have Mars Base Zero support a person (though we don't really expect it to, it may come close), we need to harvest around 2,500 calories per day. This year we had about 55% of the total area planted for about 120 days. About 40 days of this were probably not very productive (plants don't tend to do very well when nearly frozen), so our probable growth period was 80 days. 55 percent (the percent of Mars Base Zero we planted) of 80 days is 44 days; in order to support a person for 44 days we should have harvested around 110,000 calories. Our actual harvest was closer to 40,000 calories – far short.

During the summer of 2003, we will expand our crop area by around 144 square feet (11 square meters) nearly doubling the potato area and adding a number of crops. We are already experimenting with peanuts, and we hope to have a successful crop (good for vitamin E). We may also try rice. These hot weather crops will require a sheltered area and a lot of extra work, but, if successful, they will significantly improve the diet. We will also try less radical crops such as sunflower seeds (also a good source of vitamin E), barley and oats. In the fall of 2003 we hope to run Mars Base Zero closed for longer than the week we did this fall.

//

The MarsSim Project†

Carlos Echagüe, Ignacio Puig Moreno, Matías Traverso, Manual Pena and Gabriel Rshaid
Cardinal Newman College, Buenos Aires, Argentina, grshaid@marsacademy.com

Introduction

The MarsSim project consists of an on-line collaborative educational simulation of a manned mission to Mars. A Mars Simulator is being built at Cardinal Newman College, in San Isidro, Buenos Aires, Argentina. This K12 bilingual school will be coordinating the project and hosting the simulator that will be connected to the Internet. In this way, both a physical and a Virtual Mission Control (through a web-based simulation) will be carrying out the mission simulations.

The educational goals of the project are to foster the learning of math, science and other disciplines through a hands-on approach, in a natural way that allows students to interact with each other in a fun environment and using state of the art technology.

The Mission Architecture

In order to provided a working base for the mission, a baseline design was selected. It is important to emphasize that the mission design does not attempt to be original but only accurate enough to provide data for a realistic simulation.

The mission will be a Mars Direct type mission, that has been proven as the most effective and convenient. The spacecraft will be propelled by an Ares class booster, four Space Shuttle Main Engines, Advanced SRBs and a Nuclear Thermal Reactor for the TransMars Injection.

There will be four crew members and only six Mission Control stations for simplicity. The mission will also include artificial gravity of 0.38G generated by rotating the spacecraft with a tether joining the spent NTR module and the rest of the spacecraft.

Mars Descent will be accomplished by initially aerobraking with a biconic shell, parachutes and finally a powered descent phase using four RL-10 engines.

Mars Surface Rendezvous will be done with the fully fueled Earth Return Vehicle, which will take the four person crew back home.

In a long duration mission, the Life Support System's job is crucial. The basic function are to remove carbon dioxide, to generate air, to remove air contaminants and to produce vital water. Carbon dioxide will be removed with a CRS system, which uses carbon dioxide and hydrogen to produce water and methane, thus solving two problems at the same time, carbon dioxide removal and water generation, as well as urine and waste water purification. As regards oxygen generation an Oxygen Generation System will be used, which uses water and produces oxygen and hydrogen. To remove air contaminants a TCCS (Trace Contaminants Control System) shall be used.

Educational Project Synopsis

This project is based on an educational simulation, not a high fidelity one. The difference between these two lies in the complexity of the sims. A high fidelity simulation aims for the highest amount possible of details to emulate as accurately as possible the real mission. In an educational simulation, such as ours, many processes and data have been simplified to make the project understandable and enjoyable for high school students with no special background or education on the subject. The layout of the spacecraft and mission control, which are currently being built in the Robotics room in our school, consist of a spacecraft replica and mission control stations. The spacecraft itself will be about three meters high and two meters in diameter. It will have two levels, the top one for mission operations and the lower one for experiments. It will have panels connected to a PC. This PC will be connected to a Server, which will also connect to Mission Control. Mission Control will be in the same room, it will have fours stations for simplicity with one computer per station.

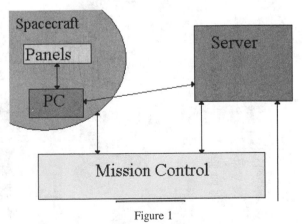

Figure 1

A unique variant on the simulation will be the Virtual Mission Control, which basically consists of people in other places or countries being able to act as the physical mission control by means of the internet and a web site specifically designed for this purpose.

Information Flow

The information flow through the whole simulation will be: from the panels of the spacecraft to the PC via parallel port. A Visual Basic application will regulate the flow. Between the PC on the spacecraft and the server, and from the server to Mission Control and vice versa, a Flash application will control data transfer. The Server application will constantly check for any changes in the panels on the spacecraft and the computers in Mission Control (either physical or virtual) and update any change taking into account the time lag due to long distances in this type of missions.

There will also be a direct audio link between Mission Control (physical) and the spacecraft to allow for direct contact with the astronauts during solution of problems. The link between the spacecraft and Virtual Mission Control will be conducted through a chat interface. The information transfer between Virtual Mission Control and the server shall be via Internet. Other features will be built-in web cams that will allow visual recording of the mission that will be updated after a certain period of time and the inclusion of the time lag in every data transfer from the spacecraft to Mission Control to emulate the actual time lags due to distance in the real missions.

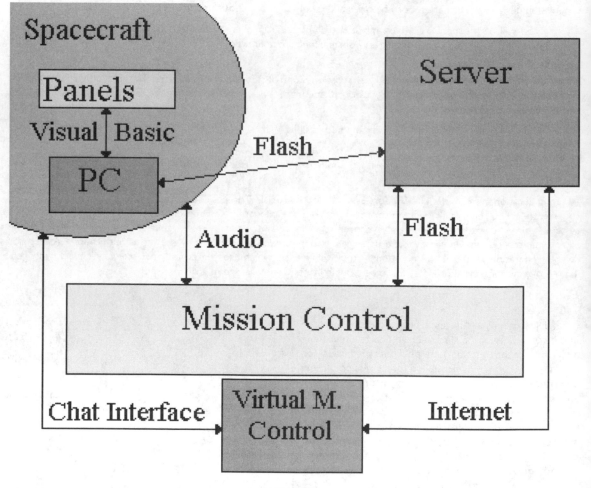

Figure 2

The Simulations

For the actual simulations previous training will be provided, both in situ for people who are interested in coming to our school, and on line for those who are far away. There will also be practice runs with the crews to allow them to familiarize themselves with the equipment. "Real" simulated missions shall be planned and implemented, using the same time schedules and real data from Mars missions. The characteristics of these simulations will be the inclusion of problems of different nature during the mission that the crew will have to solve on the march. These problems will be of non-catastrophic or time critical nature. There will be no nuclear reactor meltdowns or meteorite collisions because there is simply no way for the crew to solve them. There will be plenty of consultation among the crew and mission control (physical or virtual). Consultation and collaboration between each of these people will be crucial for the solving of these problems. This simulation is designed in such a way to allow students with little training on the subject to get a glimpse at what an actual Mars mission might be like and learn from this experience. Future developments include simulated experiments that will take place on the lower floor of the spacecraft. These experiments will be designed and implemented by actual students from other schools, thus allowing other students to bring their own ideas to this project, which will allow them to make significant contributions to the project.

//

Project Greenhab at the University of Maryland – Development† of a Research-Scale Life Support Greenhouse

David Blersch and Patrick Kangas
University of Maryland, dblersch@wam.umd.edu

Abstract

Project Greenhab is an ongoing research and development initiative to explore life support technologies suitable for the missions at the Mars Society's remote research stations. Evolving from earlier efforts of the Mars Society's life support technical task force, the Greenhab project has developed a low-cost greenhouse that is an analog of inflatable structures that might be used in an actual manned mission to Mars. The modular design of the greenhouse has allowed staging of segments at various locations around the country, providing a platform for various research initiatives within the life support context. One such segment, located at the University of Maryland's Department of Biological Resources Engineering, has been used for prototype engineering, education and R&D. Prototype engineering began with segment installation in January 2002 and is ongoing as mission support for a similar Greenhab segment at the MDRS in Utah. Research continues at the Maryland segment to develop a biologically-based waste water recycling system, based on living machine technologies, appropriate for Mars Society field simulation sites. Finally, educational opportunities created around the Maryland segment included a graduate course in ecological engineering, providing a successful model of university partnerships with Mars Society research initiatives.

Introduction

Project Greenhab is a technical task force within the Mars Society that focuses on several life support issues of possible human bases on Mars. Although many volunteers from the US and other countries have participated, one focus of effort has been by a group in the Mid-Atlantic region. Early work of this group dealt with the design of a greenhouse-based waste water treatment system for the simulated Mars habitat on Devon Island in the Canadian Arctic. Alternative designs were surveyed[1] and a specific design was offered in the form of a modified living machine.[2] In 2001 work shifted to applying the living machine design to the Mars Desert Research Station (MDRS) in south central Utah.

Several alternatives for waste water management at the MDRS are being explored but the living machine concept has been emphasized. A living machine is a waste water treatment system that combines conventional technological components (plumbing, pumps, etc.) and aquatic ecosystems that are contained in tanks connected in a flow-through pattern.[3,4] Both anaerobic and aerobic tanks are included along with a high diversity of plants, aquatic invertebrates and microbes. Treatment of waste water in a living machine occurs by physical-chemical processes (sedimentation, filtration, absorption) and by biological metabolism. This is a form of ecological engineering since constructed ecosystems are employed for a practical function.

The greenhouse-based living machine design is being implemented for the Mars Society's simulated Mars habitats. Each system consists of an external greenhouse structure and a living machine constructed inside the greenhouse with various water and power interfaces to the habitat. Gary Fisher of the Mars Society has designed the greenhouses and has supervised their construction.[5] The living machines have been designed and constructed by the co-authors of this paper. A first generation system was constructed on the University of Maryland at College Park (UMCP) campus as a prototype design. This system was tested at the MDRS in Utah and, based on experience from the test, a second generation design has recently been constructed there. The purpose of this paper is to describe the first generation system on the UMCP campus. This system is being used both to test design details for implementation at the MDRS and for educational applications by university students.

The University of Maryland Mars Greenhouse System

The overall concept of the Greenhab greenhouse is a cylindrical structure with a framework made of standard 1" PVC pipe reinforced with steel electrical conduit. To manage cost and to allow reproduction by other interested parties and hobbyists, the greenhouse is constructed with commercial off-the-shelf products normally available at a

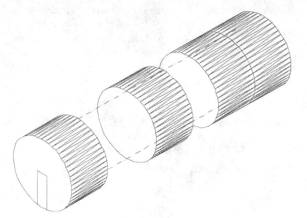

Figure 1. Rigid cylindrical segment concept to create a terrestrial greenhouse that mimics the shape and size of an inflatable Mars greenhouse.

local hardware store. The cylindrical shape simulates a possible inflatable greenhouse structure that would be attached to a habitat unit on Mars. To implement the Greenhab greenhouse modularly in a rugged terrestrial setting, however, a segmented rigid-structured greenhouse concept was pursued for modular implementation. Each segment is a cylinder on its side 8 feet (2.5 m) long and 13 feet (4 m) in diameter. Segments may be built and joined together axially as needed to simulate inflatable cylindrical greenhouse concepts (Figure 1). The outer covering of the greenhouse is composed of double-layer corrugated polyethylene sheet, which provides a low-cost translucent exterior. The end caps on the greenhouse are custom-sized translucent tarps that can be removed as necessary for maintenance work. For the Maryland segment, the entire structure is built on a large wooden pallet located in the enclosed work area of the Biological Resources Engineering Department on the UMCP campus (Figure 2).

Figure 2. The Greenhab greenhouse segment at the University of Maryland.

The living machine inside the greenhouse segment consists of three translucent polyethylene tanks connected in series (Figures 3 and 4). The tanks are elevated in such a way that water will flow by gravity from Tank 1 to Tank 3. A sump pump in Tank 3 connected to a flexible hose and sprayer manifold drives continuous recirculation from Tank 3 to Tank 1. Tank 1 is a 10-gallon (40 L) rectangular tank filled with 1" diameter plastic bioballs to create a trickling filter unit process, providing an aerobic environment for communities of attached-growth microorganisms. Tanks 2 and 3 are aerobic 30-gallon (120 L) cylindrical tanks. Tank 2 contains floating-leaved and submerged aquatic plants, dominated by water hyacinths, and Tank 3 contains submerged rocks for extra habitat space and a floating rack for terrestrial, potted plants. A fourth tank, located on a rack below Tank 1, could serve as an anaerobic tank, but it is not plumbed into the system at present. During operation, waste water would flow into the living machine either at the anaerobic tank (i.e., the presently unconnected Tank 4) or at the trickling filter. Continuous recirculation of the water from the last tank to the first increases the effective hydraulic retention time in the system, thus improving the overall treatment process. The UMCP living machine is currently a closed water loop (i.e., no system outflow, except for evaporation). In field implementation, however, outflow would occur from Tank 3, either as overflow to an underground drain field, or, following additional filtration and sterilization, to a recycled use such as for irrigation of food plants or for toilet flushing.

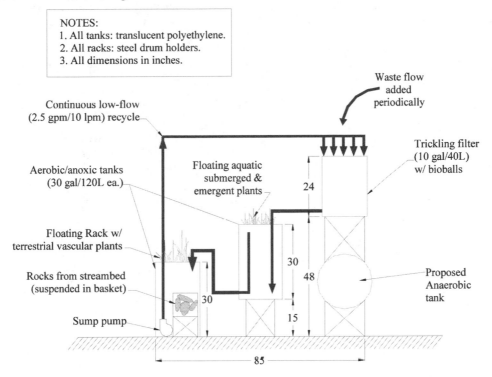

NOTES:
1. All tanks: translucent polyethylene.
2. All racks: steel drum holders.
3. All dimensions in inches.

Figure 3. Schematic flow diagram of the UMCP Greenhab living machine.

Figure 4. The living machine tanks in the UMCP Greenhab greenhouse.

Applications

The primary purpose of the Mars Society greenhouse at UMCP has been to test design features for applications at the simulated habitat field sites. Preliminary work on greenhouse sensors has been conducted[6,7] and heat and power budgets are currently being studied. Another emphasis has been on waste water treatment performance. Figure 5 shows the results of one short-term experiment in the living machine. In this case, a pulsed addition of partially-digested dairy waste water was used to simulate input of human sewage. 1.5 L of waste water at 13,800 mg/l chemical oxygen demand (COD) was added to the system, approximating the organic waste load of two crewmen per day if averaged over the total volume of the system. COD is a measure of the concentration of organic and other oxidizable compounds in the water, similar to biological oxygen demand (BOD). After the sewage was added to the system, the system was allowed to continue recirculating undisturbed for two days. One water sample was collected from each tank immediately before sewage addition, one hour after sewage addition, and two days after sewage addition. These samples were analyzed with a Hach 2000 spectrophotometer for concentrations of COD, ammonium and nitrate. The results of these measurements show that both COD (Figure 5a) and ammonium (NH_4^+)

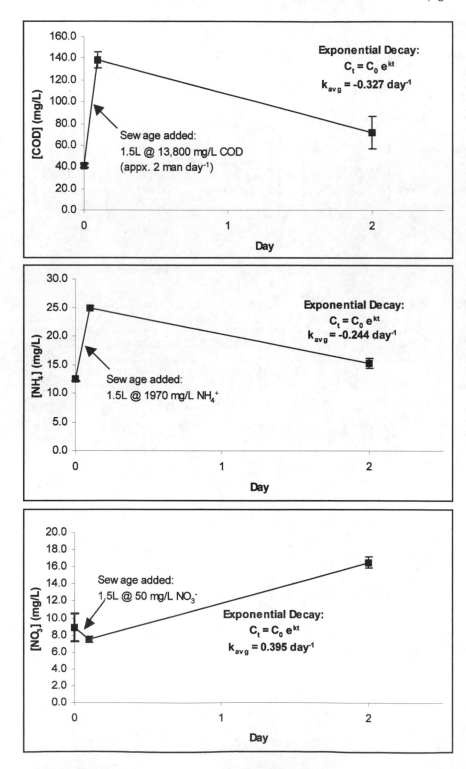

(Figure 5b) declined during the experiment while nitrate (NO_3^-) (Figure 5c) increased. The opposite pattern of ammonia versus nitrate concentrations represents nitrification, the microbial metabolic conversion of ammonia to nitrate. As a preliminary characterization of the treatment potential of the living machine, removal rate constants have been calculated for each of the chemical species assuming a first-order exponential relationship. All of the patterns shown in Figure 5 demonstrate the oxidation of the waste water by the ecosystems of the living machine, a dominant process in waste water treatment.

Figure 5a (top). UMCP living machine COD concentration vs. time.

Figure 5b. UMCP living machine ammonia concentration vs. time.

Figure 5c (bottom). UMCP living machine nitrate concentration vs. time.

In addition to effectively treating waste water with ecological processes, the UMCP living machine has proven to be amazingly robust and resilient to perturbations and intervals with little or no maintenance. The use of large-diameter plumbing connections prevents the likelihood of system clogging. The mechanical requirements are few: only one low-power recirculation pump is necessary to keep water flowing through the system. The biological components incorporated into the system organize at the ecosystem level: the high level of biological diversity introduced at the start of the system provides countless metabolic pathways for nutrient and organic removal and affords system resiliency from extreme environmental events or disease. In addition to its reliability, the living machine has been easy to build and maintain. The mechanical components of the system were assembled in a day using second-hand and available parts, and maintenance and repair operations have proven to be relatively easy. These characteristics make the living machine concept a desirable candidate waste water management technology for operations in remote locations.

Because it is located on a university campus, the UMCP Greenhab greenhouse is also being used for educational purposes. For example, research was conducted in the greenhouse by graduate students in the ecological engineering course (ENBE 688D) during spring semester 2002. As one of their course projects, the students designed a water harvesting system that collected condensation moisture inside the greenhouse. A small wetland ecosystem was added to the greenhouse in this experiment to compare plant evapotranspiration versus open water evaporation as sources of water vapor. Relationships were found with temperature and relative humidity that can form the basis for future research on water harvesting in greenhouses at the simulated Mars habitat field sites.[8] In addition to being used as the subject of course research projects, the Greenhab greenhouse has been used for tours in undergraduate courses and for recruitment of incoming freshmen during campus orientation days. All of these applications provide outreach information on the Mars Society to students, thereby advancing the society's education mission. Future plans include the offering of an undergraduate course on Biospherics, which will utilize the greenhouse for various engineering design studies, and use of the system as a starting point for an entry by undergraduates to NASA's Mars greenhouse design competition.

References

1. Blersch, D.M., E. Biermann and P. Kangas. 2000. *Preliminary Design Considerations on Biological Treatment Alternatives for a Simulated Mars Base Wastewater Treatment System*. SAE Technical Paper Series 2000-01-2467, Engineering Society for Advancing Mobility, Warrendale, Pennsylvania.
2. Blersch, D., E. Biermann, D. Calahan, J. Ives-Halperin, M. Jacobson and P. Kangas. 2001. *A Proposed Design for Wastewater Treatment and Recycling at the Flashline Mars Arctic Research Station Utilizing Living Machine Technology*. Presented at the 3rd Annual Mars Society Conference, Toronto, Ontario. Published in: Zubrin, R. and F. Crossman (eds.). 2002. *On to Mars: Colonizing a New World*. Apogee Books, Burlington, Ontario, Canada.
3. Todd, J. 1991. *Ecological Engineering, Living Machines and the Visionary Landscape*. Pp. 335-343. In: *Ecological Engineering For Wastewater Treatment*, C. Etnier and B. Guterstam (eds.). BokSkogen, Stensurd Folk College, Trosh, Sweden.
4. Todd, J. and B. Josephson. 1996. *The Design of Living Technologies for Waste Treatment. Ecological Engineering* 6:109-136.
5. Fisher, G. 2002. *Project Greenhab: The Origin, Efforts, and Future*. Presentation at the 5th Annual Mars Society Convention, August 8-11, 2002. University of Colorado, Boulder, Colorado.
6. Calahan, D. 2002. *Greenhab Sensor Net*. Presentation at the 5th Annual Mars Society Convention, August 8-11, 2002. University of Colorado, Boulder, Colorado.
7. Frederick, G. 2002. *Data Acquisition Concepts for ET Agriculture*. Presentation at the 5th Annual Mars Society Convention, August 8-11, 2002. University of Colorado, Boulder, Colorado.
8. Ballam, D., E. Hanssen, C. Nagoda, K. Phyillaier, M. Pittek, G. Seibel and E. Turner. 2002. *An Ecological Engineered Technique for Water Limited Environments*. Unpublished course report in ENBE 688D – Introduction to Ecological Engineering, Biological Resources Engineering Department, University of Maryland, College Park, MD.

//

Red Thumb's Mars Greenhouse†

Colleen Higgins, Kate Atkinson, Sara Lewandowski, Dave Klaus,
Shawn Bockstahler, Jim Clawson, Bob Gjestvang, Aaron Frey and Ryan Reis
Aerospace Engineering Sciences, University of Colorado

Introduction

NASA began a Marsport competition to elicit student involvement in the manned exploration of Mars. The Red Thumb team, comprised of students from Aerospace Engineering sciences at the University of Colorado, designed a greenhouse to be deployed on the Martian surface and meet the requirements put out by the 2002 Marsport competition while being compatible with the mission architecture set forth by the current Mars reference mission.

Intrinsic difficulties of the Martian environment are discussed in this report, including radiation, micrometeorites and dust storms. Diet augmentation and crop selection discussions address the amount of calories that will need to be produced by the greenhouse. Crop selection is also discussed, as the crops chosen will have a significant contribution to crew nutrition, greenhouse size, harvesting and processing tasks.

The final greenhouse system is comprised of seven modular, inflatable greenhouses called AGPods, a maintenance bay for the AGPods and harvesting tasks, and a 30 m² PlantHab for salad type crops.

Marsport Requirements

Marsport listed six requirements and allowed the teams to refer to the design reference mission for mission architecture and compatibility. The Marsport requirements are listed below.

- The design life of the MDG shall be 20 years.
- Crew size is 6.
- Leakage rate of the MDG should be less than 1% of the volume per day at the target internal pressure.
- MDG crops will provide a diet augmentation (i.e., will not be used to supply more than ~25% of the crew food).
- Crop lighting will be provided using incident solar radiation with or without supplemental electric lighting.
- Crew ingress / egress is not a requirement.

Mars Environment

Radiation

There are a number of forms of radiation that need to be quantified in order to complete a conceptual greenhouse design. Central to the design of the MDG is the visible portion of the spectrum ranging from 400 nm to 700 nm. This is the portion of the spectrum where plants photosynthesize. The visible spectrum varies with Mars orbital distance, eccentricity, and the change in the dust level in the atmosphere. The infrared portion of the spectrum is important because of its impact to the heating environment. The ultraviolet, X-ray and gamma ray portions of the spectrum pose a hazard to plants, humans, and equipment and must be quantified to determine their threat. High-energy particle events include solar flares, galactic cosmic rays (GCR), solar particle events (SPE), and the solar wind. Figure 1 shows the maximum, minimum, and mean spectral power density for Mars and compares them with those of Earth.

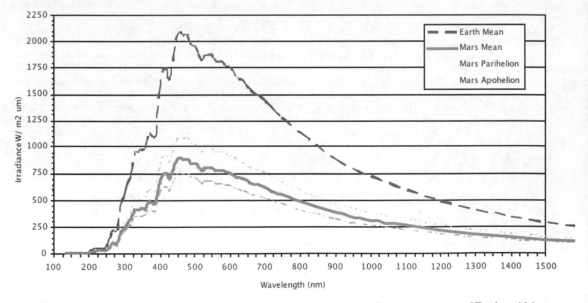

Figure 1. A comparison of the spectral power distributions at the top of the atmospheres of Earth and Mars. Mars' orbital eccentricity produces a variation in the spectral power between perihelion and apohelion (2000).

Dust Deposition / Accumulation

Measurements taken during the Materials Adherence Experiment (MAE) on *Pathfinder* indicate steady dust accumulation on the Martian surface at a rate of about 0.28% of the surface area per day (Landis and Jenkins, 1997).

The Mars Exploration Rover (MER) program has extended this analysis to account for variations in the atmospheric columnar dust amount. So, deposition rates increase with increased dust loading according to:

$$(0.0018 \ \tau) \ / \ 0.5$$

where τ is the vertical dust optical depth. Therefore, methods of dust removal must be considered.

Radiation Effects on Plants

Plants grown in a greenhouse on the Mars surface will be exposed to an increased ionizing radiation environment. The effects of this type of radiation on certain plants, and possibly humans, depending on the design configuration chosen, must be examined. In tests conducted on this subject, plants have shown greater resilience to radiation than humans do (Clawson, Hoehn et al. 1999). This evidence suggests that radiation shielding requirements of non-human-tended greenhouses would be much easier to meet than human-tended greenhouses. Table 1 lists the effects and lethal doses of radiation in Sieverts (Sv) on selected organisms.

Micrometeorites

Even though the thin Martian atmosphere provides some protection, micrometeorites pose a moderate threat to equipment and personnel on the surface of Mars.

The influx of meteorites entering Mar's atmosphere can be estimated as:

$$\log N = -0.689 \log m + 4.17$$

where N is the number of meteorites per year having masses greater than m grams incident on an area of 10^6 km^2 (Bland and Smith, 2000).

Atmospheric entry simulations indicate that particles from 10 to 1,000 μm in diameter are slowed below 1 km/s before impacting the surface of the planet (Flynn and McKay, 1990).

Organism	Observable Effects (Sv)	Lethal Dose (Sv)
Human (Annual Limit < 5 REM)	0.25	4.50
Onion	3.77	14.91
Wheat	10.17	40.22
Corn	10.61	41.97
Potato	31.87	126.08
Rice	49.74	19677
Kidney Beans	91.37	361.49
Potential Dose:	Solar Minimum: 0.40 Sv	
	Solar Maximum: 1.20 Sv	
	Proton Flare: 5.00 Sv	

Table 1. Effects if Ionizing Radiation on Selected Plants (Clawson, Hoehn et al. 1999).

Diet Augmentation / Crop Selection

The size of the growth area of the greenhouse will be determined by the requirements placed by the crew. Table 2 shows the caloric intake of the crews of previous space missions.

In addition to providing the needed caloric intake, the total food systems must provide a balanced diet. Astronaut consumption of protein is essential to offset the reduction in muscle mass that occurs in the microgravity environment [Lane 2000] and should be maintained at 12-15% of the total calories. Approximately 50% of a crew's diet should be carbohydrates, of which less than 10% should be sucrose and simple sugar. Approximately 30-35% of the total should be lipids, or fats.

	Iowa State	JSC	Average
Men	1.7*(11.6*M+879)	[66+(13.7*M)+(5*H)-(6.8*A)]	2714.34
	3436.72	1991.95	
Women	1.6*(8.7*M+829)	[655+(9.6*M)+(1.7*H)-(4.7*A)]	1505.43
	1897.12	1113.73	
Apollo			1880.20
Skylab			2832.20
Shuttle			2118.20
Average			2276.87

Table 2. Caloric requirements (crewperson/day) calculated using two different methods (M, H and A are the mass, height and age of each astronaut in kg, cm and years, respectively) and the average caloric intake for past space missions. (Lane and Schoeller, 2000; Anon., 2001).

System Architecture

During our proposal effort, research into various greenhouse technologies enabled us to develop various configurations that we could analyze. We diluted the characteristics of a number of designs into three primary configurations. These configurations were traded with consideration given to driving system parameters that included structural mass, lighting mass and power, and additional crew time requirements. Our trade study assumed that many of the components and systems would be similar across configurations; therefore, we concentrated primarily on those aspects that would be unique to each configuration.

The three primary configurations were based from various concepts proposed in the life support literature. Hublitz (Hublitz, 2000) proposed a large transparent greenhouse that could utilize artificial as well as natural lighting and is similar in concept to the greenhouse proposed by Gertner and also Sadler (Gertner, 1999; Sadler, 1999). The DRM 3.0 uses inflatable technologies, similar to the Transhab developed at JSC, for the construction of a science lab. Our second configuration is based on this technology and assumes that solar irradiance collectors provide natural lighting. Our final configuration was proposed by Clawson. Called the Autonomous Garden Pod (AGPod), it is a transparent membrane structure that is smaller in comparison to Hublitz and is non-human rated and is intended to be part of a modular system where the plant growth units are brought inside the habitat for harvest, planting, and maintenance. Each of the three configurations was evaluated at three different operating irradiance levels. Setting the required irradiance at the plant level drives the size of the resulting system and the breakdown of natural versus supplemental lighting.

Our final system architecture selection was a hybrid design combining the elements of the small modular transparent greenhouse (AGPod) with that of the larger opaque inflatable volume. The AGPod has superior mass and natural light transmittance, but limits access to the crops and requires a pressurized volume to harvest, plant, and maintain. There is no allowance in the DRM or Marsport requirements to bring the AGPods into the habitat, so we must provide that volume as part of our system. The larger opaque volume structure, which we call PlantHab, provides workstations to process the modules in a "shirt-sleeve" environment. Additionally, the PlantHab offers a place to grow short-cycle crops that benefit from more regular access. Therefore, the AGPods will focus on staple crops, such as potato cultivation, that have a long growth cycle, are amenable to the ~12 hour lighting environment at our mission locations, and do not require regular access from the crew. The PlantHab systems will focus mainly on leafy salad greens that would be accessed regularly and are amenable to lower light values that are expected with the lower efficiency of the solar collectors and/or artificial supplemental lighting. Development of automated systems will first focus on the retrieval and delivery of the modular units by remote controlled rover, which reduces crew time during EVAs.

Mechanical failures, as well as microorganism infections, can sometimes pose a threat to an advanced life support system (Schuerger, 1998). Dividing the total plant production capability into separate modules will reduce the risk of mechanical failure and crop loss due to pathogen infections. A modular system also allows for customization of atmosphere, nutrient delivery, etc. for specific crops.

AGPod

The AGPod, depicted in Figure 2, is a modular unit that resides external to the crew habitat pressurized volume to make use of natural direct solar illumination through transparent structures for all or part of the lighting needed for plant growth. This reduces the equivalent system mass (ESM) of crop production systems by eliminating the use of spacecraft internal pressurized volume and by reducing power and heat rejection resources that would otherwise be needed for total artificial lighting. By placing these structures in the surface environment, a natural difference in pressure that allows the use of mass-saving inflatable structure technology is produced. A plant-only rating on the structure and internal environment permits the use of lower pressures; further reducing mass and also leakage rates and it also lowers the required safety factors, which even further reduces mass.

 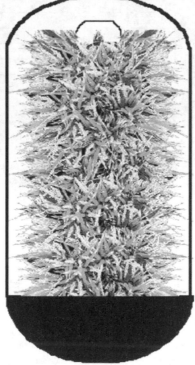

Each AGPod module must provide a suitable environment in which to grow the plants, i.e., each module must execute all the life support functions. For each of the functions we evaluated whether or not to include hardware in each unit to accomplish these functions, or to centrally handle the function and connect each module via an umbilical. For many functions there is an economy of scale (Clawson, 2000). Therefore, the solution approach was to connect the AGPods via an umbilical to allow centralization of certain services while still maintaining a capability to run autonomously for short periods to facilitate deployment and retrieval operations. The umbilical is used to supply a CO_2-rich atmosphere, collect O_2-rich atmosphere, and provides a pathway for communications to the main control computers in the

Figure 2. The AGPod.

PlantHab. Supply and collection of photosynthetic gases requires a relatively low flow rate through the umbilical and short disconnections will not adversely impact the AGPod's performance. The hardware for both thermal and humidity control is located within the module. Both utilize the entire internal recirculating flow, as well as interface with the local module environment, making it somewhat impractical to accomplish via an umbilical.

Structure

The stress of flexible membrane materials under an internal pressure load is directly proportional to the radius of curvature and pressure, while inversely proportional to the thickness of the material. Optical transmittance is directly proportional to thickness and also related to the geometry (radius of curvature). Therefore, there is a trade-off between increasing the thickness of the material and decreasing the radius of curvature when optimizing the structure for both transmittance and stress or lower mass.

To achieve transparent flexible structures capable of higher pressure involves an inflatable structure phenomenon known as pillowing, as illustrated in Figure 3. When spaces exist between restraints, the underlying bladder bulges outward in an attempt to form a spherical radius, decreasing its local radius of curvature, which decreases stress. The challenge in exploiting this phenomenon is choosing the proper type of restraint system and paying close attention to the interaction of the bladder with the restraint at the edges of the "pillow."

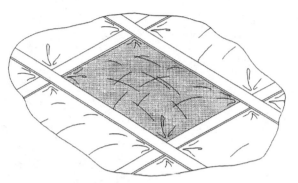

Figure 3. Pillowing of underlying fabric between spaces in the restraint (Stein, Cadogan *et al.*, 1997).

The approximate wall thickness for inflatable structures is 0.001 inches. For a micrometeorite traveling at 19 km/s, the critical particle diameter that would puncture the structure is one sixth of the wall thickness [Hyde, J., 2001, personal communication]. This results in a particle diameter of approximately 4 µm. Assuming a spherical shape, the volume of the particle can be estimated. The particles are assumed to have a density of 1 g/cm^3, which is consistent with the range of 0.7-2.2 g/cm^3 measured for micrometeorites recovered from the Earth's stratosphere (Flynn and McKay, 1990). The volume and density can then be used to estimate the mass of the meteorite particle. Therefore, 2.2×10^{11} particles can be expected to impact an area of 10^6 km^2 per year, or 0.22 particles per m^2 per year. The probability, P, of x particle impacts in t years with enough energy to puncture the inflatable structure can be estimated as

$$P = ((vx)^x /x!)\, e^{-vt}$$

where v is the rate of impacts in one year. If x is taken to be zero, in order to determine the probability that the structure would not be punctured, and t is one year, the probability of no punctures is 0.805. Over a 20-year period, the probability of no punctures becomes 0.013.

The critical particle diameter for impacts normal to the surface of the structure is 16 µm [Hyde, J., 2001, personal communication]. Using the same method as described above, the probability of zero impacts capable of puncturing the structure over a one-year period is 0.986. However, over a 20-year period, the probability is only 0.757.

Lighting and Insulation System

A transparent structure on the Martian surface is susceptible to dramatic heat loss especially at night. To counter this heat loss, flexible insulation blankets will cover the structure at night. Additionally, these blankets will double as reflectors during the day to increase the amount of light available for plant growth.

PlantHab

The PlantHab structure must be lightweight to reduce launch mass and have sufficient volume to accommodate the internal systems while using the lowest possible payload volume on the launch vehicle. At the time of this study, the required plant growth area inside the PlantHab was estimated to be at least 30 m^2. The PlantHab lower level will also provide an area for maintaining the AGPods. There should also be sufficient area for the storage of plant growth supplies such as lighting, atmospheric control, computers, nutrient delivery systems, and waste processing systems.

The structure must be able to survive the defined mission lifetime of 20 years. Since the PlantHab will be human-rated, the structure should provide sufficient protection for the crew and internal systems from radiation and micrometeorites. It must be able to maintain the necessary atmospheric pressure and constituents while reducing the system leakage. It must support the internal pressure loads as well as the equipment and crew weight. Permeability and flammability of candidate materials must be considered in the selection process. Leak-tight construction of the PlantHab is also needed to decrease system leakage. Crew ingress and egress will be necessary to maintain the plants, thus creating a requirement for an attachment to the crew habitat and/or an airlock.

Maintenance Bay

The main purpose for the lower level of the PlantHab is to harvest the AGPods and store harvested crops. However, atmospheric control for the upper level of the PlantHab is also stored here. Furthermore, inedible biomass from harvested crops is also taken care of in the waste management leaching process.

Not shown in the Figure 4 are the stairs or elevator that will allow astronaut access between levels and mechanisms that will be used to raise and lower each AGPod from the surface to the maintenance bay.

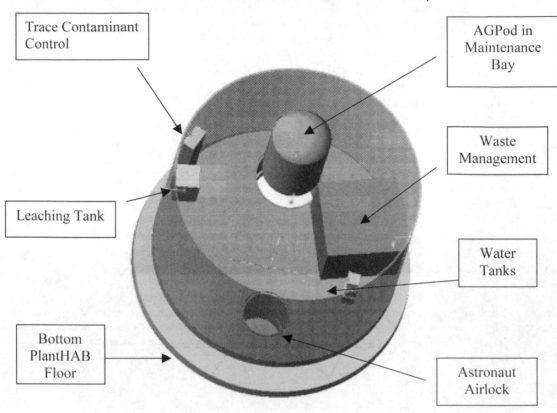

Figure 4. PlantHab lower level configuration.

Operations

The four main operations of the MDG system are crop collection, planting crops and maintenance of both the AGPods and PlantHab. Planting crops will be the first operation that occurs after deployment. The crops will be planted when the MDG arrives on Mars and the deployment process has been finished. The AGPods will have to be opened up to place seeds on the A-frame tower in the AGPods. The PlantHab trays will be planted when they are deployed. Once the system has started and the seeds planted, the next operation would be collecting the crops. Crop collection includes the actual picking of the crops, crop storage and then replanting or pruning for re-growth.

System Capabilities

The greenhouse was sized to supply 25% of the diet for a crew of six astronauts. A growing area of 100 m² is needed for this task. This was accomplished with a 30m² PlantHab and seven 9.9 m² AGPods. The maintenance bay is also included in the design.

Crop Selection

The final crop selection was made to meet the dietary requirements set by Marsport. The selected crops were selected based on productivity, lighting and environmental requirements, harvesting and post-processing requirements, psychological and dietary concerns. This final crop selection is shown in Table 3.

Physical System Mass

The physical mass of the greenhouse is the mass of the greenhouse and its components. This does not include the mass of the power generation equipment required to provide the greenhouse with the enough power to run. Table 4 shows the breakdown of the physical mass of the greenhouse. This is the launch mass but does not include the equivalent system mass of power or crew time.

System Power Consumption

The greenhouse power is broken down and shown below in Table 5. This is the maximum amount of power the greenhouse will need in order to run at full capacity at all times on the Martian surface. Power requirements will be reduced based on surface temperatures, plant maturity, etc.

PlantHab	
Crop	% of Total Growth Requirement
Lettuce, raw	6%
Red tomatoes, raw	4%
Chard, Swiss, raw	4%
Cabbage, raw	4%
Carrots, raw	2%
Strawberries, raw	2%
Spinach, raw	0%
Peanuts, all types, raw	0%
PlantHAB Total	32.31%

AGPod	
Crop	% of Total Growth Requirement
White rice, short-grain, raw	47.15%
Brown rice, long-grain, raw	23.61%
Sweet potato, raw	7.24%
Soybeans, raw	0%
Potatoes, raw, skin	0%
Wheat, hard white	0%
AGPod Total	67.25%
Greenhouse Total	99.46%

Table 3. Final Crop Selection.

AGPod Component	Mass, kg
Humidity control	50
Thermal control	162
Airflow fan	3
Transparent membrane	5
Bottom shell	25
Internal air duct	10
Nutrient delivery	50
Atmospheric control	10
Reflctor system	50
AGPod stand	10
Total AGPod Physical Mass	375

PlantHab Component	Mass, kg
Inflatable structure	508
Composite structure	276
Thermal control	1,931
Humidity control	2,012
Artificial lighting system	2,069
Waste management	150
Airflow fan	66
Nutrient delivery system	650
Atmospheric control	30
Total PlantHab Physical Mass	7,692

Total Physical System Mass – 7 AGPods + PlantHab	10316.68

Table 4. Physical Mass of Greenhouse.

AGPod	
Component	Power, kW
Thermal / humidity system	1.4
Atmospheric handling	0.11
Total AGPod Power	1.51

PlantHab	
Component	Power, kW
Thermal / humidity system	23
Atmospheric handling	3.32
Lighting ESM mass	18.6
Total PlantHab Power	18.6

Total System Power Consumption – 7 AGPods + PlantHab	55.49 kW

Table 5. Power Requirements for greenhouse system in kW.

Conclusion

The Red Thumb design meets the requirements outlined by the MarsPort competition in a unique hybrid. The hybrid design maximizes mission adaptability and environmental customization. With the AGPod, a low-power system utilizing natural lighting and a modular approach minimizes the impact of pathogen or mechanical failures. The plant rated AGPod structure minimizes structural mass. The PlantHab utilizes a man-rated structure for easy access by the crew to minimize the impact of short growth cycles and multiple harvests. The PlantHab adds safety to the mission by including another man-rated structure.

Our efforts with this year's MarsPort will continue beyond the DDR. We are currently discussing the inclusion of a system into the Mars Society's Mars Desert Research Station. This effort will be continued by a new group of students in aerospace engineering at the University of Colorado. These students will have the opportunity to apply what has been accomplished this year.

//

A Proposal for a Mars Analog Microbial Observatory†

Shannon Rupert Robles and Edward A. Martinez

Introduction

In the late 1990's, the National Science Foundation (NSF) recognized the value of studying microbial ecology in order to better understand ecological systems and instituted a program devoted to the development of microbial observatories. The program's purpose was to establish a network of sites where scientists could focus on the discovery of unique microorganisms and the study of the microbial diversity and ecological processes in various ecosystems. Currently, there are eight Microbial Observatories (MO) funded in the United States under this program, which awarded approximately 2.5 million dollars per year between 2002-2004. The average award varied from one-half million to one million dollars over five years. The major goals of a microbial observatory are to identify unknown microbes, characterize the properties and activities of newly discovered and poorly understood microorganisms and their communities, provide educational and outreach activities and to disseminate these findings using an established internet-accessible knowledge network.[1,2] In addition, microbial observatories each follow an established long-term ecological research (LTER) program, while allowing for additional environment specific research[3]. The LTER program was established in 1980 by the NSF to investigate ecological processes over long temporal and broad spatial scales and to promote synthesis and comparative research across varying sites and differing ecosystems. There are currently twenty-four LTER sites in the United States and a network office in New Mexico. The program has an annual budget of 17.8 million dollars and supports 1,100 scientists and students with an additional 44 million dollars in funding. Twenty countries other than the United States now have LTER programs of their own, including Canada and Australia[4].

The goals of these two programs are similar to many of the goals identified as being important at Mars Analog sites: to discover new microorganisms and learn more about those we have already identified, to use databases and the internet to communicate and disseminate the results of scientific research being done at a specific location, and to educate the general public and to support student researchers. For these reasons, the Mars Analog Research Station Program would benefit from adopting the procedures and protocols developed by the scientific community for the study of microbes and their ecology at microbial observatories.

What is the applicability of a Microbial Observatory to Mars? It is true that we do not know if there is or has been life on Mars. However, if there is life on Mars, studies here on Earth can focus on developing methods that will best allow us to discover it. One example is to use NASA's "Follow the Water" Hypotheses for the Mars Exploration Rovers *Spirit* and *Opportunity* and apply it to Mars analogs on Earth. Another example is to take our knowledge of latent microbial cultures and develop techniques here on Earth that will address the challenge of delayed growth in any possible Martian microbes.

A commitment to long-term research and a database that demonstrates that commitment is necessary in order to qualify for funding as a LTER / MO site. MDRS already has three years of data applicable to the development of a Microbial Observatory. However, some LTER sites with many years of data still do not receive funding. We need an ongoing directed research program, a consistently updated on-line database and web site, and collaborators from other institutions and universities. To do that we need to demonstrate to these institutions that we can support their research over the long term. Once we have developed a program that has the rigor, depth and breath of current LTER and MO sites, we can apply for NSF funding. If we were to receive NSF funding, most of the funds would need to be used for projects in the United States. As a result, MDRS would be the analog station that would benefit most. Some funding could be used to support microbial observatory programs at the remaining analog stations; however, those stations would also have to apply for additional funding from other sources, including the governments of Australia, Canada and Iceland, for their major support. Figure 1 shows a possible organizational chart for a Microbial Observatory at all four analog sites. While this chart would be most applicable once major funding has been granted, it can also serve as a model for how we should currently structure any collaborative efforts between scientists at the different stations.

Possible studies which could be used to develop a microbial observatory include, but are not limited to, phylogenetics, physiology, metabolism, genomics, growth, adaptation, survival, interactions, ecosystem processes, novel properties and modeling of systems. For example, terrestrial ecosystems can provide models for possible extinct or extant Martian ecosystems. Since development of methodologies on Mars will not be easy, it is best to develop methodologies for life detection here on Earth. Prior to our exploration of Mars, there are many things on Earth that can teach us about possible life on Mars. Therefore, hypotheses based on Earth analogs are valuable. The Mars Desert Research Station (MDRS) in the United States, Flashline Mars Arctic Research Station (FMARS) in

Canada, MARS-OZ in Australia and the future EuroMARS in Iceland are promising sites for development of a worldwide microbial observatory. We could use the same research criteria as other established microbial observatories for our research, within the framework of Mars analog environments. This would also give us the opportunity to explore more varied science questions than at most Microbial Observatories.

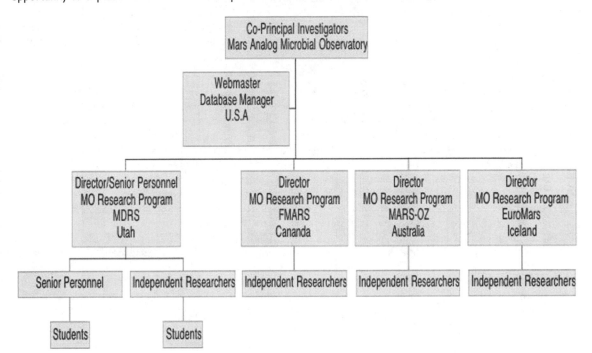

Figure 1. Proposed Organization of a Mars Analog Microbial Observatory (MO). Because NSF funding rules require that most funds for a Microbial Observatory must be used in the United States, MDRS would begin the focus of much of the research and would receive much of the funding. A small percentage may be used to fund liaisons for the other Mars analog stations but most funding for those stations would have to come from additional sources.

Most established microbial observatories are site specific. Proposing a microbial observatory based on a set of Mars analog sites has never been done. However, a recently added component to NSF's Microbial Observatory Program, new in 2004, would allow for smaller, shorter-termed projects that need not be based at a single site. This part of the program, called Microbial Interactions and Processes (MIP), will accept funding proposals through 2007, and would be ideal for Mars analog microbial ecology[5]. At up to $500,000, these awards are smaller than those awarded for Microbial Observatories, which are expected to range from one-half million to two million dollars in this funding round. Combined, the MO and MIP have a current budget of 6.5 million dollars per year[5]. This increase in funding over the last few years demonstrates the importance of microbial ecology to the scientific community at large. The flexibility of the new MIP programs points to a recognized need for examining how microbial processes differ and remain the same in varying locations.

Mars Analog Sites

The Mars Society initiated the Mars Analog Research Station Project in 1998. One of its goals was long duration geological and biological field exploration conducted in the same style and under the same constraints that will be encountered when humans first travel to Mars. The rationale for the selection of the four proposed sites for the stations – Devon Island in Nunavut Territory, Canada (where FMARS is located), Wayne County, Utah, USA (MDRS), the Australian Outback (near Arkaroola in the North Flinders Ranges of South Australia) (MARS-OZ) and Iceland (the future EuroMARS) – were that each provided excellent geological and operational analogs. The Canadian site was chosen because it has at its center an ancient impact crater and is a polar desert. Australia was chosen because it has fossil-containing deserts that date from the time when we believe the surface of Mars held water. Iceland was chosen because its basaltic and geothermal areas most closely resemble where we believe extant life may be found on Mars. Interestingly, the Utah site was selected for its ease of access and physical resemblance to Mars, and was originally slated as a test bed for equipment and isolation experiments[6]. As mentioned before, none of the four, with the possible exception of Iceland, was selected based on biological characteristics. However, there are common biological links at each site that are of great research value. Cryptobiotic crusts, found worldwide, are abundant at all sites and constitute the majority of ground cover in some areas associated with each site. Biologically, up to 80% of the living ground cover is cryptobiotic crust in nature, which consists of cyanobacteria and its associated green algae, moss and lichen. The cyanobacteria help maintain soil stability and moisture and assist in the germination and growth of the area's native and non-native plant species. These crusts are very fragile and are easily damaged by human and livestock intrusion, both of which are a problem at MDRS and possibly MARS-OZ. In addition to these cryptobiotic crusts, there are non-organic structures called desert varnish that are of great interest biologically, as they are believed to be fossilized forms created due to biological activity. Finally, all four sites can be categorized as extreme in terms of their environmental conditions. Scientific interest in extremophiles is on the rise and these sites undoubtedly contain microorganisms with unique and/or evolutionarily similar adaptations to their harsh environments.

Previous Research

Microbial studies related to the proposed project were started three years ago. During the 2002 field season at MDRS, a project studying the distribution of microbial communities based on water availability was instituted. Soil samples were classified as either wet, meaning they were collected from places where water persists (washes, run-off channels and ephemeral basins), or dry, meaning where water does not persist (escarpments and sloped terrain). Incubation of samples using soda lime as a measure of microbial respiration demonstrated a significant difference in carbon dioxide output between treatments. Wet samples appeared to contain more microbial life than dry samples, based on this measure. This suggests that it is possible to quantify microbial richness across treatments, and that more microorganisms persist during the dry season in areas where water lingers longest[7]. We applied the requirements used to assess long-term distribution patterns of microbial life at established microbial observatories to our study. This work was continued and expanded in 2003 by the science teams of Expedition One. The resulting data represent the equivalent of a four-month intensive field study. The results of this work are still being analyzed. Preliminary analyses suggest that microbial richness is dependent on water and water persistence is dependent on soil type and not microhabitat, as was the assumption in the prior year. In addition, delayed growth of up to one month in cultured samples in several microbial groups suggests that some microbes have adapted to these environmental constraints[8,9]. Due to the success of these studies we suggest that they be continued at all Mars analog stations. However, in order to use the same methodology and apply the same assumptions across sites, baseline measures of biological diversity, both spatial and temporal, must be calculated.

Current Research

Objectives of this study were to conduct baseline surveys that include transect monitoring of terrestrial plant communities, macroinvertebrate identification counts, and water quality measurements. Having quantifiable measures of biodiversity at each site will give us an indication of how similar and/or different each site is from the others, and we can design experiments accordingly. Consequently, this would allow for collaborative biological projects, such as microbial taxonomy, ecological investigations, and LTER studies at all sites. Because the locations for MDRS, FMARS, MARS-OZ and EuroMARS were selected based on their geological, and not biological, analogous characteristics, a baseline biodiversity study was needed to provide researchers information on biological richness and equitability at the macroscale level, which could then be applied to processes at the microscale level.

The development of universal methodologies for all sites must be based on measures of biological similarity between sites. Biodiversity indices developed from plant and macroinvertebrate communities, two main biological ecosystem components, are valid for the determination of biodiversity. In order to better determine the feasibility of linking all Mars analog sites into a single unified Microbial Observatory for study, we proposed that a baseline biological survey and calculation of biodiversity indices at each site be undertaken. Surveys, following the below methodology, were conducted at MDRS in May 2004, at FMARS in July 2004, and at MARS-OZ in August 2004. Research dates for EuroMARS are still to be determined.

Biodiversity Study Sites for MDRS, FMARS and MARS-OZ

Based on each Mars analog's geomorphology, stream order, elevation and environmental conditions, various watercourses were selected for our study. Plant survey sites were determined based on location of freshwater sampling sites.

At MDRS, we completed our surveys along three permanent lotic systems: Muddy Creek, Salt Creek and the Fremont River. Muddy Creek, a third order stream, was surveyed in two places: below the confluence with Salt Creek and above the confluence with the Fremont River. Salt Creek, a first order stream, was surveyed almost at its spring source in Salt Wash, and again just above the confluence with Muddy Creek. The Fremont River, also a third order stream, was sampled just east of the turnoff to MDRS at Highway 24 and again directly south of Factory Butte. Fifty-six aquatic invertebrate samples were collected from the above streams. Initial assessment indicates low benthic invertebrate diversity as well as low abundance. Taxonomic identifications of the organisms have not been completed.

At FMARS, nine streams were sampled: FMARS River (second and third orders), HMP Creek (first and second orders), Snowy Creek (first order), Hinsa River (second order), No Man River (third order), Little Comet Creek (first order) and Seven of Nine Creek (first order)[10]. A total of 75 aquatic invertebrate samples were collected.

Taxonomic identification of the organisms has not been started. A preliminary assessment of the macroinvertebrate samples indicates a low diversity. Very few organisms were found and it is believed to be due to the limited amount of carbon falling (organic matter such as leaves etc.) into the stream; however, further assessment is needed. Plant diversity was also low; a total of nine different species were found and documented. This is consistent with other studies of plant diversity in the area[11].

At MARS-OZ, our goal was to sample at least three creeks from their headwaters to the lowlands. By doing this, we would have sampled first through third order streams at various elevations. This would have allowed us to compare first order streams (highland to highland) and third order streams (lowland to lowland) and the surrounding vegetation. However, the study area was experiencing an extended drought, and no streams were flowing. Study sites consisted of various natural springs and water holes. Water holes were located where the ground water table had reached the surface, while the springs were pressurized systems. Most of the springs were slightly radioactive and had warm water while the water holes were not radioactive and had cold water. Site names did not always reflect this distinction. Survey sites included Paralana Hot Springs, Arkaroola Springs, Noodulanoodula Waterhole, Munyallana Spring, Nepouie Spring, Black Spring, Old Paralana Homestead Spring, Bolla Bollana Spring, Arkaroola Waterhole, Echo Camp Waterhole, and two water holes in Bararrana Gorge. These sites were between 100 and 500 meters in elevation. We conducted plant and macroinvertebrate studies at all sites except Old Paralana Homestead, where no plant survey was conducted because most of the vegetation surrounding the spring had been planted there by homesteaders. A low diversity is expected from these lentic sources[12].

Materials and Methods

Preliminary identification of sampling locations was conducted using topographic maps. Once actual sampling locations were established, GPS coordinates and site photo-documents were taken. In addition, water quality measurements were recorded for each sampling location prior to macroinvertebrate sample collection, and a visible physical habitat assessment was conducted.

1. Using a water quality data logger and/or water quality meters we recorded the dissolved oxygen concentration, pH, temperature and conductivity of each riffle habitat or pool. These measurements were taken at the lowest riffle sampled at each reach or, in the case of the pools, at 0.5-1 meter from the water's edge.

2. Aquatic macroinvertebrate collection and identification was conducted using the following method when sampling locations consisted of riffle habitats.
   a) A 100-meter reach of the stream was identified and three riffle habitats were randomly selected for each sampling location. The reach could be of greater length if riffle conditions dictate.
   b) From the randomly selected riffles up to three sub-samples of aquatic invertebrates, depending on the width of the stream, were collected across the stream channel[13].
   c) Samples were collected using a Surber-square stream sampler.
   d) Sample was transferred to a labeled sampling bottle and preserved with 95% ethanol.
   e) Steps 1-5 were repeated for each riffle habitat at each sampling location.
   f) Sorting and identification of aquatic macroinvertebrates will be conducted in the laboratory using a dissection microscope and the identification key of Merritt and Cummins[14].
   g) In Australia, this methodology was slightly altered to allow for better collection of marcoinvertebrates in pools. A 30 cm square area of sediment was disturbed to collect a single sample in both springs and water holes. All other methods were unchanged.

3. Plant distribution and identification was conducted using the following method and was completed within a 50 m radius of each aquatic sampling location at the same time, or within the same week, that the stream was sampled. The exception to this was MDRS, where plant surveys conducted in May 2004 were incomplete.
   a) From the middle of the lowest sampling riffle at each reach or from the single sample collection point in lentic waters, a random point, generated from 50 m radius / 360 degree random numbers tables, was identified as the starting point for the transect.
   b) Plant counts and identification were completed along transect lines run for 25 m out from each randomly selected point in the cardinal directions (N, S, E, W).
   c) Unidentifiable plants were photographed for later identification in the lab. None were collected.
   d) A plant checklist was developed for all plants within the area. All species were photographed for reference.

4. The Shannon Wiener index and Simpson's index are used to calculate biodiversity indices that are relative measures of richness and equitability for both plants and aquatic macroinvertebrates[15].

5. The % similarity between sampling locations is calculated following a protocol whereby the family biotic index (FBI) is first calculated for each sampling location and used to calculate % similarity as follows in this example: % similarity = (FBI of first order site A / FBI of first order site B) x 100[16]. Significant differences between sites is determined by calculating a T-test at $\alpha = 0.05$ using the FBI[17].

Expected Outcomes and Significance

Although we are still at the preliminary stages in the analysis of our collected data, we expect to find similar biodiversity indices for MDRS and MARS-OZ. We do not expect FMARS to be as similar to either MDRS or MARS-OZ as they are to each other. It may be that we use universal methodologies at both MDRS and MARS-OZ, but alter them slightly to reflect the biological differences at FMARS, when the same studies are conducted there. The ultimate goal of our study is to develop universal methodologies for biological research, and more specifically microbial ecology, conducted at Mars analogs worldwide. Having a baseline measure of each site's macroscale biodiversity allows us to determine whether or not we can do this with some degree of confidence. The more biologically similar two sites are, the more confidence we can have that assumptions and experiments created for one Mars analog are appropriate for another. The benefits of conducting this study, therefore, beyond creating a baseline biological survey at each site, were threefold: (1) To develop the methodology and conduct the surveys to determine the indices at these three sites, (2) the developed methodology can then be applied to other Mars analog sites and, (3) if we determine that these sites (MDRS, FMARS, MARS-OZ and EuroMARS) are biologically similar, a wide range of scientific assumptions and methodologies can be applied to all three Mars analog sites. For example, in an earlier study we determined that the soil moisture content at MDRS correlated to soil composition and not to the proximity of a water source. If our determined indices are statistically similar we can then assume with some degree of confidence that this will also be the case at MARS-OZ, FMARS and EuroMARS.

The current Mars analog sites were chosen based on their geology and extreme environments. We recognize that human activities, water quality, climate, and evolutionary adaptations of the biota at these various sites may influence differences in biodiversity. Therefore, in a scenario opposite to the one outlined above, our determined diversity indices may indicate that the sites are not biologically similar. Consequently, the methodologies developed by the proposed study may not be used as indicated in (3) above. However, if this is the case, we still believe that data provided by this study are valuable in the sense that we will have determined that one or more stations are site specific. Therefore, each site may require its own methodologies, which may also be the case on Mars. Additionally, the development of these site-specific methodologies can be used as alternatives to the proposed universal methodologies. Fortunately, the scientific community has recognized the value of looking at microbial ecology across time and space, and, just recently, in applying the same questions to more than a single physical site. Either way, studies of microbial ecology at Mars analogs should be a priority and development of a fundable Mars Analog Microbial Observatory is recommended.

Acknowledgments
The authors would like to thank Steve McDaniel, Jonathan Butler, Penny Boston, Nancy Wood, Jonathan Clarke, Rocky Persaud, MDRS Crews 4 and 21, FMARS Crew 9, the participants of Expeditions One and Two, Tony Muscatello, Robert Zubrin and Shannon's colleagues Nancy Lee, Don Robertson and Mark Yeager at MiraCosta College.

References
1. Directorate for Biological Sciences. 2002. Microbial Observatories (MO) Program Solicitation NSF-02-118. National Science Foundation.
2. The LTER Committee on Microbial Ecology. 1999. A Research Agenda for Microbial Observatories: A 1999 Perspective from Intensive LTER Projects. http://intranet.lternet.edu/archives/documents/reports/microbial_ecology/lter_me_whitepaper.html
3. LTER Microbial Ecology web site. lternet.edu/microbial_ecology/
4. Long Term Ecological Research web site www.lternet.edu
5. Directorate for Biological Sciences. 2004. Microbial Observatories (MO) and Microbial Interactions and Processes (MIP) Program Solicitation NSF-04-586. National Science Foundation.
6. R. Zubrin. 1999. Mars Analog Research Station Project. Unpublished document.
7. J. Butler, S. McDaniel, S.M. Rupert Robles. 2002. Comparative Biology of Regolith and Ephemeral Basins: A Working Test of the McDaniel's Hypothesis. Published to the web at www.marssociety.org (Marspapers)
8. R. Persaud, S.M. Rupert Robles, J.D.A. Clarke, S. Dawson, G.A. Mann, J. Waldie, S. Peichocinski, and J. Roesch. 2004. Expedition One: A Mars Analog Research Station 30-Day Mission. (AAS 03-304) Martian Expedition Planning (ed. C. Cockell). American Astronomical Society.
9. N.B. Wood and J.D.A. Clarke. 2004. Strategies for Investigating Martian Microenvironments for Evidence of Life: The Expedition One Experience. (AAS 03-305), Martian Expedition Planning (ed. C. Cockell), American Astronomical Society.
10. FMARS Crew 9. 2004. www.marssociety.org/arctic/index.asp
11. C.S. Cockell, P. Lee, A.C. Schuerger, L. Hidalgo, J.A. Jones, M.D. Stokes. 2001. Microbiology and Vegetation of Micro-Oases and Polar Desert, Haughton Impact Crater, Devon Island, Nunavut, Canada. Arctic, Antarctic, and Alpine Research. 33.3, pages 306-318.
12. Expedition Two Microbial Observatory Project Report. http://chapters.marssociety.org/canada/expedition-mars.org/ExpeditionTwo/reports/0820/
13. J. Harrington and M. Born. 1999. Measuring Health of California Streams and Rivers, AS Methods Manual for: Water Resource Professionals, Citizen Monitors, and Natural Resource Students. Sustainable Land Stewardship International Institute, Sacramento, California.
14. R.W. Merritt and K.W. Cummins. 1996. An Introduction to the Aquatic Insects of North America. 3rd Edition. Kendall / Hunt Publishing Company, Dubuque, Iowa.
15. F.R. Hauer and V.H. Resh. 1996. Benthic Macroinvertebrates. Pages 339-369 in F.R. Hauer and G.A. Lambert (Eds.) Methods in Stream Ecology. Academic Press. New York, New York.

16. J.L. Plafkin, M.T. Barbour, K.D. Porter, S.K. Gross, and R.M. Hughes. 1989. *Rapid Bioassessment Protocols for Use in Streams and Rivers: Benthic Macroinvertebrates and Fish*. EPA 444/4-89-001. US Environmental Protection Agency. Washington, D.C.
17. V.H. Resh, M.J. Myers, and M.J. Hannaford. 1996. *Macroinvertebrates as Biotic Indicators of Environmental Quality*. Pages 647-667 in F.R. Hauer and G.A. Lambert (Eds.) *Methods in Stream Ecology*. Academic Press. New York, New York.

//

Optical Dust Characterization in Manned Mars Analogue Research Stations†

Brent J. Bos
NASA Goddard Space Flight Center
brent.bos@gsfc.nasa.gov.

Abstract

Martian dust has been identified as a potentially serious hazard to any manned Mars landing mission. NASA and other organizations realize this risk and continue to support Martian dust research through the Matador project led by researchers at the University of Arizona. The Mars Society can contribute to this work by beginning a regimen of monitoring and measuring dust properties at its Mars analogue research stations. These research facilities offer the unique opportunity to study the transport and distribution of dust particles within a crewed habitat supporting active geologic exploration. Information regarding the amount, location and size of dust particles that may accumulate in a Mars habitat will be required to design a real Mars habitat and habitat equipment.

Beginning such an effort does not require a large outlay of equipment and can be accomplished using crew members experienced with station operations. Various optical techniques, such as dark-field illumination, coupled with image processing algorithms enable the collection of dust grain relative size and frequency information. Such approaches can be applied in several different zones within the research stations to evaluate the various dust reduction and isolation procedures implemented during a particular crew rotation. As the station's simulation fidelity increases, the applicability of such data to a functional Mars lander will increase. This presentation describes the optical equipment and procedures for measuring dust properties in Mars analogue research stations that can be implemented during the next field season.

Introduction

As more is learned about the conditions on the surface of the planet Mars, the more we have come to recognize the hazards that a manned mission to the planet may encounter. The planet's inhospitable atmosphere, its 0.38 g gravity field and harsh radiation environment are some of the dangers that first come to mind. Beyond these obvious concerns, however, other dangers await a manned Mars crew. Recently the threat posed by Martian dust to a manned mission has started to receive attention from mission planners as an area of concern.[1] Through crew extra-vehicular activities (EVA) and laboratory surface sample studies, Martian dust will be transported into pressurized Martian habitats through their airlocks and be distributed throughout the interior volume. The dust presence could be problematic for mechanical, electrical, thermal and human systems located inside. In fact, some have claimed dust protection and decontamination is going to be a priority consideration in designing manned Mars mission equipment.[2]

Studies related to this subject are not new but field work data on the amount of Martian dust that may come to contaminate the interior of a Mars habitat do not exist. With the ongoing operations of the Mars Society Mars analogue research stations, however, the magnitude of this threat can begin to be understood. The analogue research stations offer the unique opportunity to study a confined crew engaged in biological and geological field work under a variety of simulated Mars-like conditions. Through various means of monitoring the amount and type of dust distributed through the habitats during these simulations, the dust contamination issues can begin to be quantified.

Related Dust Research

During the Apollo Moon landings it became apparent that an extended human mission on the Moon's surface would have to contend with and mitigate dust contamination of spacecraft and habitat interiors. In 1991, National Aeronautics and Space Administration (NASA) contractors completed a study on how lunar dust may affect power system components in a manned lunar base.[3]

More recently, the Mars Surveyor 2001 lander included onboard instruments to study Martian dust and the danger it poses to a manned Mars mission. Unfortunately this mission was canceled after the loss of the *Mars Polar Lander* and so the information has not yet been acquired. There is a chance that the same instruments may fly on a Mars Scout mission in 2007. The likelihood of this occurring will be better understood as the Mars Scout mission studies mature.

Perhaps the currently most relevant research to the dust contamination issue is the Matador project led by Peter H. Smith at the University of Arizona Lunar and Planetary Laboratory. The Matador project studies terrestrial dust devils in order to further our understanding of their Martian relatives. At the core of this work is an annual field deployment in the desert southwest near Eloy, Arizona. During these deployments, not only are the dust devils themselves studied but the instruments participating in the research are critiqued as well. This produces a better understanding of Martian dust storms and the instruments we should send to Mars to study them.

Determining the level of Martian habitat dust contamination from related dust research is extremely difficult. It is a challenging exercise to quantify how much Martian soil would be displaced from the surface during EVA activity and eventually settle and adhere to an EVA team and its equipment. The calculation requires using gross estimates or assumptions such that the uncertainty in the final results is too large to be useful. In addition, there are not currently any data available to assist in constraining the results of such a calculation.

The manned Mars mission surface operation simulations currently being funded by the Mars Society are our best opportunity for determining how much dust might be brought into a Martian habitat as a result of EVA. In addition, dust transportation throughout the habitat can be monitored and studied as well. And as the rest of the research stations come on line, there will be the opportunity for obtaining results from at least three, and possibly four different Mars-like terrains: the Canadian high Arctic, the desert southwest, Iceland and the Australian Outback.

Procedure and Measurement Technique

The Mars analogue research stations' purpose is to engage in geological and biological field exploration of Martian analogue sites under manned mission constraints. Simulated space suits are required for activities outside of the habitat and egress and ingress are only allowed through the stations' airlocks. The habitats are not pressurized but they remain closed to simulate a pressurized environment. Therefore, the core requirements for studying dust contamination issues are present in the simulation.

If the incoming dust flux is found to be large enough, a useful first order measurement will be measuring the mass of floor sweepings from a defined area – most likely in the airlock and EVA room. After several EVAs have been completed, enough dust may accumulate in these areas to the point where it can be measured with a standard laboratory analytical balance (when >5 mg or so).

A more sensitive technique to potentially yield more information is to employ optical microscopy with dark-field illumination. This approach to dust measurement can be used with minimal additional infrastructure in the analogue research stations. Quality microscopes already exist in the station laboratories. The only additional items required are glass microscope slides, sample holders, a frame grabber and a dark-field illumination source if one is not already part of the microscope equipment.

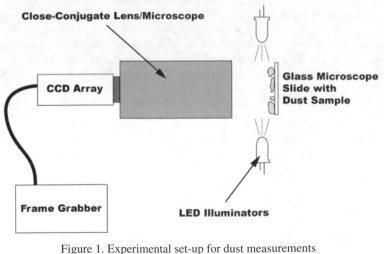

Figure 1. Experimental set-up for dust measurements using dark-field illumination.

An experimental set-up for measuring dust particles under dark-field illumination is illustrated in Figure 1. This set-up has been used successfully in the laboratory to measure and characterize dust particles. A glass microscope slide contaminated with dust is placed in the field of view of a video microscope and the lens to slide distance is adjusted until the dust particles are in focus. Then the dark-field illumination intensity is increased until the dust particles scatter a sufficient amount of light into the microscope's aperture such that all other background objects appear sufficiently dark in the image. In the set-up shown, light-emitting diodes (LED) are used as the dark-field illumination source.

After sufficient image contrast is obtained, the image from the charge-coupled device (CCD) is digitally captured using a frame grabber or similar piece of equipment. The image is then post-processed using an image-processing algorithm. Currently a very simple algorithm has been implemented in IDL (Interactive Data Language) to do the analysis. First, the user determines an appropriate image DN (digital number) threshold for the image. Image pixel values below the threshold are ignored by the algorithm. Pixel values above the DN threshold are determined to be due to dust reflections. At this point the algorithm calculates the total amount of dust detected within the microscope's field of view.

To generate further dust information the image is classified using 4-sided connectedness, region-growing logic to determine the sizes of the dust grains detected. Pixels above the threshold that are located directly above, below or next to another pixel above the threshold are determined to be the same dust grain. Pixels above the threshold that only have a diagonal connection are determined to be parts of different grains.

Once the algorithm completes the region growing, grain sizes are calculated based on equivalent spherical area. The algorithm could be changed to assume more elaborate particle shapes in the future but the spherical particle assumption is the baseline. After the equivalent grain sizes are calculated, dust grain statistics are generated and saved.

Figure 2 shows a laboratory trial of the described technique using the Mars regolith simulant JSC Mars-1.[4] Figure 3 shows the statistical information derived from the classified image.

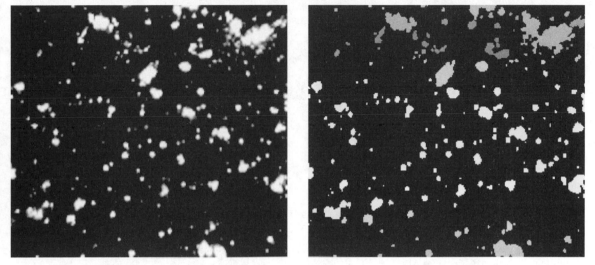

Figure 2. Dust measurements using Mars regolith simulant JSC Mars-1, on the left the captured image and on the right the classified image. Each separate dust grain is labeled with a different color or gray scale by the algorithm.

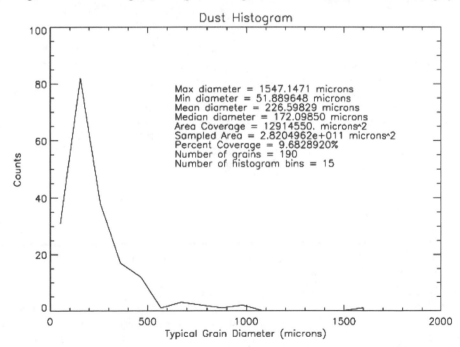

Max diameter = 1547.1471 microns
Min diameter = 51.889648 microns
Mean diameter = 226.59829 microns
Median diameter = 172.09850 microns
Area Coverage = 12914550. microns^2
Sampled Area = 2.8204962e+011 microns^2
Percent Coverage = 9.6828920%
Number of grains = 190
Number of histogram bins = 15

Figure 3. Histogram and statistics of Mars regolith simulant JSC Mars-1 dust particles.

The dust sampling procedure employed at the habitats will depend on the site location and be adjusted to produce useful results. The recommended initial approach is to position glass sample slides in a horizontal orientation beneath or at least outside of the habitat (preferably protected from the wind and rain), inside the airlock, in the EVA room, in the laboratory and in the habitat upper deck. This sampling scheme is illustrated in Figure 4. Given the space constraints, the preferred number of sample slides in each inside location is two. One slide can be removed, measured, cleaned and returned to the sample site at whatever sampling frequency is desired while the other slide can be left undisturbed for the duration of the test to accumulate the full amount of dust fall. The currently envisioned preferred sample height is as close to the floor as possible with the requirement that the same sample height be used at each site. This will facilitate comparisons between samples acquired at different locations.

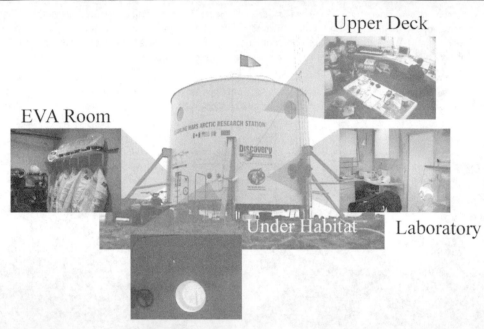

Figure 4. Mars habitat dust monitoring locations.

The various dust data acquired will undoubtedly be dependent on the types of decontamination and housekeeping procedures in place during a particular simulation. These procedures will need to be fully documented as part of the study. Very relaxed procedures are probably best implemented initially to see if the amount of dust generated is within the dynamic range of the measurements. If the more relaxed procedures produce measurable results, then more elaborate and comprehensive procedures can be implemented and compared.

Research Issues

Several issues associated with this research will require monitoring and further investigation. Arguably the most important of these issues is the simulation quality. Simulation quality will directly impact the fidelity of the dust monitoring measurements. Various simulations in the past have periodically opened habitat windows to improve air quality. Some have allowed the interior and exterior airlock hatches to remain open simultaneously to facilitate non-simulation activities. And many crews have conducted pre-EVA and post-EVA suit procedures in areas outside of the EVA room. These and similar practices will have to be eliminated to obtain dust contamination data with applicability to a real Mars mission. Along with this there is some concern that the habitats may not be sealed well enough to keep dust from blowing through cracks in the structure. This concern is certainly merited in the case of the Flashline Mars Arctic Research Station where rain has been observed leaking through the roof. Part of pre-simulation preparation may be to better seal leaks and gaps in the simulation habitat.

The dust measurements will also depend on the resolution of the optical system that captures dust images. The system's resolution affects the shape of the particle size distribution measured at the small particle end of the curve. This effect can easily be seen in the histogram of Figure 3. The number of dust particles with sizes below 100 μm should dominate the curve. But due to the spatial and radiometric resolution of the optical system that made the measurement, the number of dust particles detected at the small end tapers off. This effect is expected to be present no matter what optical system is used. But for the simulation field measurements the optical system will need to be capable of detecting particles down to a few microns (μm) in size.

Associated with the optical system resolution issue is the functional form used to fit the particle size distribution measured. Various researchers have reported using log-normal curves to fit particle size distributions,[5] while others have used gamma distributions.[6] The gamma distribution is closer to the true physical distribution while the log-normal curve can follow the measurement induced roll-off at small particle sizes. It is very likely a gamma distribution fit to only a portion of the measured particle size distribution will be the functional form used to represent dust measurements in the habitats. The portion of the measurements to fit will be determined theoretically using a mathematical model of the optical system.

Eventually additional methods for characterizing habitat dust contamination may present themselves. But optical microscopy is likely to remain the most accurate approach because it is used to calibrate most other techniques in use today.[5]

One final, important consideration for future study will be the development of a dust transportation model. Such a model will eventually be required to best relate Earth-bound dust measurements to manned Mars missions. The primary driver in this model will be gravity, but other effects, such as wind speed and atmospheric pressure, will need to be considered to address dynamics outside of the habitat. The dust contamination database will be useful in mission planning even if a model is not developed, but the results will be able to be used with better precision if a model is employed.

Conclusions

Dust contamination of a Mars habitat interior is an issue that should begin to be studied today through research conducted at Mars Society analogue research stations. Optical microscopy coupled with a segmentation algorithm can be used to accurately measure and characterize the dust that is brought into a habitat through human interactions with soil and rocks. Evaluation of these measurements will determine if such levels of contamination would be harmful to mechanical, electrical, thermal or human systems on a prolonged manned Mars mission. This information will drive engineering budgets.

The sooner this research is initiated, the sooner evaluation of decontamination procedures employed during simulations will begin, and the sooner the fidelity of the simulations will be improved. Certainly a few uncertain issues will have to be better understood as this research progresses but the core research facilities for producing meaningful results exist today. Quantitative data applicable to actual manned Mars missions can be acquired through this dust research program. To date, quantitative results from the analogue research stations have been sparse, so attention needs to be shifted to research whose results real Mars mission planners can utilize. Dust contamination is one such area of research.

References

1. Hoffman, S.J. editor, *The Mars Surface Mission: A Description of Human and Robotic Surface Activities*, NASA Technical Paper NASA/TP-2001-209371, NASA Center for AeroSpace Information (CASI), December 2001.
2. Kosmo, J.J., *Design Considerations for Future Planetary Space Suits*, SAE Technical Paper 901428, July 1990.
3. Katzan, C.M. and J.L. Edwards, *Lunar Dust Transport and Potential Interactions With Power System Components*, NASA Contractor Report 4404, National Aeronautics and Space Administration, Lewis Research Center, November 1991.
4. Allen, C.C., R.V. Morris, K.M. Jager, D.C. Golden, D.J. Lindstrom, M.M. Lindstrom and J.P. Lockwood, *Martian Regolith Simulant JSC Mars-1*, Lunar and Planetary Science XXIX, 1998.
5. Batel, W., *Dust Extraction Technology*, Technicopy Limited, England, 1976.
6. Tomasko, M.G., L.R. Doose, M. Lemmon, P.H. Smith and E. Wegryn, *Properties of Dust in the Martian Atmosphere from the Imager on Mars Pathfinder*, Journal of Geophysical Research, 104, No. E4, p. 8987, 1999.

//

Mars Habitat Dust Contamination from Simulated Extravehicular Surface Activity†

Brent J. Bos, NASA Goddard Space Flight Center, Brent.J.Bos@nasa.gov
and David Scott, The Mars Society Canada, dave.scott@utoronto.ca

Abstract

After the high radiation environment and the low gravity field on Mars, dust is arguably the next biggest hazard facing a manned mission to Mars. The seriousness of the threat depends on the specific characteristics of Martian dust and soil, which we are still trying to understand through robotic missions. At its most benign, Martian dust could cause premature failures in mechanical, electrical and thermal systems. And at its most hazardous, it could cause debilitating illness and jeopardize the health of the crew.

From April 26 to May 10, 2003, a seven-person international crew manned the Mars Society's Mars Desert Research Station (MDRS) located near Hanksville, Utah. During that two-week period the crew lived and explored the surrounding desert terrain under the constraints of NASA's Mars surface reference mission. One of the primary research objectives of the simulation was to study the amount of dust brought into the MDRS from the surrounding Mars analogue terrain during simulated extra-vehicular activity (EVA). This work characterized the soil and dust contamination brought into the Habitat through 12 out of the 14 simulated EVAs. The amount of dust, in terms of mass, and the sizes and shapes of the contaminating dust particles were measured. EVA characteristics such as type (pedestrian, all-terrain vehicle or pressurized rover), distance traveled and the work engaged were recorded to study their affect and relationship to dust contamination. We found that more than 50 g of dust and soil were transported into the MDRS during the 12 EVAs that were measured. And the amount of contamination from EVA activity was most strongly dependent on the type of terrain over which the EVA was conducted.

1. Introduction

The potential threat Martian dust poses to a manned Mars mission has been a known risk to mission planners for quite some time. The extended surface missions of the Apollo Moon program clearly demonstrated the need for the implementation of dust mitigation strategies on future long-duration planetary surface missions. Since then, various theoretical studies have been conducted to understand the nature of the threat both on the Moon[1] and Mars[2].

Perhaps the best understood, and least troubling, types of Martian dust hazards are those that might affect a Mars lander's or Mars habitat's engineering systems. For instance, Mars habitat mechanical systems could be compromised if small particles become lodged between moving parts – causing excessive wear and premature failure. Or dust could coat a lander's interior and exterior radiators and, due to the low emissivity of most soil-derived types of dust, cause premature system failures through the generation of excessive heat. A Mars habitat's electrical systems could also be damaged by Martian dust if the dust coats electrically conductive components or causes an electrical shock due to electrostatic discharge. Based on our current understanding of the Martian environment, some of these equipment failure modes may exist for a manned mission but it is expected that such risks should all be capable of being mitigated through careful engineering design and analysis.

The more troubling types of dust hazards that might confront a manned Mars mission are those that could directly affect the health of the crew. The *Viking* lander missions indicated that the Martian soil might have chemically reactive oxidation properties. And due to the high concentrations of sulfur and chlorine in the Martian soil, it is also possible that airborne dust might be acidic. Not only could this cause excessive equipment corrosion but Martian soil and dust could also be irritating to the skin or damage human organs[2]. Even chemically inactive dust poses a risk to humans. Individuals exposed to dusty environments on Earth are known to be at risk for developing silicosis, an inflammation of the lungs due to concentrated exposure to silica. Long-term astronaut health could also be jeopardized due to the presence of other hazardous materials in Martian dust and soil. Based on robotic Mars lander measurements to date, hexavalent chromium, arsenic, beryllium and cadmium are all metals that could conceivably be present in Martian dust[2]. If such particulate material were present in a Martian habitat at concentrations greater than 1 mg/m^3, then astronauts would be exposed to an unacceptably high level of risk for developing cancer during their lifetime[2]. Biological contamination of a habitat by Martian organisms is also a risk, but one that is considered extremely remote.

The primary mechanism by which Martian dust will be transported into a Mars habitat will be through human interaction with the planet's surface and atmosphere. Part of a manned Mars mission's normal activities will be for astronauts to leave the habitat and go out on extra-vehicular activity (EVA) to explore the Martian terrain, conduct scientific experiments and surveys, collect samples for further study back in the habitat or on Earth, and conduct repair and service activities on their spacecraft and equipment. This type of activity will end up coating the astronauts and their equipment in a layer of dust and soil that will be transported into the habitat through the spacecraft's airlock upon an astronaut's return.

It is generally agreed that a Mars lander will need to have some form of airlock decontamination procedure and equipment in place to keep the bulk of the airborne Martian material out of the habitat's general living space. Some have suggested a combination of air curtain and electrostatic scrubber to remove Martian dust from EVA equipment[3]. Other, more extreme engineering solutions have also been suggested whereby all EVA equipment is kept on the outside of the lander vehicle and the EVA suits are entered through small hatches in the back of the suit that are connected to the habitat[3,4]. Such "suitport" concepts are an attractive prospect for minimizing dust contamination to the inside of a lander, but are, as yet, an unproven concept. Another extreme decontamination concept is to use EVA suits with removable layers that are peeled off after each EVA before entering the lander. Most likely the first manned Mars landers will have a traditional airlock for accessing the Martian surface, and habitat dust contamination will probably be minimized through a combination of airlock decontamination procedures (such as brushing down or vacuuming), positive airflow in the airlock and habitat air filtering.

To begin to understand the scope of the dust contamination problem, we conducted a two-week Mars surface mission simulation, from April 26 – May 10, 2003, at the Mars Desert Research Station and measured the amount of soil and dust brought back into the habitat due to simulated EVA activity. Contaminating dust particles sizes and shapes were also recorded. To our knowledge, this is the first time anyone has attempted to conduct a Mars surface mission simulation for the purposes of quantifying dust contamination and understanding the important EVA characteristics that affect the amount of contamination.

2. Mars Desert Research Station

The Mars Desert Research Station (MDRS) is a Mars analogue research facility built by The Mars Society for studying the engineering and scientific challenges that the first Mars explorers will face. It is located in the American desert southwest, northwest of Hanksville, Utah, at North 38° 24' 23", West 110° 24' 31" (WGS 84) and during our experiment consisted primarily of a two-story Mars habitat; an astronomical observatory; a greenhouse for processing gray water; three all-terrain vehicles (ATVs); a large, fully-enclosed pressurized rover simulator; and simulated EVA suits. The facility began operation in January 2002 after the habitat site construction was completed and has hosted multiple simulation exercises ever since[5]. Figure I shows the facility as it existed during our simulation.

Figure 1. Mars Desert Research Station (MDRS) with the habitat, all-terrain vehicles, observatory and greenhouse visible.

The largest structure at MDRS is the Mars habitat, which is approximately 8.6 m in diameter and 10.4 m tall[5]. It is a two-deck, multi-function facility providing living and sleeping quarters, a galley, laboratory space, communications, electrical power and bathroom facilities. In addition, it has two simulated airlocks for conducting EVA in the surrounding terrain.

Geologically, the site and surrounding area are characterized by horizontal layers of relatively soft dolostone, sandstone and siltstone, with an occasional volcanic layer. The MDRS facility is located in a red Jurassic terrestrial sediment unit and the area is generally devoid of vegetation. The rugged terrain, remoteness, red rocks and lack of significant rainfall make this site a good Mars simulation area.

The differential weathering of softer siltstones, compared to the better-cemented sandstones, creates mesas capped by sandstones, with valleys filled by the mass wasting products of the softer lithologies. Due to the lack of vegetation, erosional processes dominate the area. The soft sediments weather easily, creating large amounts of unconsolidated silt and sand, usually shielded from the wind by a thin crust of desert pavement.

In terms of geology, there are some important differences between the MDRS area and Mars. For example, Mars is unlikely to have large sedimentary deposits, and is less likely to have significant lithification / cementation of these sediments. Based on known Martian rock types, land forms such as mesas are unlikely to be present on Mars. But despite these differences, the area has important similarities – little water; little or no vegetation; large amounts of dust; no development of soil; very little chemical or biological weathering of soil; and remote, rugged terrain.

To facilitate the simulation of Martian surface EVA activity, MDRS is equipped with multiple copies of the same analogue EVA suits. Each suit consists of a one-piece, heavy canvas, full-body jumpsuit that zips up in the rear from just below the waist up to the back of the neck. One piece, lace-up rubber boots are worn on the feet, with nylon gaiters added to protect the lower legs of the suit and seal off the top of the boots. A plastic helmet with a transparent hemispherical acrylic face plate is worn on the head. Two plastic air tubes and a small water hose act as umbilicals to the suit backpack which contains drinking water, a battery and two electric fans that draw in outside air to cool the head and neck of the wearer. The helmet is anchored to the backpack and canvas suit with a plastic, collar ring that is attached to the suit and backpack with nylon straps. A radio headset is also worn inside the helmet to facilitate suit-to-suit and suit-to-habitat communications. An optional head wrap or stocking cap is sometimes worn to help keep the radio headset in place and cushion the wearer's head from the helmet's hard interior. Thick nylon ski gloves are worn on the hands and are attached to the suit sleeves with Velcro that rings around the inside lining of the gloves. The entire suit weighs approximately 16 kg. Figure 2 shows how the EVA suit looks when donned.

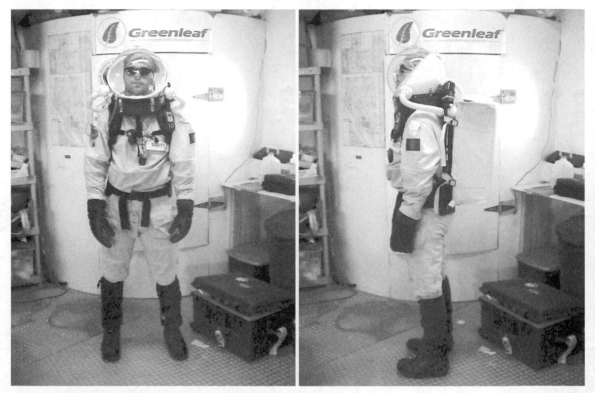

Figure 2. MDRS analogue EVA suit, as worn.

Although the EVA suits are realistic in appearance and provide some of the same functionality that real Mars surface EVA suits will provide, they are not airtight or pressurized. For instance, the backpack fans force in filtered air from the outside to provide cooling and there are a few small gaps in the helmet adapter ring where it touches the canvas jumpsuit. The positive pressure created by the air blowing into the helmets generally helps keep air and dust from entering through the adapter ring gaps, but the backpack air filters are not able to keep all outside material out of the helmet's interior.

3. Experiment Description

The goal of the MDRS dust contamination experiment was to monitor the habitat dust contamination brought into the habitat via the airlock through typical Mars surface EVA. In addition to the dust contamination experiment, several other research projects were pursued during our simulation. These projects included a general geological survey of the MDRS terrain; a biological search for halophilic microorganisms; and a drilling experiment to sample subsurface methane. Due to the activity of these other analogue research programs, there was no need to invent and design special EVAs for the dust contamination experiment. Realistic EVAs were already being carried out to satisfy the needs of the other simulation participants.

Even though the MDRS habitat has two airlocks, the simulation EVA activities were limited to entering and leaving through the main airlock in the front of the habitat (the exterior airlock door is visible in Figure 1). The second airlock was reserved for non-simulation exits and entries to the habitat for facility maintenance activities that had to be performed out of simulation. Prior to starting the dust contamination experiment, the main airlock was upgraded and cleaned. Gaps between the walls and the doors were sealed and the inside volume was washed down with water and sponges to remove as many dust particles as feasible. In addition, three small sampling platforms were added to the inside of the airlock at three different heights (305, 1,524 and 2,057 mm) to hold glass microscope slide witness samples. Figure 3 shows the interior of the airlock from above and from the inside of the habitat. In addition to upgrading the airlock, a custom-fitted cardboard mat was created to line the bottom of the airlock interior and catch any material that would fall off the EVA teams after they entered the airlock. The use of the mat facilitated the collection of contamination material and was easy to clean after each EVA.

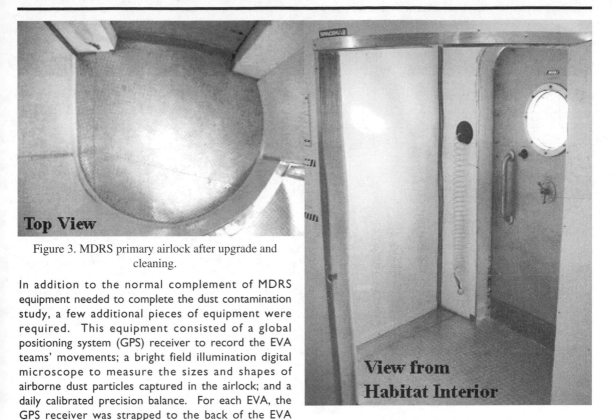

Figure 3. MDRS primary airlock after upgrade and cleaning.

In addition to the normal complement of MDRS equipment needed to complete the dust contamination study, a few additional pieces of equipment were required. This equipment consisted of a global positioning system (GPS) receiver to record the EVA teams' movements; a bright field illumination digital microscope to measure the sizes and shapes of airborne dust particles captured in the airlock; and a daily calibrated precision balance. For each EVA, the GPS receiver was strapped to the back of the EVA commander and recorded the position and speed of the EVA commander in internal memory, typically in either 5 or 10 second intervals.

Before the simulation's initial EVA, the EVA suits were cleaned and the boots scrubbed to remove contaminates resulting from the previous MDRS simulation. After that, the following procedure was followed:

1. EVA team exits airlock for EVA and closes exterior door.
2. Custom-fitted mat is placed on the airlock floor.
3. Glass microscope witness samples are placed at 3 heights inside the airlock.
4. EVA team enters the airlock.
5. EVA team brushes down suits and equipment during 5 minute simulated airlock repressurization.
6. Airlock interior door is opened and EVA team is helped out of EVA gear inside airlock.
7. EVA gear goes through another brush down and cleaning in airlock.
8. Airlock mat is swept and contaminating material mass measured.
9. Microscope slides are removed and imaged.
10. After removal from the airlock, EVA equipment and airlock mat are washed down.

The brush down procedure generated a small amount of airborne particles, some fraction of which adhered to the airlock walls. The glass witness samples were intended to help understand the amount and types of dust particles that did this. But such particles were not able to be included in the contamination mass measurement and were simply recognized as once source of experimental error.

4. Results

In all, 14 EVAs were conducted during the course of our simulation, but dust contamination was only measured for 12 of the 14. EVA 3 contamination was not measured because it was purely a greenhouse maintenance activity involving liquids and particles with no Mars analogue. Similarly, EVA 12 contamination was not measured due to a small amount of rainfall during the excursion. Table 1 summarizes the simulation's EVAs and the amount of contamination in terms of mass. A comprehensive analysis of the dust that fell on the airlock glass microscope slide witness samples has not yet been completed but a representative of the 305 mm height data from EVA 9 is shown in Figure 4.

The most significant factors affecting the amount of habitat dust contamination were not those we originally suspected. Our analysis did not find very strong correlations between dust contamination and EVA duration or the type of work activity. Before completing the experiment, we had assumed the digging and drilling activities would produce significant contamination and that the total time spent outdoors would be an extremely significant variable. Although our experiment cannot rule out the importance of these factors, our data indicates that other variables were more important.

EVA	Date	EVA Team Size	Team (leader first)	Travel Mode	Time (min.)	Distance, (km)	Moving Avgerage Speed	Activity Summary	Dust and Dirt, (g)
1	4/28/03	3	1, 2, 3	walking	106	2.1	<3.9 km/hr	Deploy biology exp. around Hab	0.57
2	4/29/03	3	4, 5, 6	walking	136	3.3	2.8 km/hr	Dig with shovels, retrieve bio samples, maintenance	1.16
3	4/29/03	3	2, 4, 5	walking	115	—	—	Greenhouse maintenance	—
4	4/30/03	3	4, 3, 5	ATV	167	20.5	11.6 km/hr	Geology survey, bio sampling	0.52
5	5/1/03	3	4, 2, 1	ATV	251	30.1	11.0 km/hr	General survey, retrieve bio samples, maintenance	4.4
6	5/2/03	3	7, 5, 3	ATV	221	29.1	11.5 km/hr	General survey and sample deploy	0.90
7	5/3/03	3	6, 1, 4	rover	325	24.7	8.5 km/hr	Drilling, retrieve samples	1.54
8	5/4/03	2	2, 3	walking	58	2.2	2.3 km/hr	Retrieve bio samples	0.23
9	5/5/03	3	4, 2, 3	ATV	294	65.8	18.5 km/hr	Bio sampling	6.85
10	5/6/03	3	4, 6, 7	ATV	294	46.1	14.0 km/hr	Sample deploy, general survey	13.45
11	5/7/03	3	4, 3, 5	ATV	282	70.9	21.4 km/hr	General survey, retrieve samples	9.67
12	5/8/03	3	4, 1, 6	ATV	169	23.2	11.0 km/hr	Drilling	—
13	5/9/03	3	4, 2, 3	rover	492	107.7	29.5 km/hr	Bio sampling	7.09
14	5/10/03	3	2, 1, 6	2 ATV, 1 walking	124	22.5	15.4 km/hr	Bio survey, drilling	5.03

Table 1. Summary of simulated extra-vehicular activity.

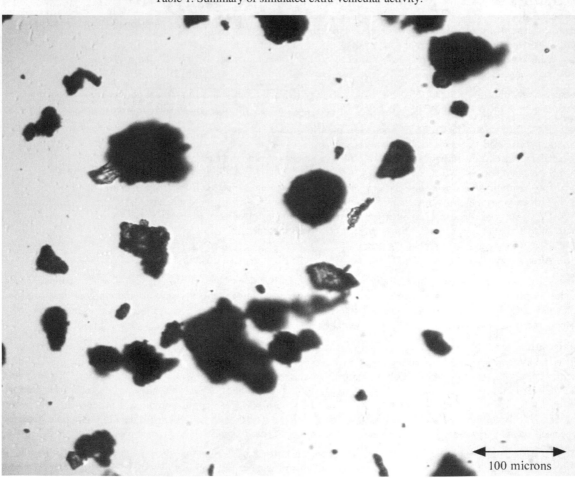

100 microns

Figure 4. Dust particles captured in the airlock at a height of 305 mm after EVA 9.

Figure 5 is a plot of habitat dust contamination versus EVA. The error bars are unilateral and represent our worst-case estimate on the amount of dust that was not measured, either because the dust adhered to the inside of the airlock or made it's way into the inside of the EVA suits through the air filters and the gaps between the helmet collar rings and the canvas suits. The mass measurements did not include any manmade material or pieces of vegetation visible to the naked eye. Such particles were removed from the samples before weighing. The EVA contamination results can be classified into three types: low, medium and high contamination.

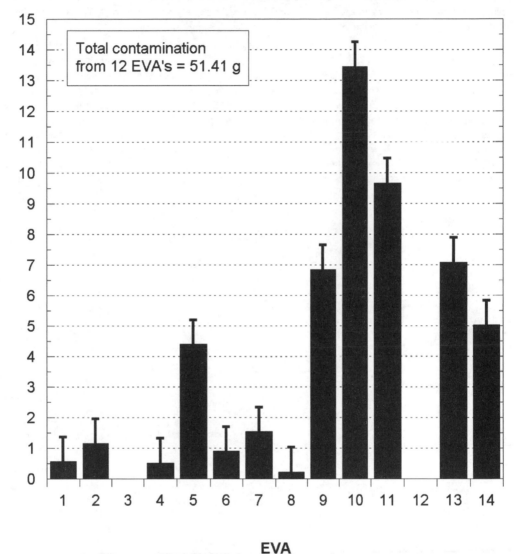

Figure 5. Dust contamination from simulated Mars EVAs.

The six low contamination EVAs (EVAs 1, 2, 4, 6, 7 and 8) only generated about 1-2 g of dust and soil contamination per EVA. Three of those were walking-only EVAs and one utilized a pressurized rover. Two of the low contamination EVAs used ATVs at rather moderate speeds (~11.5 km/hr). EVAs 2 and 7 consisted of significant amounts of digging and drilling activity, but generated very little contamination.

The two medium contamination EVAs, EVAs 5 and 14, were both conducted on ATVs at average moving speeds of 11.0 km/hr and 15.4 km/hr respectively. They both generated around 5 g of contamination. EVA 14 only had two team members on ATV and it is conceivable (assuming a simple proportional relationship between the amount of dust contamination and the two ATV riders, and no dust from the team member on foot) that had the third individual been riding an ATV as well, that the EVA 14 dust contamination would have been closer to 8 g. That would have made that EVA the third worst contaminator overall. So it is difficult to understand from that EVA data how the high rate of travel affected the results.

The high contamination EVA group consisted of EVAs 9, 10, 11 and 13, but we believe that the results for EVA 13 may be misleading. EVA 13 was a long duration, long distance pressurized rover excursion primarily searching for biological samples. Upon return to the habitat the EVA team reported that they might have walked through wet and damp areas to obtain their samples. And inspection of their boots indicated that some of the contaminating material might have actually been dried up mud. So the utility of the EVA 13 results for a Mars analogue simulation is in some doubt. The EVA 9, 10 and 11 results, however, are not questionable and are intriguing. Together those three EVAs alone accounted for over 58% of the total 51.41 g of dust contamination. EVAs 9, 10 and 11 were completed on ATVs and all had relatively high average moving speeds in common: 18.5 km/hr, 14.0 km/hr and 21.4 km/hr, respectively. In addition, they also all traveled through the same unique MDRS terrain. And we contend it is the combination of the terrain and the high rate of ATV travel that made those three EVAs produce the extraordinarily large amount of dust contamination.

The first time our Mars mission simulation found the fine-soiled MDRS terrain was on EVA 9. The EVA 9 team (crew members 4, 2 and 3) reported that approximately 10 km northwest of the habitat they found a relatively flat valley full of fine material – almost as fine as talcum powder – that was easily lofted into the air by the moving ATVs. In addition, it was thought that this area could be used as a corridor for reaching Factory Butte, the most impressive geologic formation in the MDRS vicinity. Even though it is only 11 km away from the MDRS habitat via a direct route, the most direct route is impassable riding ATVs with space suits due to the topography. EVAs 10 and 11 were subsequently sent out to scout an easily traversed route to Factory Butte through the valley of fines and were successful in doing so. Figure 6 shows the path that EVA 11 took to Factory Butte and the area of fine material. Figure 7 shows an aerial view of the fines area and the surrounding terrain northwest of the habitat. Figure 8 is a view to the west, near the beginning of the fine-soiled terrain.

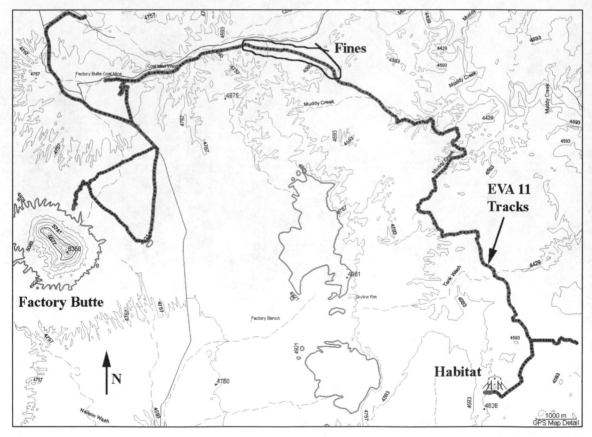

Figure 6. GPS tracks of EVA 11 indicating area of fine soil material.

The fine soil that seemed to play a major role in dust contamination is due to mass wasting off the higher topography to the south. The mesa to the south is capped by a thin layer of more resistant sandstone overlying a thick (up to 50 m) layer of very soft, poorly cemented siltstone. Due to the sediment load from the mass wasting of the siltstone cliff and the lack of vegetation for ground cover, the unprotected siltstone plain below the mesa is a high-erosion area.

The sporadic rainfall the area receives is not enough to compact the loose dust, but does manage to form a thin, weak crust that caps the 10 to 15 cm of loose, uncompacted silt that overlies this area and around the base of the mesa. This crust is strong enough to protect the underlying dust from wind transport, though once this cover is broken by ATV wheels, the silt is transported by the wind and by turbulence from the ATVs. This silty sediment is more easily entrained by the wind (and picked up by EVA crews) due to the silt mineralogy, which contains an abundance of soft, platy minerals including micas, gypsum and chlorite. The platy minerals have a higher surface area to weight ratio and more jagged grain edges than minerals such as quartz, making these grains stick to fabrics as well. In addition, the grain shape inhibits compaction, which also aids in dust transportation. We contend the following factors made this particular area the highest dust loading terrain we explored: high sediment load from the weathering cliff to the south of the trail; sediment type; protection of loose underlying dust by thin crust; and the greater speed at which ATVs could traverse this flat area.

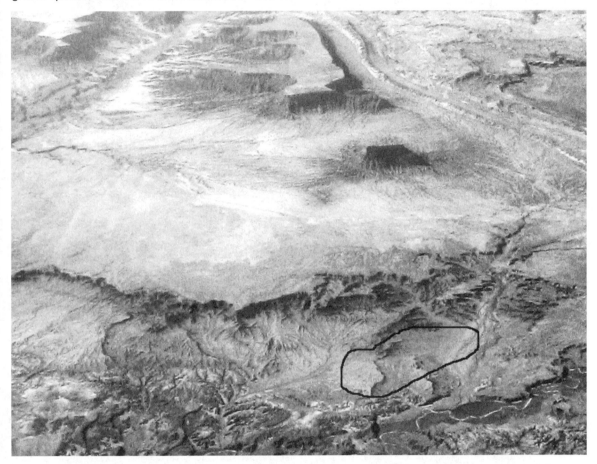

Figure 7. Aerial view of MDRS terrain toward the southwest: the habitat is outside of the picture to the left, Factory Butte is visible in the upper right center and the area of fine soil material has been circled (courtesy of R.D. "Gus" Frederick).

Figure 8. View of fines terrain looking toward the west, taken during EVA 10.

5. General Conclusions and Implications

The dust contamination we measured needs to be put into context. Most importantly, we need to note that no particular dust removal protocol was implemented prior to the EVA teams entering the airlock and sealing it shut. So the amount of dust contamination we were measuring was worst-case for this particular area in Utah. We expect that on Mars there will be some type of dust removal protocol that will be executed before EVA teams are even allowed back into the spacecraft's airlock and, indeed, we hope to eventually measure the efficacy of such procedures with a simulation. But this initial experiment was intended to develop a baseline for dust contamination at Mars analogues on Earth, and to begin an understanding of the major factors affecting it.

One subjective observation made during the course of sweeping down the EVA equipment was that dust contamination would have been substantially reduced if one of the design goals for the equipment had been the reduction of contamination. Holes, divots, voids, folds and pockets in the EVA gear acted as dust traps and when these items were brushed down and shaken out, they released a good deal of dust and soil material into the airlock. Folding pocket tools suffered from this type of problem as did most sample bags and tool pouches. To minimize dust contamination, Mars EVA equipment must be designed to be free of dust and soil traps, similar to good beach equipment on Earth. For instance, bags and pouches should be made out of netting material so that they cannot hold particulate matter. Folds, pockets and areas of overlapping material should be minimized on the outside of EVA suits and other EVA equipment. The introduction of such design requirements on EVA gear, early in the design process, will significantly reduce the amount of material an EVA team will transport into a Mars habitat. Just how much of a reduction is a subject for future work.

Another effective dust contamination strategy for a real Mars mission would be for EVA teams to leave a majority of their EVA equipment in lockers or storage containers on the outside of the lander. The equipment could originally come from the inside of the habitat but once exposed to the Martian surface, be prohibited from being brought back in. A separate, smaller lander airlock for samples that would open directly to a sealed glove box would help the efficacy of such a protocol. Then scientific samples would never have to be exposed to the cabin environment, reducing the likelihood of cross-contamination.

Arguably one of the most important conclusions of this work is that Mars habitat dust contamination experiments need to be conducted with large swaths of terrain available, which mimics that on Mars. Our experimental results indicate that traveling with a medium speed, open cockpit, high-ground-clearance vehicle will be the predominant source of Mars habitat dust contamination if such a transportation system is used for exploration on Mars, which is likely. The speeds of some of our ATV EVAs may have been slightly higher than what will be achievable with early Mars rovers, but they are not out of the realm of possibility. In 1972, *Apollo 16* set a lunar rover speed record of 17.70 km/h[6]. And the distances over which our ATV EVAs were conducted are not without precedent either. For instance, *Apollo 17* roamed a total distance of 33.80 km during the course of three lunar surface EVAs[6]. Although more work is needed to further quantify the affect of EVA work types on dust contamination, based on this experiment it does not seem likely that currently envisioned EVA work activities could become primary sources of dust contamination. And so we contend that any dust contamination experiment attempting to understand the important Mars contamination mechanisms needs to be able to simulate mechanized travel over Mars-like terrain.

If the early Mars explorers use open cockpit vehicles, like we primarily used in our simulation, then vehicle design requirements and travel protocols could be instituted to reduce the amount of dust contamination generated by travel. Comprehensive vehicle fenders and dust shields could be designed to minimize the amount of dust kicked up. These types of features limit the ground clearance and navigable terrain of vehicles. So a trade study would need to be conducted to determine the maximum amount of dust shielding possible that would not restrict the vehicles' necessary range of operation. EVA speed limits could also be implemented to help keep down the dust contamination and ease the burden on dust shielding hardware.

If more than one vehicle is used on EVA, then an EVA travel formation might also be able to be designed to minimize dust contamination but more study is needed in this area. For instance, maintaining a minimum distance between vehicles when traveling might help cut down on the dust contamination from leading vehicles. Or traveling with the vehicles in a line perpendicular to the direction of travel might be a feasible strategy when the trail allows it. But preliminary, non-turbulent air modeling of the ATV-generated dust dynamics indicates that the majority of the dust that falls onto an EVA team member when traveling via ATV comes from the team member's own ATV. So implementing restrictive travel formations may not be worth the effort. But more work needs to be done to understand the results and how the results scale to Mars.

The 51.41 g of total dust contamination this experiment measured entering the habitat would be large enough to require air filtration or decontamination in an actual Mars habitat. For example, if we assume all of it could go airborne and it was confined to just the habitat airlock volume, then the particulate concentration would be approximately 10,000 mg/m^3. Or if all the material went airborne and uniformly filled the entire enclosed habitat volume, then the concentration would be approximately 80 mg/m^3.

The current recommended maximum particulate concentration for a Mars habitat is 1 mg/m^3, which is approximately equivalent to industrial city air[2,7]. And 1 mg/m^3 is 20 times greater than the 0.05 mg/m^3 average requirement set for the International Space Station[2,8]. So, even if only 2% of the contaminating material was lifted into the habitat air, the air would still need to be filtered.

Another way to put the dust contamination measurements into perspective is to calculate the worst case amount of material that would need to be removed from the air per three-person EVA. Referring to the results for EVAs 9, 10 and 11, the amount of material brought into the habitat per three-person EVA on Mars could be around 10 g. If all this material eventually went airborne into a 600 m^3 cabin air volume, then the cabin filtration and decontamination protocols would need to remove approximately 9.4 g per EVA to meet the 1 mg/m^3 air quality recommendation.

Drawing any further conclusions from this research for the amount of dust contamination to expect on Mars is probably not warranted. This experiment was conducted at a Mars analogue on Earth where the force of gravity is 2.65 times stronger and the atmospheric pressure is about 143 times greater than on Mars. In addition, it is difficult to predict if the ambient sky dust fall at a landing site on Mars will be greater than our simulation experienced in Utah. If it is, then the total amount of time simply spent outside of the habitat will be a more important factor than this experiment found it to be. Quite surprisingly, the United States Geological Survey has measured non-organic matter dust settling rates as high as 0.03 g/cm^2/yr (and typical values of 0.001 g/cm^2/yr) in the United States desert southwest, whereas Pollack et al. estimated an average global sedimentation rate of approximately 0.002 g/cm^2/yr after the 1977 dust storms[9,10]. On Mars we expect this value to depend on latitude, but it is certainly conceivable that ambient dust fall may be less of an issue on Mars than we experienced in Utah.

To be able to further accurately extrapolate the results of this experiment to a Mars mission, a fairly sophisticated dust contamination model will need to be developed with variables that can be adjusted for Earth and Mars terrain, gravity and atmospheric pressure. The results of this experiment could then be used to calibrate that model for Earth terrain, pressure and gravity and then be used to predict Mars contamination by changing the model inputs for Mars. But due to the extremely dry conditions on Mars and the loose-soiled Martian terrain observed at the *Viking 1*, *Viking 2*, *Mars Pathfinder* and Mars Exploration Rover landing sites, we expect that disturbing the Martian surface will inevitably produce greater amounts of dust contamination than we measured in Utah. And therefore our measurements probably should be considered the lower bounds for Mars habitat dust contamination.

6. Summary and Recommendations

The habitat dust contamination experiment found that the speed, transportation mode and type of terrain traversed were the most important factors affecting the amount of dust contamination that EVA crews bring into a Mars habitat. The type of EVA work activity undertaken, once the EVA travel ended, did not significantly affect the amount of dust contamination. The results of this experiment also indicate that the early manned Mars landers will need to implement dust mitigation strategies and air filtration systems to keep Mars particulates down to an acceptable level. In addition, we found that three-person ATV EVAs generated approximately 10 g of dust and soil contamination per EVA. We expect that after the Martian atmosphere, gravity and terrain effects are considered, a dust contamination model may predict an even larger amount of dust contamination per EVA.

Building on this work, several different paths of inquiry should be pursued to further our understanding of the issue. It is clear that a dust contamination model needs to be created to help us understand how Mars analogue dust contamination measurements on Earth should be extrapolated to an actual Mars mission. This will be a largely theoretical effort but the model will need to be calibrated and checked with the results from simulations on Earth. So the experimental work needs to continue as well. For instance, at MDRS, the area of fine, loose soil 10 km northwest of the habitat should be further studied to understand its relevance to Mars. Additional EVA simulations should be conducted to simply verify this experiment's limited number of ATV EVA results. Three-person ATV EVAs could be sent out across all different types of MDRS terrain under the constraint that they travel the same distance, at the same speed and carry out the same type of work. This would allow the experimental determination of the relative importance of terrain type on dust contamination. Another two-week simulation at MDRS could also investigate the effectiveness of various decontamination protocols and strategies by sending EVA teams out to do identical work and then randomly determining which ones have to implement decontamination procedures before entering the habitat. There are many possibilities and it is clear that there are still many dust contamination issues that must be addressed before we send the first crew to Mars.

Acknowledgments

We express our thanks to our fellow Mars mission simulation crewmates: Elia Husiatynski, Simone Kosol, Mark Moran, Petra Rettberg and Joan Roach, for participating in our dust contamination experiment. Without your dedication and effort this work would not have been possible. We would also like to thank the Mars Society volunteers who helped run our Mars mission simulation and granted us research time at the MDRS. In particular we acknowledge the tireless efforts of Tony Muscatello in getting us to MDRS, supporting us while we were there and getting us home safely.

References

1. C.M. Katzan and J.L. Edwards, *Lunar Dust Transport and Potential Interactions With Power System Components*, NASA Contractor Report 4404, NASA Lewis Research Center, November 1991.
2. F.H. Hauck, H.Y. McSween, C. Breazeal, B.C. Clark, V.R. Eshleman, J. Haas, J.B. Reid, J. Richmond, R.E. Turner and W.L. Whittaker, *Safe on Mars: Precursor Measurements Necessary to Support Human Operations on the Martian Surface*, National Academy Press, 2002.
3. G.L. Harris, *The Origins and Technology of the Advanced Extravehicular Space Suit*, Univelt, Incorporated, 2001.
4. M.M. Cohen and S. Bussolari, *Human Factors in Space Station Architecture II: EVA Access Facility: A Comparative Analysis of Four Concepts for On-Orbit Space Suit Servicing*, NASA Technical Memorandum 86856, April 1987.
5. R. Zubrin, *Mars On Earth*, Tarcher / Penguin, 2003.
6. R.D. Launius, *Apollo – A Retrospective Analysis*, Monographs in Aerospace History Number 3, NASA SP-2004-4503, July 2004.
7. H.L. Green and W.R. Lane, *Particulate Clouds: Dusts, Smokes and Mists*, E. & F. N. Spon, London, 1964.
8. NASA, *System Specification for the International Space Station*, Revision W, Section 3.2.1.1.1.15, Part F, SSP 41000, December 20, 2000.
9. M.C. Reheis and R. Kihl, *Dust Deposition in Southern Nevada and California, 1984-1989: Relations to Climate, Source Area, and Lithology*, Journal of Geophysical Research, V. 100D5, pp. 8893-8918, 1995.
10. J.B. Pollack, D.S. Colburn, F.M. Flasar, R. Kahn, C.E. Carlston and D.C. Pidek, *Properties and Effects of Dust Particles Suspended in the Martian Atmosphere*, Journal of Geophysical Research, V. 84, pp. 2929-2945, 1979.

//

In-Situ Martian Construction – MDRS Crew 22 Masonry Construction Simulation†

Georgi Petrov, Laguarda.Low Architects, gpetrov@alum.mit.edu
and James Harris, Austin Community College, james@james.harris.name

Abstract

As part of the Mars Society's continuing operational research, this project aimed to demonstrate that masonry construction is a viable building method that will help establish a permanent human presence on Mars. It has been proposed that bricks can be manufactured from Martian regolith. Using pitched-brick vaults and self-supporting domes a wide range of spaces can be constructed using no scaffolding, thus greatly simplifying construction.

To explore this possibility Crew 22 aimed to manually construct a barrel vault with a one meter inner radius, using local stone and sand under simulation constraints. Portland cement and hydrated lime were the only imported materials. Construction lasted for 64.5 man-hours in simulation, and six man-hours out of simulation for comparison. Working in the Mars suits was difficult, but not overwhelming. The biggest constraints were decreased visibility and communication had a bigger effect than the weight of the backpack and suit. The use of irregular stone also proved to be a major obstacle. Time and mortar can be reduced by using masonry units of the same shape and size.

Overview of Masonry Construction

If the decision to establish a permanent presence on Mars is made and the construction of a large, permanent habitat begins, then the construction methods must be carefully considered. Relying on habitats brought entirely from Earth is an unsustainable strategy. A more realistic approach would be to maximize the use of Martian materials and to implement simple, well understood and tested building techniques.

Masonry has been proposed as one possible construction method that might be employed by early settlers on Mars because: it is the only readily available resource on the Martian surface; it is simple to produce; and it is extremely durable. The most abundant materials on the surface of Mars are regolith and rocks. In fact, the whole planet, except for the polar caps, is covered with nothing but regolith and rocks. Bruce Mackenzie has proposed that the first settlers can manufacture bricks using the regolith [Mackenzie 1987]. Using pitched-brick vaults and self-supporting domes one can construct a wide range of spaces using no centering, thus greatly simplifying construction [Richards 1985], [Robinson 1993], [Petrov 2004].

Masonry's low tensile strength however poses a challenge that must be overcome. In order to balance the interior pressure, the only option is to cover the masonry structures with as much as 10 m of regolith. However, turning the settlers into cave dwellers is highly undesirable. Therefore, in order to give the settlers the ability to view outside the habitat and to facilitate access to the surface, it is necessary to use masonry in combination with another system that can resist the pressure through tension [Kennedy 2002].

The history of masonry arches and vaults can be traced back to about 3000 BC in Lower Egypt and Mesopotamia. One of the oldest brick vaults covers the storehouses at the Ramesseum, the tomb complex of Ramses II, who reigned to about 1224 BC. There are three distinct methods for constructing vaults that were developed in antiquity. The most common is the radial vault, where the space between the side walls is filled with loose bricks that serve as temporary support for the vault. Then, bricks are laid in successive courses, with mud and small stones placed at the outer edges to cant the bricks in until the vault closes at the crown. After the mud dries the centering is removed. The most rare method involves leaning two long, slightly curved bricks against one another over the center of the space. Though they are the simplest to construct, ribbed vaults are the least strong [van Beek 1987].

Pitched-brick vaults, also known as leaning arches, are the most useful technique for work on Mars, because they require no centering. The first bricks on each side are laid at an angle against a side wall; the second set of bricks is placed on top of the first, also leaning against the wall and so on until the arc of the vault closes at the top. Successive arcs are leaned against the first one until the end of the vault is reached. The remaining triangular space is filled in with smaller arcs. Often, more courses are laid on top of the first one leaning in opposite directions [van Beek 1987].

Domes might be even more advantageous because they only need support at four points, unlike vaults that require continuous support along two walls. Domes can also be constructed without centering using a similar idea to the pitched-brick vaults. First, four piers are erected and connected with four arches. The first set of arches will require some centering, however, subsequent domes can be built by leaning the arches on the previous domes. Next the spaces between the arches are filled in concentric courses. The construction can be stopped after the completion of any course and the structure will remain stable. Brunelleschi most famously and spectacularly applied this principle at the dome of the Duomo in Florence.

Masonry Construction Simulation

In order to gain some practical experience in the feasibility and challenges of constructing masonry structures, Crew 22 undertook a project to build a masonry dome during our rotation at the Mars Desert Research Station (MDRS).

We decided to construct a pitched-brick vault which obviated the use of centering. Due to the MDRS command's concern for safety, and in order to minimize the impact on the land, the project was restricted to a vault with a one meter radius, erected directly on the ground.

Three main considerations determined the choice of a site for the project. The primary requirement was to be at the base of a steeply inclined slope, which we could use as a support for the first row of pitched masonry. Second, we wanted to be close to the Hab in order to facilitate the work and to be able to take pictures of the project with the rest of the "base." The final objective was to locate the vault in such a way that if the project failed it would not be visible from most vantage points in the area. We found a site to match all of the objectives about 50 meters north of the Hab on the opposite side of a small protrusion of the ridge next to MDRS (Figure 1).

Figure 1. Site for the masonry vault project near MDRS.

An investigation of the subsurface conditions determined that most of the vicinity of MDRS is covered by a crumbly layer of weathered stone. Relatively hard shale is located 20-30 cm below the top crust. This bedrock can still be fractured easily with a mason's hammer to a depth of another half meter. Balancing the need for stability of the vault and the need for expediency in our tight schedule, it was decided to remove the top crust and any loose rock that was knocked out in the process and to lay the foundations on the exposed bedrock (Figure 2).

Figure 2. Excavation of the top crumbly layer exposes relatively solid rock on which the vault was constructed.

Material for the masonry units was collected from an outcrop about one kilometer away from the Hab. The location was chosen near an existing road in order to minimize off-road travel. The raw material was loaded manually on the ATVs and transported to the pressurized rover (a crew member's SUV), which remained on the main road. The pressurized rover then brought the material to the construction site. Five crew members brought two rover-full shipments of rock to the site in two hours (Figure 3). The other local ingredient was sand, which was sifted from the surface of the main road by one crew member. Cement and lime were the only materials that were imported from outside the local area. The mixture that was used for mortar was two parts cement, one part hydrated lime and six parts sand. This mixture constitutes Type II mortar and is recommended for southern Utah.

The crew planned the work schedule in two EVA slots. In the morning when temperatures were still below freezing, the EVAs concentrated on projects by other crew members. On most days the temperature rose past the freezing point in late morning and the afternoon EVAs were dedicated to the masonry project. A summary of the man-hours for each phase of the project is covered in Figure 4. The surveying and excavation tasks were accomplished by one crew member and did not require many man-hours. Gathering of rocks involved the coordinated efforts of the whole crew and was less efficient in terms of man-hours, even though in absolute time it was accomplished expediently. Sand, on the other hand, was gathered and sifted by one crew member, often on the way to an unrelated EVA. We experienced a steep learning curve once we began the actual construction. The first day five crew members went to the construction site, leaving one in the Hab. In one and a half hours we managed to lay only four stones for one side of the foundations. Most of the time there were people standing around without a task. When we began the construction of the first arch we used a team of four. Finally, we determined that the best labor division for the construction of a vault of similar size is a team of three, with one crew member mixing the mortar and two crew members laying the masonry (Figure 5). This configuration allowed us to mix many small batches of mortar which could be used up before hardening past workability.

One caveat that bears mentioning is that manual mixing of mortar is a dusty process – and possibly would be even more so on the Martian surface due to decreased atmospheric pressure and increased wind velocity. On a particularly windy day one crew member was inadvertently exposed to airborne lime during the construction process. It is recommended that this experiment not be recreated or extended in windy or closely enclosed locations due to the potential dangers of chemical inhalation.

Figure 3. Two rover loads of rocks were collected in two hours by five crew members.

A schedule of all of the materials used for the project is given in Table 1. Clearly the level of efficiency needs to be greatly improved. The first step that we recommend is to use regular building blocks instead of rough stone. A large portion of the time was spent sorting through the pile of rocks and chiseling them into usable shapes. Additionally, a lot of mortar was wasted filling gaps between rocks that did not fit together well. We estimate that the amount of mortar can be reduced by half or more. Additionally, much of the water can be recovered if the operation is conducted inside a closed construction tent where the atmosphere can be controlled and moisture can be condensed.

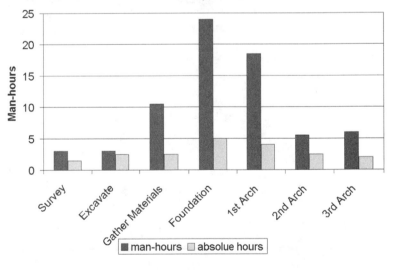

Figure 4. Summary of the man-hours for each phase of the project.

The foundation and the first two arches of the vault were completed in 64.5 man-hours. On the last day of Crew 22's rotation we completed a third arch working for six man-hours out of simulation. This gave us an opportunity to compare the experience. The main negative effects of the simulation space suits were a decrease in visibility and in communications. Working vigorously in the suits made the visors steam up much more than during science observations. This problem was exacerbated in the afternoon when the Sun was low behind the hill next to which we were working and most activities required facing it. Additionally, communication was accomplished mainly though gesturing or not at all. Both problems reduced one's awareness of the activities of other crew members, thus greatly reducing teamwork. Both of these adverse effects can be eliminated if work is done inside a pressurized construction tent. The weight of the life support backpacks had a secondary effect on work efficiency, though its role will probably increase if longer work shifts are employed.

It is useful to estimate how many man-hours would be required to complete a vault of useful dimensions. A vault with a radius of 3.5 meters and a length of 4.5 meters can cover a space of roughly 16 m². By extrapolating from the efficiency which we achieved at the end of the project of 5.5 man-hours per arch we estimate that such a vault can be completed using 353 man-hours including surveying, excavation, collection of materials and placing of 30 arches necessary to cover 4.5 m (see Table 2). Many of the required tasks can be automated in order to improve efficiency. The robots and machinery will still need supervision and

Figure 5. Optimum labor division for the construction of a vault of similar size is a team of three, with one crew member mixing the mortar and two crew members laying the masonry.

Materials Schedule		
Water	36 gal = 137 litres	(local)
Sand	540 lb = 246 kg	(local)
Portland Cement	180 lb = 82 kg	(imported)
Hydrated Lime	70 lb = 32 kg	(imported)

Table 1

Projected Man-hours for Full Vault	
Survey	3.0
Excavate by Hand	10.0
Collect Materials	160.0
Build 30 Arches	180.0
TOTAL	353.0

Table 2

Figure 6. The tools necessary for masonry constructions are simple – mason's hammer, spades, trowels, mixing bucket, and a rope to guide the geometry of the vault.

maintenance, however, this can be accomplished by fewer crew members. We estimate that by excavating and collecting materials using teleoperated robots the same vault can be built using 240 man-hours.

This experiment provides the first attempt to construct masonry structures in a simulated space environment. It is obvious that before making plans to equip the Martian settlers with trowels and shovels there are a great number of problems that will need to be solved. One of the most important challenges is to find an appropriate mortar. Mortar is used to even out stress concentrations by filling in the irregularities between bricks, as well as to hold the individual bricks in place during construction. Initially, mortar will be made by mixing dust and water. If more strength is needed, some additives might be acquired from plant products or will have to be imported from Earth. Eventually the mortar may be entirely from plant-derived polymer extracts. Shaping the brick so that successive courses interlock can obviate the second function of mortar. Additionally, a temporary inflatable tent will have to be developed so that work can be performed in a controlled environment without the need for bulky and cumbersome space suits. Once the enclosures are completed they can be glazed to make them airtight. Any remaining cracks can be patched up with plant products. A series of plastic sheets can be laid inside the cover material to trap and recover air that might still escape. If further leaks occur, the moisture in the air will quickly freeze, thus sealing the crack [Mackenzie 1987], [Petrov 2004].

Figure 7. The first three arches of a vault completed at the end of Crew 22's rotation.

Figure 8. Mars Base with a deployed habitation module, greenhouse, observatory and the first step in the construction of a permanent habitat using local resources.

Future projects in this area can include repeating our experiment using regular masonry units, developing automation processes for as many tasks as possible, and designing and testing a reusable pressurized construction tent, which will allow work in a shirt sleeve environment and control of the temperature and pressure during construction.

All of these challenges are awaiting solutions and will provide ample ground for further study by enthusiasts interested in helping to develop the knowledge necessary for humans to survive on Mars.

Acknowledgments

The researchers would like to thank the additional members of MDRS Crew 22. They are John Burgener – Commander, Sandy Muscalow – Chief Geologist, Sanjiv Bhattacharya – Journalist, and Richard Thieltges – Agronomist / Stromatolite Specialist. Additionally we owe a great debt to Professor John Ocshendorf from MIT and Bruce Mackenzie for their guidance in setting up and completing the project.

References

1. Gertsch, Leslie and Richard Gertsch, *Excavating on the Moon and Mars*, Chapter 16 in *Shielding Strategies for Human Space Exploration* J.W. Wilson, J. Miller, A. Konradi, and F.A. Cucinott. Ed. NASA Conference Publication 3360, December 1997.
2. Kennedy, Kriss, *Lessons from TransHab: An Architects Experience*, AIAA Space Architecture Symposium, 10-11 October 2002, Houston, Texas, AIAA 2002-6105.
3. Kennedy, Kriss, *The Vernacular of Space Architecture*, AIAA Space Architecture Symposium, 10-11 October 2002, Houston, Texas, AIAA 2002-6102.
4. Mackenzie, Bruce, *Building Mars Habitats Using Local Materials*, pg 575 in *The Case for Mars III: Strategies for Exploration*. Stoker, Carol ed., American Astronautical Society: Science and Technology Series v74, 1987.
5. Meyer, Thomas and Christopher McKay, *Using the Resources of Mars for Human Settlement*, pg 393 in *Strategies for Mars: A Guide to Human Exploration*, ed. by Stoker, C.R., and Emmart, C., American Astronautical Society: Science and Technology Series v86, San Diego, California, 1996.
6. Petrov, Georgi, *A Permanent Settlement on Mars: The First Cut in the Land of a New Frontier*, Thesis (M. Arch.) Massachusetts Institute of Technology, Dept. of Architecture, 2004.
7. Richards, J., Ismail Serageldin, Darl Rastorfer, Hassan Fathy. *Concept Media*, 1985. ISBN 9971-84-125-8
8. van Beek, Gus, *Arches and Vaults in the Ancient Near East*, Scientific American, July 1987, vol 257. pg 96-103.

How Would a Landing Party Sample Life on Mars? – Methods Testing at the Mars Desert Research Station, April 7-20, 2002†

Nancy B. Wood, Ph.D, Microbiologist, Crew Five.
Njbwood@hotmail.com

Abstract

Considerable evidence exists to suggest that conditions permissive for some type of microbial life might exist on Mars. Strategies for detecting putative life forms present considerable challenges, and are necessarily based on assumptions regarding their chemistry and ecology. The Mars Desert Research Station (MDRS) provides an analog setting for testing methods and hypotheses under simulated Martian operational conditions. We had three main objectives during our rotation; first, to test sample collection methods under simulation; second, to evaluate the human impact on the near-habitat microflora; and finally, to integrate the entire crew into biology activities. In the future, there should be a broad science mission for MDRS involving both biology and geology, in order to provide for sustained scientific achievement.

Introduction

What sort of life might we look for on Mars? The Viking Lander missions were designed to look for metabolic activity, such as photosynthesis and respiration, based on assumptions regarding Earth organisms.[1,2] This mission was a remarkable technical achievement, despite the negative but somewhat ambiguous conclusions. It is more reasonable to start with only the most general assumptions, namely that any organism would be carbon-based and would operate under chemical and thermodynamic principles.

If life on Mars and Earth shared a common origin, by somehow seeding each other during early planetary history, then Martian organisms, fossil or extant, may well be based on some form of nucleic acids and proteins familiar on Earth. Also, it is possible that the chemical options for functional life are restricted, and an independent origin would yield the same result. On the other hand, Martian life may have evolved a completely novel biochemistry, with different molecules for genetic storage and expression, and different structural and catalytic polymers. In any case, some type of cell surface or membrane must bound such organisms, in order to enclose the living chemistry from the outside environment. Energy-yielding metabolism might be chemically exotic, based on available compounds for oxidation / reduction.[3] Organisms might be dormant for millions of years, and only metabolize and reproduce when conditions of temperature, water and nutrient availability are permissive.[4,5]

Regardless of Martian biochemistry, there must be more than one kind, and the principle of natural selection would apply. If life originated on Mars around the same time that it did on Earth, then genetic variants must have arisen; those most successfully adapted to its particular niche would have persisted. Given the massive geological changes and the resulting selection pressures which have taken place since planet formation, numerous lineages would have arisen in any extant life. Probably these various types exist in communities, and together participate in an ecosystem. All types must be able to colonize new areas, in order to escape frequent habitat destruction.

These more general assumptions can be used to guide strategies for the next generation of Martian life searches. Analog testing of any strategy in representative Earth environments is essential to developing a mission that can answer questions regarding the presence of past or present Martian life. MDRS provides such an opportunity in a desert setting. The biology goals during the two-week Rotation 5 were designed to take advantage of the unique possibilities of this environment:

1. To test sample collection and detection methods under closed simulation.
2. To evaluate the human impact on the near-habitat microflora.
3. To integrate all crew members into biological research activities.

Sample Collection Strategies

All samples were collected by crew members wearing simulated pressure suits, which included bulky gloves. Therefore, manipulation was difficult and required preparation of all tools and sampling devices prior to leaving the "Hab." In all, three different detection and sampling strategies were explored.

Detection Direct: Attachment of Microorganisms to a Sampling Unit Under Dry Conditions

While the chemical nature of the outer surface of any Martian organism is unknown, it likely includes a complex mixture of polymers bearing regions of positive and negative charges, and non-polar, hydrophobic regions. Earth microorganisms are a particular example of this structural arrangement. Attachment of an organism to a sampling device makes no assumptions, other than charge interactions, and does not require metabolic activity of any kind.

Use of a glass microscope slide has been a common technique for retrieving microbes from aquatic environments, and it is of considerable interest to determine whether attachment can take place under completely dry conditions. The area chosen for this determination was waypoint 105, an area of rocky potholes about 500 m south of the Hab, which is covered by a luxuriant growth of unidentified bronze-yellow lichen (Figure 1). A sample of this lichen eluted from a small pebble revealed both rounded algal cells and fungal mycelium under the microscope.

Figure 1. View of waypoint 105 showing lichen-coated rock and a telltale from an inserted microscope slide sampler. The backpack and GPS unit provide scale.

Methods
Prior to the EVA to this site, microscope slides were attached to conspicuous bright pink plastic tape telltales and alcohol sterilized. They were placed in a Ziploc plastic bag, likewise sterilized with alcohol, for transport. At the site, several locations were chosen for placement in direct contact with rock or sand. The slides were imbedded in the sand or, if against rock, covered with another rock to keep them in place. The pink telltales were visible to identify the sites. After five days, the slides were retrieved for microscopic examination by pulling on the telltales with a large forceps. They were then placed in alcohol-washed collection bags. Unfortunately, many of the slides were broken, including all of the ones in a sandy area, perhaps as a consequence of a severe windstorm that had occurred during the period of attachment. No rain fell during this period. However, some slides were available for inspection.

Results
Under wet mount, slides that had been in direct contact with the lichen-coated rock showed the same algal and mycelial cells originally observed at that site. Clearly it is possible for at least some microorganisms to attach to a solid substrate under completely dry conditions. In the future, this method should be improved by design of a holder to insert fragile slides into compacted regolith, and by use of plastic slides or a flexible, transparent membrane that can be mounted on a microscope slide in the laboratory.

Sample Collection: Windblown Dust
If life exists on Mars today, there must be some sort of dispersal mechanism to allow transfer of viable forms to new habitats. An obvious candidate for this is windblown dust. Major storms lasting months and covering significant portions of the planet are known to exist. On Earth, dust is known to transfer microorganisms thousands of miles, such as between the Sahara Desert and the Caribbean.[6,7] The goals for our rotation were first, to establish a method for collecting windblown dust at MDRS, and second, to profile any microorganisms present in the sample.

Methods
The collection devices used consisted of cylindrical, clear plastic sample vials attached to 60 cm bamboo garden stakes with duct tape. Holes (~4mm) were drilled around the rim to allow dust entry from any direction. The

interiors of the vials were sterilized with alcohol, holes were covered with tape, and plastic snap lids were placed on top. The collectors were bundled for transport by pedestrian and ATV EVA teams. The collectors were installed at previously established windy sites, and retrieved for inspection.

Collected material was suspended in 2 ml of sterile, distilled H_2O and transferred to a sterile conical centrifuge tube with a transfer pipette. After tabletop settling of the suspension, 10 µl of the aqueous slurry was streaked onto LB agar plates. After incubation at 30° C for 1-2 days, colonies were observed and counted.

Results
The first attempts at collecting dust were not successful. Wind conditions in the MDRS area are quite variable; the first installation of collectors at waypoint 102 contained no visible sediment after two days, probably due to the dead calm conditions. The second attempt at the same site, a small, exposed hill, suffered the opposite problem. A severe windstorm blew over the collectors, which were found to be empty. One problem with the collector design is the difficulty in placing the bottom in compacted regolith, especially by pressure suited crew members. Even pounding the stakes with a geology hammer, and bracing the bottom with nearby rocks and sand, did not provide adequate stability. Finally, a collector was taped to the flagpole immediately outside the Hab in a second windstorm. While this technique did succeed in collecting a sizable sample, the collector design should be improved so that they can be left in remote areas, and collectors should be left in place long enough to ensure adequate wind transfer.

The material collected settled out into three obvious zones: at the bottom was a coarse (0.5-2 mm) gray layer. Above that was a medium- to fine-grained deep rusty red layer. A zone of non-settling, gray-white fines blended with the aqueous supernatant. This layer was sampled for plating. After incubation, many colony types appeared, as distinguished by color and morphology. One type was unique to this sample, and not found in samples taken directly from the soil around the Hab. These results and their limitations will be discussed in greater detail below.

Detection by Growth: Winogradsky (Ecosystem) Columns Prepared with MDRS Soil Samples
Since most microorganisms in soil samples are undescribed and not readily culturable, a column growth arrangement provides a setting for enrichment growth under natural conditions. In general, the sample is placed in a closed glass or plastic tube, which is then allowed to incubate under specified conditions of temperature, light, etc.[8] Organisms that require oxygen will proliferate at or near the top air space, while anaerobes will only grow near the bottom. If the organisms are pigmented, colored layers can be observed in the column. The value of this technique is that nothing need be known about the nutritional requirements of the organisms, since the environment in which they are found presumably supplies what is required. Modern versions of this technique have ports through which gasses and other metabolic products can be sampled.

In a desert environment, the types of organisms present in a sample may depend on the availability of water. Therefore we collected samples from areas which were always dry (any rain would immediately run off), intermittently wet (from the dry bottom of an obvious rivulet), and always wet (mud from a stream bed).

Methods
The columns used were flat-bottomed glass tubes (2 cm x 15 cm) with plastic caps. The soil samples had been collected in alcohol sterilized plastic sample vials with snap-cap lids. Samples of each (10 g) were placed in the tubes and sufficient sterile, deionized water was placed in each to thoroughly wet the soil, with a substantial aqueous layer above the stream bed sample. The columns were then incubated on a sunny windowsill in Chicago, Illinois, for three months.

Results
The samples differed in color and texture, probably because water plays a major role in flushing soluble and finely pulverized minerals. The dry sample was pale, pinkish gray, with a fines layer at the top. The wet sample was gray-brown and coarse. The intermittently wet sample was very finely textured and deep rusty red.

After undisturbed incubation under natural light / dark daily cycles, different types of presumed biological activity were observed in each column. The dry sample showed three colored layers in the top cm of the soil. Just below the water, a yellowish-white layer appeared over time, with a pinkish layer just below. Below that a dark gray-brown layer appeared. No gas bubbles were ever observed. The bottom portion of the column showed no visible changes. A microbial mat was seen floating on the aqueous layer.

The wet sample is quite different. A purple-black zone appeared near the bottom, under somewhat anaerobic conditions. A cluster of bubbles 1-2 mm in diameter appeared in the upper half of the column. The surface of the soil was greenish, and floating clumps of green material were dispersed throughout the aqueous layer.

The intermittently wet sample showed several types of apparent growth at the top 2 mm of the column. Just below the liquid surface, separate streaks of reddish-purple and green-black appeared after two months. Several kinds of colonies, mostly viscous and mucoid, grew on the glass wall above the liquid surface. Some of these colonies were pale pink.

While none of the presumed biological growth regions has been further characterized, it seems likely that each sample region contains different types of microbial populations. Selection for different organisms may well be a consequence of the differing chemical compositions present in the soil. This point should be explored further and a more sophisticated column that permits direct sampling should be employed to analyze metabolic activity in situ.

Human Impact on the Near-Hab Microflora

Humans obviously disturb the microbial environment, both by altering soil and water chemistry and by releasing human-associated microbes. While it is definitely not clear that any Earth microorganism would survive and reproduce if released on Mars, the question is of considerable interest for the ultimate development of agriculture. The objective for this study was to determine the extent of variation in the kinds of microorganisms that could be cultured adjacent to the Hab following construction and five rotations. It was assumed that the extent of disturbance varies with distance from the Hab.

Methods

Two directions were chosen for sampling: relatively undisturbed (U), southwest of the main airlock (197° from geographic north), and contaminated (C), across the leach field (111° from geographic north). Surface soil samples (0.5 ml) were taken at 1, 5, and 20 meters from the Hab. Distances were measured with a steel tape. Prior to sampling, alcohol sterilized plastic snap-cap microcentrifuge vials with volume markings were labeled and placed in Ziploc bags. At the designated distance the appropriate vial was removed from the bag with a large forceps and the sample collected from the surface.

For analysis, each sample was resuspended in 1 ml sterile H_2O, vortexed for 15 seconds, and allowed to settle on the workbench for 30 minutes. Soil was then pelleted by centrifuging in an Eppendorf microcentrifuge at maximum speed for 15 seconds. A 20 µl portion of each supernatant was transferred to 190 µl sterile H_2O, from which serial 10-fold dilutions were prepared. Each dilution was streaked onto LB plates with a 10 µl plastic inoculation loop. All plates were transferred to an incubator at 33° C for 1-3 days. Colonies differing in color and morphology were counted separately from the dilution providing the most accurate count. The pH of the original supernatants was estimated with narrow spectrum pH paper.

Results

Collection of these samples was nontrivial, due to the major windstorm taking place on the day this was carried out. Sustained wind speeds of greater than 40 mph with gusts up to 56 mph were recorded while we were on EVA. Dr. Vladimir Pletser assisted with the collection and measurements, and Dr. William Clancey made still and video recordings. While it was frequently difficult to stand during wind gusts, it was essential to collect samples on that day to allow time for microbial growth before the end of our rotation. Exploration parties on Mars may well be under the same kind of constraints.

Soil pH varied somewhat among the samples, as shown in Table 1. The highest pH was recorded in the sample presumably least affected by Hab occupation, the 20 meter sample in the undisturbed direction. Samples in the contaminated direction were somewhat more acidic.

After incubation of the plates, several colony types were observed whose distribution varied among the samples. Some were evident in all samples, and some were unique to particular samples or directions. The colony types observed are as follows:

	Distance From Hab, meters		
	1	5	20
Undisturbed	7.8	7.5	>8.4
Contaminated	7.5	7.5	7.2

Table 1. Estimated pH Values of Near-Hab Soil Samples.

A. Round, bright orange
B. Small, white, cusped, like a molar tooth
C. White, glossy
D. Large, white, puckered
E. Flat, bright yellow
F. Flat, translucent
G. Rounded, yellow
H. Large, white opaque

The distribution of colony types is shown in Table 2. The most abundant type is shown in boldface, and any unique to that sample is underlined. Orange colonies (A) were present in all samples, while C, E, and F types were only found in the undisturbed direction. The distinctive cusped colonies, B, were only present in the 20 m undisturbed sample, where they were the dominant form. The contaminated direction samples uniquely produced colonies of types D and G. The windblown dust sample, discussed above, produced colonies of types A, C, F, G and H, which was unique to this sample. None of these organisms has been identified at present.

Distance From Hab, meters		
	1	5	20

	1	5	20
Undisturbed	A, C, E, F	A, C, F, G	A, B
Contaminated	A, D, G	A, C, D	A, C, D

Table 2. Summary of Colony Types
in Near-Hab Soil Samples.

While these results suggest varying patterns of microbial populations surrounding the Hab, the methods employed in this study are crude and by no means identify all types present. It is to be expected that most microorganisms in a sample cannot, in fact, be cultured. LB plates are rich a medium which no doubt excludes many autotrophic forms. Many microbes may well remain trapped within crevices in individual soil particles.

Integration of the Crew into Biological Research

All members of Crew 5 participated in this work, which could not have been accomplished without their help. They helped with sample collection, detector installation, field measurements, and photography. In order for biology to succeed in an analog mission, all crew members should feel connected to a successful outcome. The biologist is obligated to explain procedures carefully, and to provide properly packaged materials so that other crew members can carry out their assignments without confusion or difficulty. Crew 5 members, who came from different professional backgrounds, made many valuable suggestions for improving both equipment and procedures. Finally, scientific results are interesting and exciting, and everyone should share in the pleasure as well as the work.

Conclusions

Analog field testing is essential for the design of an effective experimental strategy for searching for life on Mars. The work done during Rotation 5 provides encouragement for the approaches to sampling methods attempted, and suggests improvements for further methods and equipment design. Some improvements could be made in the Hab itself. While every attempt was made to carry out the culturing procedures under sterile conditions, construction aspects in the building left holes open to the outside through which dust readily penetrated. A simple and inexpensive solution would be a "clean workstation" or glove box apparatus in which to carry out sterile operations.

Further improvements in equipment and procedures are suggested by the experience of the conditions encountered in this environment, such as high winds, compacted soil, and limited power availability. For example, the initial attempt at collecting soil samples near the Hab was aborted when one of the sample collecting bags was blown away in the very high wind. Dr. Pletser, who had astronaut training, suggested that all bags and tools should be tethered to the person who was not using them; it is much easier for a person to remove objects from a partner's pockets than from his own. After this procedure was implemented, collection took place without further problems.

In future missions, there should be a broad, overall science mission that takes advantage of the uniqueness of Martian analog research and of the remote desert site. Possible components of such a program might include:

• Development and testing of life detection methods.
• Study of soil chemistry and mineral formations in relation to microbial communities.
• Miniature instrumentation development and testing.
• Development of field data recording methods.

A general program would not preclude other initiatives, but provide a framework for recruitment of collaborators. Consistent research progress would also attract funding and, best of all, contribute to successful exploration of Mars.

Acknowledgments
I would like to thank all of my crew mates, Dr. Bill Clancey, Andrea Fori, Dr. Jan Osburg, Dr. Vladimir Pletser and David Real, for their help, suggestions, enthusiasm and companionship during our rotation. The camaraderie and good will of all made the work easy and the problems small.

References
1. Horowitz, N.H., and Hobby, G.L., Viking on Mars: The Carbon Assimilation Experiments, J. Geophys. Res., 82, 4659-4662, 1977.
2. Levin, G.V., and Straat, P.A., Recent Results from the Viking Labeled Release Experiment on Mars, J. Geophys. Res., 82, 4553-4667, 1977.
3. Boston, P.B., Ivanov, M.V., and McKay, C.P., On the Possibility of Chemosynthetic Ecosystems in Subsurface Habitats on Mars, Icarus, 95, 300-308, 1992.
4. Mattimore, V., and Battista, J.R., Radioresistance of Deinococcus Radiodurans: Functions Necessary to Survive Ionizing Radiation are Also Necessary to Survive Prolonged Desiccation, J. Bacteriol., 178, 633-637, 1996.
5. Stan-Lotter, H., Pfaffenhuemer, M., Legat, A., Busse, H.-J., Radax, C., and Gruber, C., Halococcus Dombrowskii sp. Nov., An Archaeal Isolate from a Permian Alpine Salt Deposit, Int. J. Syst. And Evolution. Microbio., 52, 1807-1814, 2002.
6. Griffin, D.W., Garrison, V.H., Herman, J.R., and Shinn, E.A., African Desert Dust in the Caribbean Atmosphere: Microbiology and Public Health, Aerobiologia, 17, 203-213, 2001.
7. Perry, K.D., Long Range Transport of North African Dust to the Eastern United States, J. Geophys. Res., 102, 11,225-11,238, 1997.
8. Lehman, R.M., Colwell, F.S., and Bala, G.A., Attached and Unattached Microbial Communities in a Simulated Basalt Aquifer Under Fracture- and Porous-Flow Conditions, Appl. Environ. Microbiol., 67, 2799-2809, 2001.

The Geophysical Study of an Earth Impact Crater as an Analogue for Studying Martian Impact Craters†

Louise Wynn, Jason Held, Akos Kereszturi and Judd Reed

Abstract

Devon Island's Haughton meteorite impact crater is a focus of Mars analog research because of its similarity to impact craters on Mars. However, its size (20 km diameter) and age (about 23 million years) make close-up geologic examination in a limited field season difficult, because of erosion of the wall surface, as well as backfall from the original impact, which has substantially refilled the crater. We hoped to confirm previous work at Haughton and similar craters using hand-held geophysical instruments to show the validity and practicality of this type of research in the initial exploration of Mars. Our research achieved that objective and additionally modeled a possible way to collate the results of several data-mapping methods to draw inferences about the approach and angle of Haughton and other impact objects.

Introduction

The Haughton meteorite impact crater is located on Devon Island in the Canadian Arctic, at 75° 22' north longitude and 89° 41' west latitude. The Arctic desert provides a Mars-like location for research into the formation of the crater and subsequent geologic processes of such structures on Mars.

The object that struck Earth on Devon Island was too big to be stopped by the atmosphere. It maintained its coherency until impact, when it was deformed as it plowed into the ground and, because of that deformation, penetrated not much more than its diameter. The result was an explosion perhaps the size of a 1,000+ megaton hydrogen bomb going off about a kilometer below the surface.

The Haughton impact crater has been the object of several earlier studies by geologists, seismologists and geophysicists. We hoped to compare the results of our geophysical surveys, including magnetic susceptibility and total count radiation, conducted at the crater during the Mars Society's Flashline Mars Arctic Research Station field season in July 2004, with other evidence, such as aeromagnetic and seismic surveys and satellite photography, in order to paint a more complete picture of the impact event.

In other meteorite impact craters that have been studied extensively the materials thrown out of the Earth by the impact (the ejecta) are scattered around the outside of the crater in a pattern that reflects, in roughly reverse order, the stratigraphy of the Earth beneath where the meteorite fell. In general, the dense mafic and ultramafic rocks from lower in the Earth (Precambrian crystalline basement rocks), which fall closer to the rim of the crater, should show relatively high readings on the magnetic susceptibility meter, while the younger felsic and sedimentary layers from closer to the surface, which are lighter and can be carried farther by the original explosion and wind sorting, should give lower magnetic readings. Conversely, higher radioactivity readings are often found in felsic than mafic and ultramafic rocks. Our research was designed with this model in mind; we conducted magnetic susceptibility and radiation surveys in the northwestern quadrant of the crater's rim.

Geophysical studies are by nature inferential; the geophysicist typically arrives at a site of interest millions of years after the event has occurred and must use instruments that measure indirectly the results of the event. Yet much can be learned from such studies, including the size and composition upon impact and depth of penetration of a bolide and perhaps even information about the impactor's direction and angle of approach to Earth.

Previous seismic evidence suggests that the impactor penetrated no more than 1.7 km of Paleozoic sediments, leaving a crater about 20 km in diameter and excavating substantial amounts of the local Paleozoic carbonate sediments. Chunks of limestone and dolomitic sediments up to 1 m² were thrown into the air and deposited back into the crater and beyond its rim.

Research Plan

The 23-million-year-old Haughton impact crater located near the west end of Devon Island has been studied through remote sensing methods, including seismic studies,[1,2] aeromagnetic surveys,[3] ground-penetrating radar,[4] and geochemical analysis of samples.[5] However, until this summer, it appears that no one has done a systematic walking geophysical survey for radioactivity and magnetic susceptibility of the debris thrown outward from the blast.

We conducted surveys from the rim of the crater to the north and to the west in five kilometer lines extending radially in the direction from the center of the crater as it is marked on generalized geologic maps (and apparent from the morphology of the feature). In addition, wherever possible, we took readings while participating with other research groups in the area around the crater.

Discussion

We began with the hypothesis that basement rock was excavated and deposited beyond the crater rim. Further, we anticipated that there would be a rough sorting distribution in the ejecta field, e.g., denser and deeper crustal blocks lying closer and shallower, and lower density carbonate rocks farther from the center of the impact.

We conducted three surveys, two to the north and one to the west, sampling every 50 to 200 meters, depending on terrain, for magnetic susceptibility and radioactivity. These areas, most accessible to our site, effectively covered the entire northwest sector of Haughton crater for a distance of five kilometers from the identified rim.

Magnetic susceptibility readings would detect the heavier metals (such as iron and nickel), expected in mafic and ultramafic rocks of the lower crust. A deeper impact would presumably excavate the denser rocks hosting these minerals, with higher magnetic readings, settling closer to the center of the crater, while lighter felsic and carbonate rocks, with lower magnetic readings, should be thrown further out. Any positive magnetic readings would strongly indicate an impact depth greater than two kilometers and would additionally manifest the presence of metals such as iron or nickel in the impactor itself.

Radioactivity readings, taken at the same points, would indicate the presence of potassium 40, which is common in felsic layers of the upper crust (less than two kilometers). Thus these readings would show a gradient of lighter materials and higher radiation readings the farther we traveled from the crater's center. Heavier materials would show higher magnetic readings the closer we were to the center.

Instead, we found no measurable magnetic susceptibility or radioactivity in any of our profiles to a distance of five kilometers from the crater rim, an approximate distance of 10 km from the center. Testing of the instruments on site verified that they were functioning correctly; in addition, the instruments were tested successfully in July at Columbia River Basalt sites along SR14 east of Vancouver, Washington. In addition, in other area scouting excursions along the way, we took measurements in interim points along our routes to expand upon our data. These interim points also showed no positive radiation or magnetic susceptibility readings.

While not excluding excavated crystalline bedrock, these consistent readings suggest that: (1) No mafic or ultramafic bedrock can be found in the ejecta field, and (2) the incoming bolide was not an iron-nickel body. An iron-nickel bolide would show readings of scattered fragments on the magnetic susceptibility meter. This information gives us information about the composition of the bolide and the upper crust around Devon Island, as well as a depth limit of the impact.

As a rough approximation, the incoming bolide would likely have an effective diameter of about 1/20th the size of the resulting crater. Earlier studies at Haughton suggest that the depth of penetration is considerably less than the radius of the crater; in other words, the excavated crater was not hemispherical, but instead shallow and circular. One explanation for this is that the incoming bolide deformed as it penetrated, dispersing itself and its kinetic energy in the less consolidated and more easily mobilized upper (younger) geologic units.

The fact that there were no significant magnetic susceptibility readings outside the crater, where due to kinetic energy-driven detonation much of it might be expected to lie, supports this inference and would seem to constrain the bolide to a stony or chondrite body. With satellite photography showing a crater rim, possibly elliptical, with a 12-kilometer semi-major axis,[1] we can estimate the impactor to be about one kilometer in diameter.

An apparent elliptic crater shape is implied by both visual analysis of satellite imagery and seismic data. While it is tempting to speculate on the incoming bolide's trajectory from the shape of the crater morphology, there is not enough evidence to finalize the bolide's angle of approach. In order to do this, we will need the following:

1. An ejecta blanket model around the south side of the crater. We haven't been able to map the complete ejecta blanket in any detail around all sides of the crater, and therefore do not have statistics on its distribution as a function of azimuth. Although satellite photos and walking surveys provide a big-picture view, it is still incomplete. Landsat 29-meter-per-pixel imagery will only show larger concentrations of ejecta, but one-meter multispectral data are required for adequate analysis. The walking surveys on the West and North sides provided a clear picture of ejecta on those sides.
2. More information and evidence of pre-impact faults and geology are needed. There are several other possible reasons for crater ellipticity other than impact trajectory (subsequent tectonic deformation, faults and inhomogeneity of the pre-impact geology, etc.).
3. Model studies by Shoemaker et al.[5] state clearly that the impact angle would have to be shallower than 15 degrees from the horizontal to overcome the normal physics of a hypervelocity impact – detonation from beneath masks any trajectory characteristics with a much larger circular crater than the incoming body.

Conclusions

Based on the evidence from satellite photography and seismic data we can estimate the size of the incoming bolide at just over one kilometer in diameter. Our magnetic susceptibility data of consistently zero readings of up to 10 kilometers outside the crater rim strongly support the hypothesis that it was not an iron-nickel asteroid, but deformable and likely a stony (chondrite) or hybrid body.

Very low radiation readings taken during our survey also strongly support the hypothesis that the bolide did not penetrate to the crystalline basement and point to an impact of no more than 1.6 to 1.7 km in depth. Since all of the breccia discovered from the impact is consistently made of sandstone and dolomite, we can conclude that the carbonate sequence on Devon Island is oceanic in origin and at least 1.7 km thick.

We conclude that hard work and careful thought, using simple hand-held instruments which can be realistically used by individuals wearing pressure suits, can still provide substantive information about an impact crater, its causative body, and the geology of the underlying target rocks. This shows, first hand, the viability of this type of study for similar locations on Mars. This was a realistic analogue, and of the type that would be extremely difficult for a robot to accomplish. Since magnetic susceptibility and radiation readings require surface contact for an accurate reading, and since the rugged terrain is considered extremely hazardous for a robot to traverse, we conclude that this study can only be accomplished with human astronaut researchers rather than robots.

Finally, evidence for the angle of descent is conflicting and warrants further study. From Landsat 10-meter panchromatic imagery we measure an apparent elliptical crater angled approximately 40 (±5) degrees from True North. Topography shows buckling near the crater rim on that side, supporting this hypothesis. This is in conflict with seismic data, however, which show the longest fault lines on the same side but also many shorter faults due east of the crater rim. Therefore, we cannot fix an incident azimuth angle with confidence.

There are several possibilities, however, which should be addressed. One is the direction of erosion, another is a bolide which partially melted and deformed prior to impact. Erosion due to rain and melt water, as well as 23 million years of glacial activity, could potentially warp the crater walls. Glacial activity would produce a north-south warping (explaining deformation of surface features), although not necessarily enough for a half-kilometer extension of the semi-major axis. This still does not explain the off-centered impact point, which would not be significantly moved from erosion, supporting an elliptical shape. Nevertheless, study of glacial effects on the crater rim is needed for more evidence.

Bolide deformation prior and during impact is another likely explanation for conflicting evidence. With a partially melted body the crater will show some evidence of both solid and liquid impacts, as well as explain the difference between surface and subsurface evidence. This is another area warranting further study.

Acknowledgment
This work was supported by the Mars Society.

References
1. D. Scott and Z. Hainal (1998) *Seismic Signature of Haughton Structure*, Department of Geological Sciences, University of Saskachewan, Saskatoon, Saskatchewan S7N 0W0, Canada
2. Vladimir Pletser, Philippe Lognonne, Michel Diament, Valerie Ballu, Veronique Dehant, Pascal Lee, Robert Zubrin (2001), *Subsurface Water Detection On Mars By Active Seismology: Simulation At The Mars Society Arctic Research Station*, The Mars Society, 11111 West 8th Avenue, Unit A, Lakewood, Colorado 80215, USA. Tel: ++1/303/9800890; Fax: ++1/303/9800753; Email: Zubrin@aol.com
3. Glass *et al*, NASA, Aeromag Survey
4. C.E.I. Nieto, R.R. Stewart (year?), *Geophysical Investigations at a Mars Analog Site: Devon Island, Nunavut*, Dept. of Geology and Geophysics, The University of Calgary (GLGP, 2500 University Dr., N.W., Calgary, Alberta, Canada T2N 1N4. cenieto@ucalgary.ca, stewart@ucalgary.ca).
5. T.E. Bunch, R.A.F. Grieve, P. Lee, C.P. McKay, J.W. Rice, Jr., J.W. Schutt and Z. Zent, *Preliminary Observations on Highly Shocked Crystalline Basement Rocks from the Haughton Impact Crater* (1997), Geological Survey of Canada, K1A OE9, Ottawa, Canada; Space Science Division. NASA / ARC, Moffett Field CA USA 94035
6. G.R. Osinski and J.G. Spray (2001) *Highly Shocked Low Density Sedimentary Rocks from the Haughton Impact Structure*, Planetary and Space Science Centre, Geology Dept., University of New Brunswick, 2. Bailey Drive, Fredericton, NB, E3B 5A3, Canada
7. Michael C. Malin and Kenneth S. Edgett (2000) *Sedimentary Rocks of Early Mars*, Malin Space Science Systems, Post Office Box 910148, San Diego, CA 92191-0148, USA
8. E.M. Shoemaker, Dave Roddy

//

Using Geophysical Field Methods in the MDRS Area†

Louise Wynn

Abstract
One of my goals at the Mars Desert Research Station (February and December 2004) has been to use geophysical methods to learn more about geologic field methods while studying the geology of the Morrison Formation in southern Utah. As a US Geological Survey volunteer working with field crews in southeast Alaska and northern

Mexico, I realized that I would learn more when I was on my own, getting hands-on experience and being forced to use my own deductive skills rather than relying on the knowledge of coworkers. My geophysical research at the MDRS and the Flashline Mars Arctic Research Station has given me the opportunity to learn fast and apply my knowledge in the field; in addition, it has prompted me to return to school to gain the background knowledge essential to doing better research.

Introduction

I wanted to see if geophysical measurements of radiation and iron content of rocks and sediments would correlate with geological maps already prepared of the area. To my knowledge, this was the first geophysical survey undertaken in the area.

Theoretically, geophysical readings would add to geological observations of the sedimentary deposits of the area to distinguish between layers from a more oxidized (arid, dry) paleo-environment and layers from wetter and more anoxic environments.

Guided by geologic maps prepared by previous MDRS crews, and with help from crew geologist and journalist Kyoichi Sasazawa, commander Digby Tarvin and crew mates Bob McNally, Celeste Gale, and Diego Casa, I used two hand-held instruments to take geophysical readings of rocks and sediments in the area around the MDRS Hab in February 2004.

A major constraint on my research was the need to accommodate research goals of other crew members. The result was that I did not conduct a regular grid survey of the area, but rather took my instruments with me on every EVA so I could make readings at every possible opportunity. Only toward the end of the two-week Crew 24 rotation was I able to plan an EVA to make readings in a more regular pattern. However, the apparent randomness of my surveys did not make the results any less valid or helpful.

Upon my return to the MDRS in December 2004, I hope to take more geophysical readings and samples in areas I missed in February.

Equipment and Procedures

I used two hand-held instruments for some very basic geophysical research: A radioactivity meter (which the manufacturer markets variously as a Geiger counter and a scintillometer), to measure radiation; and a magnetic susceptibility meter, to measure iron content. Ideally, I would take readings with both instruments at regular intervals around the Hab and in a grid pattern in any area where the geologic map or my own observation indicates possible metals. The geophysical instruments should find a pattern of mineralization corresponding to the known geologic features.

In fact, as mentioned above, my sampling was not regular; but when I compiled data points I could see where I needed to take more readings and was able to go to one of those locations before the end of Crew 24's rotation.

Narrative and Lessons Learned

In spite of time and logistical constraints, I developed a data set of some 40 points, reaching outward around the Hab as far as about 1.5 km. In addition, I collected rock samples and took photos of some of the distinctive features where I made readings.

I realized that my goal of learning to do field geology was being accomplished spectacularly when I began to be able to predict what readings I might find on my instruments based on sediment / rock color, clast size, and location.

In addition, I learned how to deal with the hour-to-hour, minute-to-minute, let's face it, downright constant tedium of walking through a cold desert while wearing a cumbersome space suit and trying to push little tiny buttons with fat clumsy gloved fingers. This is what doing geophysics is going to be like on Mars, but what the hey, it's what it's like on Earth.

The tedium is compounded by occasional frustration. For example, why do those reddish rocks show no evidence of iron? Is the apparent oxidized iron in a form not detectable by the magnetic susceptibility meter? Do the rocks not really contain any iron, after all? What other kinds of rocks could they be? And where's the radiation I expected to find in the darker layers I thought were shales? Are my meters broken?

I learned that getting a lot of "zero" readings, while discouraging, is the norm – unless you're in an auto junkyard or a uranium mine. I learned to look for rocks that were definitely shale to get the radioactivity meter to start beeping again; and for old rusty cans or other metal to make the magnetic susceptibility meter soar.

Conclusions

In Table 1, I've charted the geophysical data with descriptions of some of the locations where they were measured. I recorded readings in the order I made them, not according to their positions on the map. Waypoint numbers correspond with my GPS readings and MDRS Crew 24's waypoint numbering system; those with no number are copied from another person's GPS unit. The waypoint numbers have no further significance.

The geophysical maps in Figures 1 and 2 show the pattern of mineralization detected by the two meters. These maps must be taken with a grain of salt – in addition to some major gaps in data collected, indicated by the white areas on the maps, the software used to collate the data points filled in some areas where coverage is not as complete as the maps make it appear.

Future Plans

I am returning to the MDRS in December 2004 to collect more geophysical data to complete these maps. I also hope to conduct lab studies of some of the rock samples collected in order to more fully relate the geophysical and geochemical data with the geological descriptions made by Crew 24's geologist Kyoichi Sasazawa and geologists from previous crews.

Acknowledgments and Some Final Comments

Thanks to Dr. Robert Zubrin and The Mars Society for giving me the chance to do this work. The effects of my experiences at the MDRS in February 2004 reach farther than I ever suspected when I applied to work there. Realizing that my years of hanging out with geologists and geophysicists hadn't given me the education I needed (by osmosis? – what was I thinking!) to do competent field work, I went back to school, auditing Geology 101 at Clark College in Vancouver, Washington. This in turn prepared me to do better work at FMARS in July 2004. That experience inspired me to return to school full time to study geology and physics "for the fun of it." (Yes, I realize this shows just how ignorant I was, to think there could be anything "fun" about studying physics.) At any rate, my full-time return to school is the beginning of studying more of the sciences, helping me prepare for a future I can only dream of at this point.

Wpt.	UTM Coordinates	Mag.	Rad.	Comments
41	4250767 N 0518144 E	N/A	N/A	HAB
42	4250855 N 0518286 E	82	0	
43	4250667 N 0518446 E	50	0	At Juras.-Cretac. Boundary
44	4250632 N 0518537 E	4	0	White feldspar or plag.
45	4250655 N 0518661 E	0	0	"Martian mushrooms"
46	4250655 N 0518799 E	0	0	
47	4250400 N 0519102 E	0	0	
49	4250692 N 0528675 E	0	0	
101	4250929 N 0518169 E	0	0	Sandy wash near HAB
—	4251750 N 0518595 E	55	0	ATV road
104	4253670 N 0518620 E	15	0	Streambed at ATV crossing
107	4250320 N 0519090 E	0	0	Sedimentary layers
108	4250520 N 0521838 E	13	0	Red sandstone hill
111	4253736 N 0527894 E	7	0	Sed. wall at streambed
112	4253719 N 0517487 E	16	0	Purple mud sediment
113	4253716 N 0517492 E	0	0	Red mud face of strm wall
114	4253775 N 0517424 E	0	0	Orange hard sed. layer
115	4253834 N 0517424 E	0	0	Orange sed-conglom.
116	4253864 N 0517406 E	0	0	Near Gryphaea in stream
117	4253853 N 0517375 E	0	0	Black crumbly sediment
—	4253248 N 0516984 E	0	0	Brown mud
119	4254520 N 0516970 E	0	0	2.61 km from HAB
—	4250790 N 0518160 E	0	0	Weather station near HUB
—	4250876 N 0518219 E	0	0	Rebar Hill: Red mud
122	4250931 N 0518166 E	0	0	
125	4252447 N 0514804 E	0	0.05	Black silt from bluffs
126	4252817 N 0515396 E	0	0.05	Stream bed below bluffs
127	4252841 N 0515658 E	0	0	
128	4252841 N 0515658 E	0	0	
130	4250767 N 0518144 E	0	0.02	
131	4250611 N 0518178 E	0	0.04	(Same as Bob's # 137)
132	4250451 N 0518063 E	0	0.04	
135	4250452 N 0518064 E	0	0.09	
136	4250669 N 0518131 E	0	0.04	
137	4249225 N 0515355 E	0	0.06	
138	4249032 N 0515101 E	0	0.08	
141	4251214 N 0515415 E	0	0.10	Black dirt cliff bottom
142	4251237 N 0515379 E	0	0.08	Brown-red sandstone cliff
143	4250942 N 0518125 E	0	0.08	Behind HAB to west
144	4250942 N 0518119 E	0	0.08	Just beyond # 143

Table 1. Readings are as follows, all at UTM 12 N, NAD27 datum.

Thanks to Crew 24 Commander Digby Tarvin for giving me so many opportunities to work on this project and encouraging me in every way.

Thanks to all the members of Crew 24 for accommodating me as we went on EVAs, stopping frequently and waiting for me to kneel in the red soil of Mars to take my geophysical readings.

Thanks to the members of the Remote Science Team for their support. I am looking forward to working with them in the future to further our understanding of the geologic history of the area.

Thanks to FMARS 9 Commander Jason Held and all my FMARS 9 crewmates for their support and encouragement.

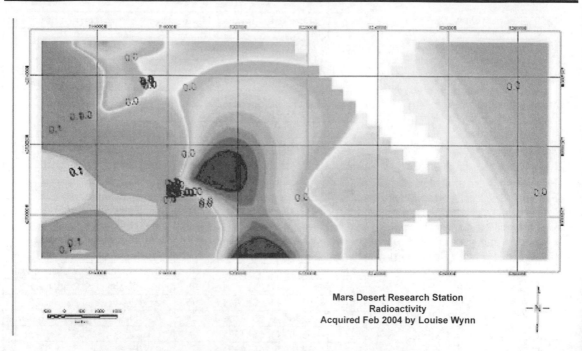

Mars Desert Research Station
Radioactivity
Acquired Feb 2004 by Louise Wynn

Figure 1. MDRS Radioactivity – Acquired February 2004 by Louise Wynn.

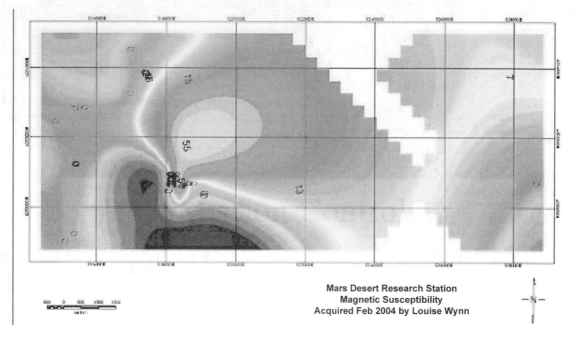

Mars Desert Research Station
Magnetic Susceptibility
Acquired Feb 2004 by Louise Wynn

Figure 2. MDRS Magnetic Susceptibility – Acquired February 2004 by Louise Wynn.

Martian Literature: Educating Generation Mars for the Voyage Ahead†

John M. Heasley, Richland Center High School
Richland Center, WI, heaj@richland.k12.wi.us

"It is the scientists who have thought hardest and best about the realities of Mars . . . But there are artists in here too, and writers, and poets, and people whose dreams take no such articulated form, but still focus themselves on the same rock in the sky. They illuminate Mars; Mars illuminates them."
— Oliver Morton, *Mapping Mars*

Mars has long attracted our gaze and our imagination. This fascination has been well documented, but there are a few stories that stand out for me. I read of Robert Goddard, climbing a cherry tree as teenager in Massachusetts in the 1890s and imagining a voyage to Mars. He learned to keep this dream to himself, but he did go on to pioneer the use of liquid-fueled rockets. In his audio reading of *The Martian Chronicles*, Ray Bradbury recounts how he stretched out his arms in 1930s Illinois and tried to will himself to Mars. In Cosmos, Carl Sagan tells how he attempted to repeat this experiment in 1940s New York. I found myself wondering if the young people in school today still have Martian dreams.

In the closing minutes of class, one day last February, I read to them a passage from Michael Benson's *Beyond*, about the images sent back by our interplanetary probes.

Sifting through a self-congratulatory final press release archived at the Mars Pathfinder site, I was suddenly, unexpectedly, moved. Contact with the lander was lost, it said, in early October of 1997. That was after nearly three months of continuous operation – much longer than expected. The loss of communication was attributed to the failure of the lander's battery, which in turn cut power to the heater. "After that," the text read, "the lander would begin getting colder at night and go through much deeper day-night thermal cycles. Eventually, the cold or the cycling would probably render the lander inoperable."

But little *Sojourner* is almost entirely solar-powered. It was just as animated as ever when all contact with Earth was lost. I came across the following sentence: "The health and status of the rover is . . . unknown, but . . . it is probably circling the vicinity of the lander, attempting to communicate with it."

The poignancy of it! The pathos! Powered forever by the inexhaustible Sun, impervious to the cold, *Sojourner* may to this day be wearing grooves in that ocherous desert floor. And we've forgotten our cybernetic creation, literally leaving it to its own devices. Having chipped, hammered, glued, and then welded and screwed together the matter we're surrounded with, we've finally endowed it with eyes, ears, and a capacity for self-direction – something like early life itself. We've propelled it at extreme velocities to distances that redefine how far human artifacts can go. And we've left it to circle, or even to beeline out of the solar system – still seeking orders, still trying to communicate with us.

My students were unexpectedly moved. They remembered *Sojourner*. It roved across Mars when they were in fourth grade. They had played with the Hot Wheels and Lego models. With parents, they visited the web site. There was some anger at NASA's seeming indifference. "Why doesn't *Sojourner* fly home?" "Is NASA going to send another probe to get *Sojourner*?" "Why didn't NASA include extra batteries?" I found myself wondering if they saw in *Sojourner* something of themselves. Is this what it is like to be a teenager? Roving into some new world, trying courageously to report back what you see? They seem to have adopted a policy of "No Rover Left Behind."

I had this experience in mind as I designed and taught a three-week summer school course in Martian Literature at our rural high school in the Driftless Region of southwest Wisconsin. There are a number of advantages to teaching in summer school. I was able to team-teach with a Social Studies colleague. There is more room for experimentation. You can teach outside your content area. You can create blocks that extend beyond the normal forty-five minute class. As an elective, students have volunteered for the course and you can create a space where it is safe to be smart.

It is important to understand who these students are. Neil Howe and William Strauss in *Millennials Rising: The Next Great Generation* identify students born 1982-2000 as Millennials who "will correct what they perceive to be the mistakes . . . of boomers by placing positivism over negativism, science over spiritualism, team over self, duties over rights, honor over feeling, action over words." The reliability of this prediction is still being determined, but it does provide students with a sense of identity and positive mission for the future.

It is the grandparents of these students who remember *Sputnik* and the parents who remember the Apollo landings. They have no memory of *Viking* or *Challenger*. They do recall the success of *Pathfinder*, *Spirit* and *Opportunity*, and the failure of *Columbia*. In a Mars Society paper entitled *Preparing for the Journey: An Introduction to Mars Education*, Donald M. Scott identifies the work to be done: "A major Mars preparation task needs more attention. This is the education of the 'Mars Kids': the children now in school who will design, fund, and conduct the human missions to Mars."

The success of Martian Literature depended on an integrated, multidisciplinary approach. Too often, a young person's experience of high school is one of fragmentation and disconnection. John Locke wrote, "The mark of genius is the ability to discern not this thing or that thing, but rather the connection between the two." More recently, Jesse von Puttkamer argues that, "We need a new frame of mind that shifts the emphasis from individual subjects to the interactions and relationships between them" (*Spaceflight and the New Enlightenment*).

Martian Literature follows this vision. Students read science fiction novels and viewed movies to discover how Mars has been imagined. They used planetarium software (Starry Night) and an inflatable StarLab to learn how Mars moves through the heavens. They built model rockets to learn the basics of propulsion. They viewed space and landscape art to see how Mars has been pictured. They studied history to place Mars in context. They constructed a web site and made public presentations at MarsFest 2004 to share the results of their research. They heard a guest lecture from propulsion researcher Dr. Jordan Maclay and made a field trip to Chicago to tour the Adler Planetarium and the Museum of Science and Industry.

The course considered several questions. When it comes to Mars, do we see what is there or what we want to be there? Will we terraform Mars, or will Mars areoform us? Is the frontier thesis applicable to Mars? Are aliens wise or demonic? Do ecosystems and planets have rights? Is Mars a promised Utopia? Does gender matter when voyaging to Mars?

You do not need an astrologer to know that Mars is very much a planet for our times. In 1999, young people were looking forward to a fine new millennium. There was every indication that we were moving closer to a vision of "one world, one people, one future." Within two years, students and teachers watched together as fundamentalists attacked the Twin Towers. It became a time of war, fear, and terror. Students were presented with a future in which they would be called upon to wage an interminable war against an emotion. Martian Literature is an attempt to end fear and regain hope by allowing students to imagine a more peaceful and cooperative future as we leave our home and travel to our new world.

Reference
Lists of novels, movies, and resources can be found at the web site we created. Go to www.richland.k12.wi.us and select Martian Literature under student projects.

Space Propulsion Systems for Mars Missions

Half Way to Anywhere – On-Orbit Electrolysis to Cut the Cost of Traveling to Low-Earth Orbit and Beyond[†]

Tom Hill
The Aerospace Corporation
hillkid@earthlink.net

Abstract
Robert Heinlein was quoted as saying "Once you're in low-Earth orbit (LEO), you're halfway to anywhere." This is due to the mechanics of space launch, where accelerating into LEO is a large portion of your journey. As a corollary, storing mass in LEO is a way to make trips beyond LEO easier. This paper discusses a project that, for on the order of $1 Billion, creates a flexible cache of rocket propellants (hydrogen and oxygen) and human consumption supplies (oxygen and water) in low-Earth orbit. Part of the project involves increasing launch vehicle flight rates through open competition, which will lower the per-kilogram cost of launching payloads into LEO. Exploiting this cache will cut the launch weights of interplanetary spacecraft by up to two thirds. This material, stored in orbit for years, would serve any space mission. The plan is modeled after historical cases that jump-started the airline industry, and calls for the best of governmental and/or commercial efforts to get us half way to anywhere.

Historical Context

Before a proper discussion about a solution can take place, it pays to review the historical events that led to its need. While many of the discussions here are tired, there are some essential points to take from them. The majority of this discussion relates to the United States' experience in space exploration, although many of the same lessons apply internationally.

Space Launch

Space launch got its start as an outgrowth of ballistic missiles, both intermediate-range ballistic missiles (IRBMs) and intercontinental ballistic missiles (ICBMs). Ballistic missiles were built on the premise that after much preparation, they could be stored for long periods of time and then be ready to go on short notice to destroy enemy targets. Because of the preparation time followed by storage time, rapid change-out of missiles (or payloads, in this case, the warheads) was not a priority. Long maintenance cycles with individual missiles out of service were the norm, and large fleets of missiles kept an acceptable launch readiness rate.

At about the same time, the United States developed a keen interest in what was going on inside the Soviet Union. The secretive adversary was growing more technologically adept, which made it difficult to rely on traditional spy planes to carry out reconnaissance.

At about the same time, the United States developed a keen interest in what was going on inside The Soviet Union. The secretive adversary was growing more technologically adept, which made it difficult to rely on traditional spy planes to carry out reconnaissance. The US government decided to move into the realm of satellite reconnaissance, and was willing to pay a lot of money for the information the satellites would gather. Since IRBMs and ICBMs already traveled to the edge of space in their travels, adapting them to launch these satellites was the simplest, fastest way to achieve the desired result.

As satellites grew in mass and complexity, and the new National Aeronautics and Space Administration (NASA) started to gain interest in crewed flights to orbit, missiles (mostly renamed rockets or boosters by this time) grew in size, largely through upgrades to their missile cousins. As new boosters were developed, many of the same people who worked on missiles worked on rockets, so many of the same design strengths (reliability grew as time went on) and weaknesses (the price for launching a kilogram into orbit continued, as a rule, to go up, and timelines between launches were not significantly shortened). Something that hadn't been tried yet was reusability.

After its stunning victory in the cold war of landing a man on the Moon, NASA looked for another mission. Since a crewed journey to Mars was too expensive for the seemingly war-ravaged country at that time, the space agency built a space truck, known as the space shuttle. The shuttle was conceived to make space travel routine, and in 1971 launch costs were estimated to be on the order of $10.5 million[1] per flight, or a cost of $330/kg. As design and budget realities hit the program, however, NASA found itself building a vehicle that wasn't designed to operate on short timelines. Meticulous construction led to difficult maintenance and preparation for launch. Architectural compromise led to several design flaws, one of which cost a crew their lives. The space shuttle was not the answer for routine space flight.

Commercial interests, at times, seemed like the answer to driving the cost of space launch down. In the late nineties, the number of satellites scheduled for launch outstripped the supply of launch vehicles by an order of magnitude, and the number of commercial launches consistently exceeded the number of government launches. New commercial ventures sprung up to meet this demand, including many of the old, familiar names in rocketry (now Lockheed-Martin and Boeing) as well as some unlikely partnerships (Russian and US companies) and some new startups, taking their own approach to cut the cost of launch to orbit (Kistler Aerospace and Pioneer Rocketplane). The recession of 2000, as well as the very public failures of some of the satellite industries driving this new revolution (Iridium, Globalstar), caused the market that the new space launch industries were to support to collapse.

The Existing Impasse

Today, there are several launch services providers chasing after a small (and steady or declining) market. In most industries this type of situation leads to a decrease in costs, but the US government still has a need for reliable rides into orbit, and they've grown used to paying a lot of money for the service. While two hybrid governmental / commercial rockets are nearly ready for launch as of this writing (the Evolved Expendable Launch Vehicle, or EELV, from Boeing and Lockheed Martin), many of the market assumptions that were made to make the systems cheaper are no longer valid. To maintain the programs (and a viable method of achieving orbit) the US government has increased its sponsorship of the programs, perpetuating the high-cost launch business.

It is an industrial fact that production volume (or flight rate, in the case of a resource used repeatedly) leads to efficiency and cost savings in an un-tampered-with market. While predictions of the late 1990's showed that launch rates would reach a sufficient number to provide production volume for space launch, the foundations of the predictions were shaky at best. Even with a fully functional Iridium (72 satellites) and Teledesic (numbers varied, but some early market research showed a requirement for 288 large satellites) constellations, once the original satellites were launched, the number of missions required to replace aging satellites would not be enough to sustain a viable launch industry. Communications satellites are not the only possibility, however.

Some advocates, including Buzz Aldrin, maintain that space tourism is the only market driver that will create enough demand for launches to make launch cost-effective. While this is possible, space tourism faces a lot of challenges as a startup business. For example, one factor playing a major role in how much (and how often) people are willing to pay to ride on a vehicle is its safety record. The safety record for commercial airliners is on the order of 99.99%, and some people will still not fly. (The safety record for automobile travel is significantly worse, and many of the people who won't fly don't mind riding in a car, but that's a discussion for another day.) Safety records are very difficult to prove without a large sample size, in this case the number of flights. In the case of a new launch vehicle, 14 launches must be successful in order to claim a 95% success rate, which translates directly into a safety factor if your cargo is paying customers. The investment required to build a new launch vehicle and demonstrate its safety through that number of launches (likely without any paying customers until an FAA certification takes place, which doesn't exist for orbital vehicles, by the way) is prohibitive.

So, following capitalistic principles, the launch rate must go up in order to decrease costs, but the most talked about market driver is unrealistic because of startup costs for a new system. The uncertainty is multiplied in the risky business of rocketry, where a 95-97% success rate is excellent. What does that leave us?

A Success Story

The airline industry was cited above as a safe way to travel. The industry was not always as safe as it is today, nor was it a viable industry for many years. Early airlines got their start from airmail routes created by the post office. The Postal Service guaranteed an amount of cargo (or, lacking the cargo amount, would pay for a flight anyway) and made regularly scheduled routes worthwhile for businesses to maintain. These routes, augmented by larger and more capable aircraft, became the first air travel routes in the nation. While the weaning process is by no means complete, (note the fragility of the airlines' financial situation in light of the change in world politics in September, 2001) the airlines are a much more viable industry than most space operations today. This comparison is not completely fair, due to the different timelines: air travel is approaching its 100th anniversary, while space travel is a bit over 40 years old. Plus there's been very little effort expended by agencies charged with space exploration to make it pay for itself.

Project Overview

The HaWaTa project is designed as a hybrid government / commercial project, although it has potential as a purely commercial venture. Either way, at project end there will be two solid results:

1. A multi-ton cache of rocket propellants and other supplies in low-Earth orbit (LEO).
2. A launch system capable of rapid flight rates and a much lower cost per kilogram to LEO compared to today's launchers.

For this discussion, a hybrid government / commercial project is assumed, meaning that government contracts build some portions of the system, while the government pays for launches on delivery, with minimal, or preferably no, interference with a capitalistic process. It could just as easily work with a corporation sponsoring the project, although until a marketable product or service (beyond what currently exists) is obvious in space, this possibility is slim. Throughout the rest of this paper, whatever entity runs the project will be referred to as the Agency.

The project begins with a decision as to a cost per kilogram that will allow a thriving space industry. Several studies have taken place to find this "magic" dollar-per-kilogram-into-orbit number, and the values vary from each. The purpose of this paper is not to come up with this number, so for discussion's sake, a value of $1,000/kg will be chosen.

Once this dollar figure is determined, another factor must be set, that being the minimum useful cargo weight. Here again, a lot of theories abound, and these theories are not the topic for this paper. For this analysis, a value of 7,500 kg will be chosen. This puts the required lift vehicle between a light and medium EELV, and should still allow some company to build a craft to carry passengers or some other useful commodity into LEO. Note: This throw mass is not enough to launch a modern-day communications satellite into geosynchronous transfer orbit, a common make-or-break throw mass for rockets today. Market research will have to determine if such a goal is required. A larger payload capability may be good, as a vehicle that could later carry more passengers could cut its cost per passenger, much like the 747 does by carrying so many people today.

With these values determined, the government builds and launches (or contracts through normal procurement channels) an electrolysis station and a number (on the order of 100) of cargo vehicles. The electrolysis station will serve as a target for the cargo vehicles, which will carry water up to it. Once they dock, the station will use solar power arrays to electrolyze the water into its constituent hydrogen and oxygen elements. Because it is much more difficult to store liquid hydrogen and liquid oxygen compared to water, the option exists to leave the cargo vehicles' payload intact, simply keeping it from freezing, until its constituents are required. Any hydrogen and oxygen produced will be refrigerated and stored in orbit, awaiting its use by another vehicle.

Any mission desiring to use the cache of fuel and supplies provided by HaWaTa will have to design a docking port in accordance with design specs. After launch into LEO and docking with the station, the user can take on a mix of liquid hydrogen, liquid oxygen, and/or water in whatever proportions they prefer. An interplanetary craft using

HaWaTa's supplies will be able to get a much larger payload to its destination, either through launch of an unfueled upper stage or the reuse of a common upper stage used today (examples include Centaur and the unnamed liquid hydrogen / liquid oxygen Delta IV upper stage).

Components

The components of the HaWaTa system are relatively simple, being designed for long life and multiple users.

Electrolysis Station

The electrolysis station is envisioned as what is now considered a normal government procurement. The Agency requests bids from contractors and monitors the construction. When construction is complete, the station is lifted into orbit using one of the well-known launchers of the time. While the space shuttle (and its requisite human involvement) may be required if the station contains several complicated deployments, the job should be feasible as a payload on board the heavy version of an evolved expendable launch vehicle. The station is designed to be automated, with the possibility of an occasional crew visit to inspect the system and/or fix any problems.

The station is made up of many systems common to other uncrewed satellites (attitude control, power, thermal control, etc.) but has some important differences:

1. Storage: The electrolysis station will have to store large quantities of liquid hydrogen and liquid oxygen. Liquid hydrogen is particularly difficult to store because of its low density (requiring large storage tanks) and tendency to leak (requiring meticulously constructed tanks, piping and valves) Cryogenic liquids are difficult to store because of their temperature extremes, and this storage will require some form of active refrigeration. Storage methods require study, using either multiple tanks (preventing a single meteor strike from puncturing the only oxygen tank, but making storage of large quantities difficult) or single tanks (requiring multiple layers of protective material to prevent punctures, but allowing much greater volume storage.)
2. Two Types of Docking Adapters: Because the station will have to receive cargo (the water) and pass on products (the hydrogen, oxygen and water) it will require two types of docking systems. Both require a mechanical locking mechanism to hold the docked spacecraft to the station. The adapters need to be located to provide a consistent flow (from the delivery tanks, into the processor / storage tanks, into the receiver craft).

Figure I shows a diagram of the station. Descriptions of components proceed from the "top" of the spacecraft (the side with the solar arrays) to the "bottom."

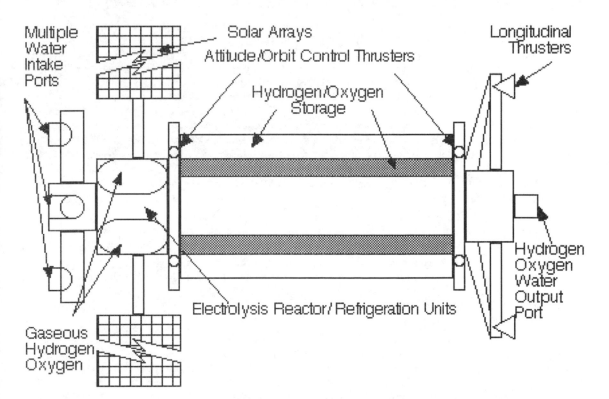

Figure 1. The electrolysis station.

Multiple Water Intake Docking Ports – These ports would receive the water cargo ships for long-term storage. They also include electrical connections that allow the station to control temperature on board the cargo craft. Multiple ports prevent a failure in one from causing a mission loss. Each port requires a valve which, when opened, allows water to flow into the electrolysis reactor.

Solar Arrays – The solar arrays need to be large to meet the huge power requirements for electrolysis, but for this mission have no particular requirements other than separate steering (each array being able to move independently) that is common on communications satellites today. This capability allows the arrays to be used to steer the station along some axes using solar pressure. The size and number of arrays depends on the production rate of hydrogen and oxygen desired, and the power requirements for cryogenic storage. Further analysis is necessary in this area. One possible source of solar arrays for the satellite would be the International Space Station – some backup arrays or components should exist, although their functionality, depending on their design for the station, may be limiting.

Electrolysis Reactor – Most water delivered on orbit feeds from the cargo craft into the reactor for separation, although some water is split into a separate channel to flow the length of the station. The electrolysis reactor's primary driver is reliability, as it will likely need to function for several years. Since the satellite itself will be rather light (it is launched almost devoid of fuel), it will likely pay to build a heavier electrolysis reactor if the added weight will add reliability.

Refrigeration Units – Redundant refrigeration units liquefy the byproducts of the electrolysis reaction (hydrogen and oxygen) and maintain the liquids in the cryogenic storage tanks. They should be reliable and able to stand long periods of time without running.

Attitude / Orbit Control Thrusters – The station will likely rely on gravity gradient stabilization (a natural condition experienced by any spacecraft which doesn't have its mass distributed uniformly throughout), but may require some form of attitude control thrusters. Orbit control thrusters will be necessary to maintain the station's orbit. Any thrusters should run on gaseous hydrogen and oxygen, two materials readily available on board the craft. Gaseous hydrogen and oxygen are stored in pressure bottles located around the electrolysis reactor.

Liquid Hydrogen / Liquid Oxygen / Water Output Docking Port – This is where a visiting spacecraft will dock to "fuel up." The pipe feeding each liquid to the docking port can be opened or closed, allowing the user to take any mixture of liquids on board. One design issue with this port is the fact that flowing water near the cryogenic liquids (hydrogen and oxygen) will freeze. A way around this is to avoid flowing all the fluids at the same time.

Longitudinal Thrusters – One great concern in orbital propellant transfer is how to get a liquid to flow into the new vessel when desired. While liquid oxygen is paramagnetic (meaning that a nearby magnet will induce a magnetism within the oxygen, and get it to flow),[2] hydrogen does not have such a useful property. In order to keep things moving, longitudinal thrusters invoke a small acceleration when needed. Pressure within the tanks will force the fluids out once the fluid is located at the drain, although some thrusting may be required to keep the flow uniform. When water is required in the electrolysis chamber, valves open between the cargo ships and the station, and thrusters fire – the water then flows into the station. Water flow within the station is likely to be handled with a bladder and external pressure system, while the dynamics of refrigeration and storage of the cryogens requires more work. Smitherman, *et al.*, describe gravity gradients (used to keep the spacecraft stable) as keeping the hydrogen and oxygen together in their tanks.[3] When a visiting craft docks with the station and requires supplies, the necessary valves are opened and the longitudinal thrusters fire. Originally, it was thought that the longitudinal thruster firing would suffice for orbit maintenance, but aligning the longitudinal axis with the orbital velocity vector would be difficult for a (hopefully) simplistic attitude control system. More study will show if the trade-offs between attitude control and orbit maintenance are worthwhile.

Cargo Vehicles

The cargo vehicles may rank as the simplest, most mass-produceable spacecraft in history. A schematic of one appears in Figure 2. Their design requires them to carry water and remain active for a couple days at most in order to dock with the electrolysis station. Depending on the amortization required to make the delivery-on-orbit contract feasible (see the next section), and whether the idea of on-orbit refueling catches on, the number of cargo vehicles ordered could range into the hundreds, making it a unique group of spacecraft. There are two components: the water tank and the orbit assist ring.

Water Tank – As its name implies, this portion of the cargo vehicle holds water. Design simplicity is important, although required units include a fore and aft docking port (so that it can attach to the electrolysis station or the tank ahead of it in line, and another can attach to it) and heating elements throughout to prevent the water from freezing. The heating elements must be able to be powered by the orbit assist ring or the electrolysis station. The tank may be made of composite materials, since it will not have to face temperature extremes brought on by exposure to cryogenic propellants, and it should be outfitted to last in orbit for some time without leaking its contents. This requirement will likely drive the need for some form of meteorite shield.

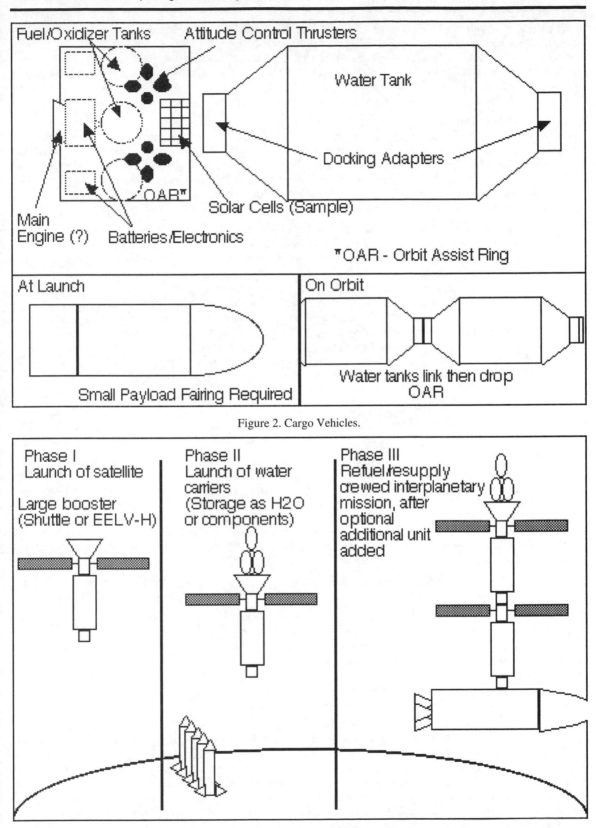

Figure 2. Cargo Vehicles.

Figure 3. An overview of the HaWaTa program.

Orbit Assist Ring (OAR) – The OAR's purpose is to deliver the tank to the electrolysis station. Once the cargo vehicle is delivered into orbit (within the parameters specified in the delivery contract) the OAR takes over, making final orbital adjustments and the precision maneuvers required to dock. This type of procedure has been automated in the past, such as with the Progress cargo vehicles' deliveries to the International Space Station, so the technology is not new. Once the tank is docked to the station, the OAR's job is complete. A small pyrotechnic charge will separate the OAR from its tank, and the OAR will maneuver itself to a safe burn-up in Earth's atmosphere. The internal components of the OAR include a guidance computer, telemetry equipment, batteries, and an orbit / attitude control system. The solar cells on the outside of the OAR are not fully researched. It is possible that given the craft's short life span in orbit, batteries alone may do the job, but solar cells would provide a longer life in case it was necessary.

Delivery-on-Orbit Contract
The primary difference between this program and others is the delivery-on-orbit contract. This option for space services, while it's been used in some cases such as the Navy's Ultra-High Frequency Follow-on (UFO) program and the Geostationary Operational Environmental Satellite (GOES), has never been used to pay for mass delivered to orbit. Of course, mass on-orbit has never been a stated goal of a space system before. After the mass and volume requirements for the cargo vehicles are established, the price desired, and the orbit where the electrolysis station will be is determined, an announcement in the Commerce Business Daily states the following (or something similar):

> "Z Agency will pay Y dollars to place X kilograms of water into orbit at W degrees inclination +/- w degrees. The orbital altitude must be V kilometers +/-v. Payment will take place when the orbit is verified by independent sources. The Agency will provide the cargo craft to carry the water, and will provide interface control information to anyone who contacts the Agency. Current launch services providers are eligible to make this effort, but upon a successful delivery, their launch costs to the Agency (and any of its parent organizations) for X kilograms will be Y from now on, in any contract. Any entity working to meet this contract should keep the Agency informed of its flight schedule. Bonuses will be paid to companies that demonstrate a high success rate (> U%), and high sortie (flight) rate (>1 flight every T days). Any launchers will adhere to the flight rules established at their chosen launch site. Entities seeking this delivery contract should consider other uses for their vehicle to make their launcher viable after this contract runs its course."

The contract is designed to be simple. Anyone who delivers the water into an orbit with the required parameters receives payment (only inclination and altitude are specified because other parameters, such as phasing, can be adjusted over time using the OAR. If solar cells aren't used in the OAR, then the phasing will be important to minimize flight time to the station).

Since this has the potential of being a government contract, and the government has been accused of stifling free competition by cutting development costs for some companies by feeding development work, two clauses are designed to either keep existing players out, or change their ways of doing business. By requiring any existing launch service provider to make the cost of launching water cargo vehicles their new standard launch cost, current launch providers will likely stay out of the business all together. Bonuses for high sortie rate will also likely keep the big players out, because the current demonstrated launch rate for a United States company off one launch pad is 28 days (Boeing, launching its Delta II vehicle during the Iridium launch campaign).

It must be noted that international launches are not prohibited. If an entity can establish a launch presence on some island, the cargo vehicles will be shipped to that island and payment will be honored upon successful delivery.

The method of lifting to orbit is left out on purpose as well, along with a required technical review of any proposed methods. The idea of the first concept is to allow any idea to be tried, so long as the entity in question can raise the cash to do so. At this writing, it's believed that the lift method would have to be reusable by a large percentage in order to be viable at the proposed launch costs per kilogram. The second concept may be a little outrageous, as current space activities require oversight beyond most other industries' imagination. It's been introduced, however, to try and cut down on the Agency stating that design A is viable and design B is not, in which case the entity producing design A is now at a tremendous advantage over entity B.

The contract will not specify a number of flights, other than stating it will be more than 50. A recent article in *Space News*[4] cites two reports describing 50 flights/year as the critical flight rate to make a reusable vehicle cost effective. The idea here is to prove that there's enough business for more than one entity to make the effort. Amortization of a launch vehicle from development through profitability has never been accomplished with any degree of accuracy (the space shuttle serves as one example, and Ariane's repeated requests to its parent agency for more money is another), but the more guaranteed flights there are for a launch vehicle, the easier such an amortization is. It is hoped that the launch capacity and flight rate created by this project would open up new space industries, space tourism being one of them.

Cache Effects on Space Flight
Immediate Applications
Once the Agency starts paying a contractor for regular deliveries of water to orbit, any other user desiring the new low-cost launch service would be free to negotiate their own flight. It's possible that the launching contractor will

negotiate a higher cost to low-use customers (such as a single flight on board their launcher), but this type of price work is best left to the newly created market, and by definition would be much more interesting and worthwhile if there are two launch service contractors available.

Assuming a launch cost of $1,000/kg and a payload mass of 7,500 kg, a small company or university could contract to launch a large satellite into low-Earth orbit for $7.5 million. Depending on the dynamics chosen for delivery to LEO, (disposable vs. reused second stage, etc.) that same university or small company could have the option to send a smaller payload (on the order of 2,000 kg) to the Moon or Mars. Unfortunately, since launch costs are largely the same no matter what the rocket is carrying, a small company or university who wanted to launch a 2,000 kg satellite into LEO would have to pay $7.5 million, unless they combined their payload with another and launched more satellites at once.

When compared to today's rates for launching small payloads, these numbers are extremely favorable. The Ariane 5 launch vehicle has the option of launching small payloads along with its primary, but the costs per kilogram are actually greater than the primary. A 120 kg microsatellite will run a customer $3 million to launch, costing a whopping $25,000/kg. Ariane also offers flights for 300 kg payloads at a cost of $6 million, translating to a cost per kilogram of $20,000. These numbers are negotiable through Arianespace. A niche market may always exist for small payloads such as these, but these launches will not open space to large-scale use by the industrial or public sectors.

Short-Term Future Applications

The Atlas IIAS rocket is a two-stage launcher in use today. Its payloads consist of military, civil and commercial spacecraft, most of which are destined for geosynchronous orbit. The rocket uses a first stage powered by kerosene and liquid oxygen (boosted by two or four solid rockets strapped to its side) and a second stage, called a Centaur, powered by liquid hydrogen and liquid oxygen. Atlas is listed as having a LEO launch capability of 8,610 kg.[5] The same source lists the Atlas' Earth escape throw mass as 2,680 kg. Though it is not specified, for this analysis this value is assumed to be the mass that the booster can push to a hyperbolic excess speed of zero launched from Cape Canaveral Air Force Station. To achieve this mass to Earth escape, the first stage burns to depletion, followed by the second stage burning to depletion. Having an orbiting fuel depot in orbit allows a different flight plan and much greater capability.

By definition, a vehicle that can push 8,610 kg to Earth escape can push that same amount of mass to geosynchronous transfer, with a little propellant left over to raise the orbit a little closer to today's goal of geosynchronous orbit. This fact will almost double the size of current communications satellites, or allow a current design to achieve geosynchronous orbit with much more fuel, translating to a much longer life.

In order to exploit this capability, the Centaur upper stage would have to be modified from its current configuration. For this discussion, it is assumed that the modifications increase the weight of a Centaur by 500 kg. Additional equipment required includes:

1. Docking Adapter – Designed to mate with the orbiting electrolysis station, this adapter will likely be mounted between the hydrogen and oxygen tank. The feeds to both tanks will need to be valved.
2. Additional Life – The current mission timeline of the Centaur is extremely short. Most missions last 6 hours or less. In order to take on an orbital refueling mission, the stage will need to be active for a much longer period of time, and may need to have rechargeable batteries or solar panels.
3. Precision maneuvering / guidance – The Centaur has never been required to maneuver in any close quarters with another spacecraft other than an avoidance maneuver from the satellite it just dropped off in orbit.

In a possible scenario during the HaWaTa project, an Atlas IIAS with a modified Centaur lifts a payload of 8,610 kg into LEO, and burns the Centaur to depletion, then docks with the electrolysis station. The longitudinal rockets on the station fire, pushing fuel from the station into the Centaur. When the Centaur takes on a full load of propellants, it will be able to push itself and the original 8,610 kilograms launched to Earth escape, with an additional delta-v available of 1.2 km/sec. This method multiplies the Atlas IIAS throw mass to Earth escape by 3.2 times, with propellant to spare. It should be noted that the *Cassini* spacecraft launch to Saturn in 1997 had a mass of 5,712kg.[6]

A Centaur-derived upper stage is used as the second stage of the Atlas V version of the EELV, and the logic spelled out above applies. Boeing also uses a hydrogen-oxygen booster for the second stage of their EELV, and could use the same approach.

Long-term Future Applications

As humankind moves beyond low-Earth orbit, fuel and/or oxidizer will be necessary to make such trips possible. No matter what form of propulsion is chosen (such as chemical, which uses hydrogen and oxygen, or nuclear, which would likely use hydrogen), a cache of propellants in low orbit will decrease the amount of mass a particular mission will need to lift off the surface and accelerate to orbital velocity. The question comes in as how much mass is saved in such a launch?

Zubrin argues that a mission to Mars can be accomplished using a Saturn V (the rocket used to take humans to the Moon) class booster.[7] Such a booster does not exist operationally today, but could lift 140 tonnes into low-Earth orbit. For a trip to Mars, of the 140 tonnes in LEO, fully 100 tonnes is fuel. Since Zubrin is a proponent of getting to Mars first and letting the newly created need for nuclear propulsion drive the development of the technology, chemical propulsion (hydrogen / oxygen) is assumed. By this argument, in order to reach Mars a new heavy launch vehicle, capable of launching 140 tonnes at once or 2 launches of 80 tonnes (the proposed lift weight for a booster called Magnum, proposed by NASA), is required for a crewed Mars mission.

A refueling station in low-Earth orbit changes the situation significantly. Assuming that a HaWaTa station could hold 100 tonnes of propellant (additional would be required for chill-down fueling losses), it would be possible to launch a crewed Mars mission with a launch mass from Earth surface of 40 tonnes. Two launches of an EELV-H would achieve this mass to LEO. This doesn't make such a launch easy, since the current diameter limitation for an EELV-H is 4.8 meters (estimated, based on a 5-meter outer diameter payload shroud).[8] Launching a hydrogen tank that could hold 86 tonnes of extremely low-density fuel would make a 4.8 meter tank very long or require a much larger diameter tank.

If such difficulties were overcome, the HaWaTa project could support routine missions to Mars. If and when nuclear propulsion becomes an option, HaWaTa can still provide a useful service. A nuclear rocket destined for Mars launched with empty tanks will weigh 40-50% less than a nuclear rocket launched with full tanks. Once again, the weight savings can be used to decrease the size of the booster required to start the mission.

Dollar Value

During or after the HaWaTa project, the Agency that operates the station will have a large supply (on the order of hundreds of tones) of material in low-Earth orbit. The material is usable by various other agencies, but how much will they be willing to pay for it? Or, more importantly, how much money could the Agency expect to make in profit from the sale of its commodity?

The simplest (and most flawed) way to look at the project is to assume an initial cost (we'll say $1 billion for the station, cargo vehicles, and delivery-on-orbit contract), an amount delivered to orbit (for discussion, 50 flights of 7,500 kg of water each), and a current cost to low-Earth orbit (dollar figures go as high as $10,000 a pound, or $22,000 a kilogram, but we'll cut that by 25% to be conservative, and give everyone a discount). With 375,000 kg of supplies on orbit, multiplied by the "going" rate of $16,500/kg, the Agency has $6,187,500,000 of commodity available for sale.

For the first launches, before the concept proves out that launching fuel and oxidizer in to orbit separately from the payload is a good idea, this sale price will be reasonable. Before the cargo launch vehicle proves itself to be reliable, mainstream missions such as those launched by NASA or the US Air Force will likely rely on existing launch vehicles. The spacecraft launched by these vehicles will then dock with the electrolysis station for a fill-up before traveling on to their final destination.

It is possible that this project will become a victim of its own success. Assuming that the cargo launch vehicle becomes a successful method of achieving low-Earth orbit, the cost for one of its launches will be much lower than the going rate for other launches. Odds are, there will still be a core of government customers who'll desire the "old" way of launching, where ultra-high maintenance satellites are babied in their cradle right up until the rocket is lit sending them on their way. Many users, including commercial interests that exist today and others that have not even been imagined yet, will use the cheaper service with reliable schedules. Depending on the payload support team for this commercial business (the cargo vehicle for HaWaTa is designed to be extremely low-maintenance, requiring very few crew personnel for preparation, so additional personnel will be required to support any other payload) the cost will go up slightly, but on a per-kilogram basis, the price for this new launch vehicle or method cannot be beat.

Expansion Possibilities

With a docking adapter designed to allow all products (hydrogen, oxygen and water) flow through, the electrolysis station can be expanded quite easily if the demand for on-orbit fueling becomes more than one unit can handle. A second, near duplicate station could connect to the first via the docking port normally used to fuel customer vehicles. If the original station is fully functional, but simply needs a greater storage area, the additional station will be able to hold the extra hydrogen and oxygen, and use its own solar arrays to cool the fluids. If the original station is having difficulty, or operating at a reduced efficiency (without the need to be completely replaced), the auxiliary station can accept water flowing through the first station and process the feedstock into hydrogen and oxygen on its own. This same approach could be used with two stations operating at or near peak efficiency – together they would produce hydrogen and oxygen at double the rate of the first station.

If the idea of on-orbit fueling catches on, it may be necessary to place electrolysis stations in orbit around Earth in different orbits. This would allow different launch sites to use the service without paying the penalties associated with a drastic change in orbital inclination.

Remaining Issues
The following list is not exhaustive, but discusses some of the issues this project could face as it moved from concept to reality.

Power
Electrolysis is an extremely power-intensive process. Research found one commercially[9] available electrolysis unit that produces 8.2 kg of hydrogen and oxygen per hour with a power feed of 100 kW. At this production rate, it would take 38 days to process one 7,500 kg tank of water into its hydrogen and oxygen components. Plus, to maintain a power supply of 100 kW in orbit, much more power must be available to charge batteries that will supply energy during eclipse time, when the Earth shades the orbiting vehicle from the sun. Another power draw will be refrigeration of the hydrogen and oxygen. While storing water on orbit is not difficult, and could still support a one per week flight rate, that storage will take power.

For comparison, one set of solar arrays for the International Space Station provides 60 kW, clearly not enough to provide an acceptable production rate. This vehicle will require multiple solar arrays of space station design, or larger arrays.

Volume
The design shown here for the electrolysis station shows multiple tanks holding hydrogen and oxygen. This design provides redundancy for a single-tank failure, but may hamstring the project because of the need for large tanks to store hydrogen. Large diameter tanks provide the most volume per unit length, but a single tank subjects the system to a single-point failure.

Reliance on One Type of Propulsion
In the current research dollar driven world of space programs, an idea such as this can generate as much negative interest based on its perceived threat to other programs as much, if not more than based on any technical flaws. For instance, one argument against this system is that it would provide an excuse to keep using liquid hydrogen and liquid oxygen to travel beyond low-Earth orbit, instead of focusing research dollars on the more efficient, though more controversial, nuclear-powered propulsion. This argument has some validity, but much of it is diffused through the ability of HaWaTa to support nuclear engines. Even if nuclear propulsion makes a debut in the next 10 years, it will not become the mainstream method of propulsion for many years after that, so there will be plenty of hydrogen / oxygen burning rockets available to use the HaWaTa propellants.

In the case where a nuclear spacecraft fuels purely with hydrogen, leaving a store of oxygen on board the station, someone in low-Earth orbit will find a use for it. Any developing space interest in orbit would not turn down such a supply, especially considering that the Agency would likely be willing to sell it at a discount. The alternative would be venting the precious fluid / gas into space.

It should also be noted that electrolyzing water provides oxygen and hydrogen at a mass ratio of approximately 8 to 1, that is 8 kilograms of oxygen for 1 kilogram of hydrogen. The best ratio of these propellants' rocket engines are operating on right now is 6 to 1, because of the troubles maintaining a stoichiometric (fully-balanced) reaction in a combustion chamber. Because of this imbalance, any station used to simply provide liquid hydrogen and liquid oxygen propellants for chemical engines will have a supply of leftover oxygen.

Orbital Location
The best initial orbital location for the first electrolysis station will likely be hotly debated, assuming it moves beyond the concept phase. A primary use for the fueling service is expected to be interplanetary missions, and the optimal orbit to leave Earth from varies from one launch opportunity to the next. A station in the 28-degree inclined orbit would serve NASA and US Air Force launches from Cape Kennedy and Cape Canaveral, but would be unreachable (well, reachable, but at such a fuel cost the advantage gained by refueling is likely to be greatly diminished) from Russian launch sites. Depending on market growth, any second station should likely be placed in a 57 degree inclined orbit, to allow service for users from Russia. Users closer to the equator than 28 degrees would be able to reach the station rather easily, but will have to trade their natural "boost" received by launching close to the equator as some of that advantage is lost by launching into a higher inclination.

Attitude Control
The electrolysis satellite is unlike any other previous spacecraft in its constantly changing center of mass. When first launched, the satellite will be largely devoid of fluids, and will have attitude characteristics based on its layout. When the first cargo craft is launched to it, the 7,500 kg of water carried on board will be a significant increase to the original craft's mass, and will shift the center of mass towards the cargo vehicles' docking ports. Further cargo deliveries will compound the problem. Once the electrolysis process begins, the center of mass of the vehicle will shift again, only the total mass will remain steady.

These changing conditions make selection of an attitude control system problematic. A common attitude control method, reaction or momentum wheels, work well when a spacecraft is nearly balanced, that is, distributed evenly around its center of mass (a materially uniform sphere meets such balance perfectly). This spacecraft, however, will likely not be evenly distributed about its center of mass. Even if such a design were possible at the start, the above mentioned shifts in center of mass would force changes in attitude control. Thrusters are another option, and though the station will have plenty of fuel on board, relying on thrusters for full-time attitude control is not an elegant solution.

One of the simplest ways to control the attitude of a spacecraft is to not do anything. When left alone, the natural "lop-sidedness" of the satellite will cause it to orient its long axis pointing towards Earth (the physics are a little more involved than that, but the description will suffice for now) as evidenced by images of the Long Duration Exposure Facility. The LDEF was deployed to test materials for their response to long periods in space. When STS-32 approached the facility in 1990, it was very stable, and allowed easy grappling and retrieval.[10] The same principal can be applied to the electrolysis satellite. When the satellite is first launched, it will orient in one direction, depending on its mass properties. If such a direction requires some active control, the solar arrays can be used to rotate the spacecraft along its primary axis. As cargo vehicles dock with the station and the center of mass changes, the craft will rotate 180 degrees in a slow yaw (or roll, or pitch as the case may be). This motion, while not common in spacecraft, is manageable, and is worthwhile considering the added complexity that other attitude control strategies would bring. If such a rotation is undesirable, water could be processed into hydrogen and oxygen, shifting the center of mass and managing any undesired changes. More analysis is required in this area, as the size of the cryogenic storage tanks, and possible shifting of hydrogen and oxygen within them, will make it difficult to predict exactly where the center of mass will move to.

Valving / Leakage
As mentioned before, the handling, transfer and storage of cryogenic propellants are not simple on the ground. Doing so in orbit is only going to be more difficult. As this paper is written, the space shuttle is grounded due to flaws in cryogenic propellant lines, with a launch date listed as "indefinite." Hydrogen is a particularly difficult commodity. The frigid temperatures (only liquid helium stores at a lower temperature) and tiny molecular size provide challenges to ground operations involving the liquid. Some technology development is required in the automated transfer, valving, and leakage detection / control of hydrogen before this project is feasible.

Similar Research
Smitherman, *et al.* described a system similar to this in a paper presented at the Space Resources Utilization Roundtable III at the Colorado School of Mines. Their research showed an increasing need for hydrogen and oxygen in LEO, both to resupply craft in LEO and to fuel missions beyond. In their paper, they discuss how this type of electrolysis system must wait for some exotic future transportation to LEO, not how this system could bring about such a new form of transportation.

Conclusion
The implementation of an orbiting electrolysis facility was discussed. While challenges remain in the production and control of on-board cryogens, the payoff in both common access to low-Earth orbit and leverages for exploration beyond are immense. Using the low-tech mass of water as a guaranteed payload also has the capability of jump-starting a low-cost transportation option to low-Earth orbit, and could be run by any government or large corporate entity. Immediate payoffs include increasing the interplanetary throw mass of a currently medium-sized Atlas IIAS booster to greater than that of the accepted heavy-weight Titan IV. Immediate payoff in communications satellite size and lifetime are worthwhile, and future applications include allowing a Mars Direct style mission to Mars using this plan and existing EELV launch technology. A business model showing the value of hydrogen and oxygen stored in orbit is unclear, as the lowered cost of lifting the material may decrease the material's value, but the end payoff in decreased cost to orbit is likely worthwhile.

References
1. Jenkins, Dennis R., *The History of Developing the National Space Transportation System: The Beginning through STS-50*. Page 115. Walsworth Publishing Co, 1992 ISBN 0-9633974-1-9
2. Zubrin, R and Wagner, R., *The Case for Mars*, 1996, The Free Press, NY, page 109
3. Smitherman, D and Fikes J (Marshall Space Flight Center), Roy, S (Futron Corporation), Henley, M and Potter, S (The Boeing Company), *Space Resource Requirements for Future In-Space Propellant Production Depots*. Space Resources Utilization Roundtable III, October 24-26 2001, Colorado School of Mines, Golden Colorado
4. Elias, Antonio, *No Time for RLV's?*. Space News, August 26, 2002 Page 13
5. Jane's Spaceflight Directory 1995-96
6. www.jpl.nasa.gov/missions/current/cassini.html, accessed 7 July 02
7. Zubrin, page 89
8. The Boeing Corporation, Delta-IV Payload Planning Guide, Page 1-8.
9. www.hsssi.com/applications/echem/oxygen/ogp.html, accessed 13 October 2002
10. http://images.jsc.nasa.gov/images/pao/STS32/10063428.jpg, accessed September 3, 2002

Mars X – Mars in 10 Years†

J.E. Brandenburg, Research Support Instruments, Princeton, NJ,
John Kline, Research Support Instruments, Princeton, NJ,
Ron Cohen, The Aerospace Corporation, El Segundo CA,
and Kevin Diamante, Florida Space Institute, Kennedy Space Center

1. Introduction: Mars Direct and its Offspring.

> "Better is the enemy of good enough." — old Russian proverb

Mars Direct by Robert Zubrin[1] was a watershed event in the exploration of space and most probably, in human history. It demonstrated that the very things that make Mars desirable – its atmosphere and near terrestrial surface conditions – allow a Human Mars Mission (HMM) to be mounted to at much lower cost and Mass In Low-Earth orbit (MIL) than previously conceived. Prior concepts for a Mars mission were based on the experience of the Apollo missions and assumed that Mars was a Moon-like planet, bare of resources. However, the Mars Direct concept of using the richness of Mars resources themselves to assist in the mission has given rise to many other ideas, all focused on the goal of reducing the cost of a HMM further. It is the purpose of this article to briefly summarize a new proposed architecture for a HMM, that builds on the foundation of Mars Direct with several new concepts to hasten the day of the first human footsteps on Mars.

Cost, not technology, has been the major barrier to a HMM since the days of the Apollo missions, when human space flight between major space bodies was demonstrated repeatedly. Therefore, any mission architecture that reduces cost without sacrificing crew safety can hasten the day of a human landing on Mars. In this study, the key concepts of Mars Direct, the technological legacy of the Apollo effort, and the advance of key electric propulsion technologies are used to attempt to lower mission costs to $20 billion and the time of development of an HMM effort to ten years.

Controllable factors which can impact Mars mission cost can be traced to four principal areas:

1. MIL – the Mass In Low Earth Orbit (LEO), consisting mostly of fuel, that must be placed in LEO in order to ultimately place human team on Mars and return them safely. This mass can cost 20 million per metric ton (MT).
2. Research and development costs for new technologies, engines and vehicles – the tendency of HMM planners to reinvent capabilities and vehicles.
3. Controversy – the launching of such a mission without strong public support being fostered beforehand or incorporating mission elements that provoke strong opposition.
4. Boondoggle – the tendency for large projects to become larger and more costly as they are seen as a source of congressional largess rather than as a goal-oriented program.

How does one build on Mars Direct to lower cost further and minimize the above listed factors?

1. Use Mars resources optimally with Mars nuclear site power, as proposed in Mars Direct.
2. Use the Shuttle and ISS (international Space Station) as LEO infrastructure.
3. Use an MOR (Mars Orbit Rendezvous) mission architecture to minimize mass required on the Mars surface.
4. Keep crew size at three.
5. Avoid nuclear propulsion for the initial Mars mission unless broad public support is evident for it. The provoking of controversy always ends up raising costs. Often this occurs by forcing the mission to become a boondoggle to shore up congressional support in the face of determined opposition.
6. Use aerocapture and aerobrake on planetfall to lower propellant needs.
7. Make the program a ten year program like Apollo. Crash programs actually save money by forcing people to make decisions.
8. Make the HMM joint with the Russians and other ISS partners to take advantage of cheap Russian LEO access and long duration flight experience.
9. Minimize R&D by using legacy technologies from Apollo wherever possible. This seems counterintuitive at first, but rather than being anti-new technology, this approach seeks to channel R&D into focused areas such as propulsion. This will minimize costs; the goal is to get to Mars at low cost rather than funding R&D for R&D's sake.
10. Use new technologies – electric propulsion with large solar arrays rather than nuclear to lower mass in LEO. Nuclear power would require substantial R&D, and bring with it the controversy that accompanies nuclear propulsion in LEO. The key technology for this mission is the MET (Microwave Electro-Thermal) thruster using water propellant with 800 seconds I_{SP}. Such a system, using 500 kW of solar electric power generation, will be able to use 80 tonnes of water to boost 30 tonnes of payload to Mars on a Hohmann-like transfer orbit.

2. Mars X: Mars in Ten Years

Mars X basic architecture is to use water-fueled MET (Microwave Electro-thermal) thrusters for interplanetary transfer to lower LEO by high I_{SP} and use of water fuel to lower fuel costs. See Table I. We use a Hohmann-like transfer orbit. In this the Solar Aqueous MET (SAMET) propulsion functions much like a chemical burn. We assemble vehicles in LEO at the ISS, boost water propellant into orbit, and accumulate it using the Progress resupply vehicle. We use aerocapture and aerobraking at Mars and Earth. We build on Mars Direct by making LOX-kerosene on Mars using a nuclear powered chemical plant. This would involve man-rating a "pony" version of the Atlas 5 for the Mars ascent vehicle – an earlier Atlas lifted John Glenn into orbit, and a similar rocket will be able to lift three persons into Mars orbit from its surface. We will use an Apollo-derived command service module for a three-person Crew Transfer Vehicle (CTV).

Vehicle	Payload	Fuel Mass	Type of Propulsion	Role
SL-4 Progress	4 MT	N/A	LO_2-Kerosene	Boosts cargo and fuel to ISS
Space shuttle	30 MT	N/A	LO_2-H_2/solid fuel	Boosts crew and cargo to ISS
MOTV (Mars orbital TransferVehicle) – Pioneer	30 MT 4 MT H_2	80 MT H_2O	SAMET (Solar Aqueous MET)	Boosts MLV Pioneer (uncrewed precursor) to Mars
MLV (Mars Landing Vehicle) – Pioneer	28 MT	Makes 100 MT LO_2-RPI	Aerobrake	Lands and makes fuel sets up habitat and MAV
MOTV (Mars Orbital Transfer Vehicle) – Command			Aerobrake	Takes crew to Mars Orbit and back again
CTV (Apollo Command-Service Module-derived vehicle)			Nitrogen tetroxide Unsymmetrical Dimethyl-hydrazine	Apollo Command-Service Module derived vehicle. Takes crew from ISS to join MOTV in high orbit
MDV (Mars Decent Vehicle)	4 MT		aerobrake	Apollo Command-Service Module derived vehicle. Takes crew to Mars surface
MAV (Mars Ascent Vehicle)	10 MT		LOX-RPI	Atlas 5 (pony version) with Apollo Command Module as payload) Brings crew and return fuel water back to Mars orbit

Table 1. Vehicles and Appropriate Masses.

LEO Operations: Assembly of Vehicles and Mission Elements in LEO at ISS

The HMM will consist of two parts using the same MIPS (Mars Interplanetary Propulsion Stage) (Figure I) with SAMET propulsion as a booster from LEO and later from LMO (Low-Mars Orbit). The first portion will be the MPV (Mars Pioneer Vehicle) and the second will be the MCV (Mars Command Vehicle) carrying crew and supplies. The MPL (Mars Pioneer Lander) will contain the Mars in situ fuel plant (MIFP), the MCH (Mars Crew Habitat), the Mars Ascent Vehicle (MAV) and a diesel- (and oxygen-) powered robot tractor for deployment of the nuclear reactor on site.

Some have complained that the ISS orbit is too highly inclined to allow efficient mass accumulation from Kennedy. This is an important objection, since even with SAMET electric propulsion most of the mass for a Mars mission is dedicated to propulsion. However, the ISS orbit is perfect for an international mission anchored on a US partnership with the Russians. Such a mission can be supplied from Baikonur where most of mass and bulk – water propellant in this case – will be lifted using heavy and inexpensive Russian boosters and delivered using modified versions of the Progress resupply vehicle. The shuttle, due to its high reliability, will be reserved for both high value cargo and crew deliveries.

The mission preparation will begin with the delivery into ISS orbit of a large, empty propellant tank of approximately 100 cubic meters capacity (2 meters in diameter, 8 meters long). This will be parked at the station and modified in place as a mission fuel tank. A special docking port on the station will allow dedicated Progress resupply vehicles to bring up water and off load it into pipes connected to the mission tank. Off-loads of water from shuttle missions will also be used. Assuming 8 MT of water brought up per month, two years will allow the accumulation of the approximately 80 MT of water required.

The MOTV (Mars Orbit Transfer Vehicle), on the SAMET booster, is boosted into orbit in pieces and assembled by astronauts at the station. It will consist of six large space station solar panels mounted around a central hub that contains power conditioning units and approximately 20 MET thruster modules (Figure I). This will occur concurrently with the accumulation of water fuel. The MET thruster uses a vortex stabilized electrodeless microwave discharge to heat gases – in this case water vapor – to very high temperatures at high thermal efficiency.

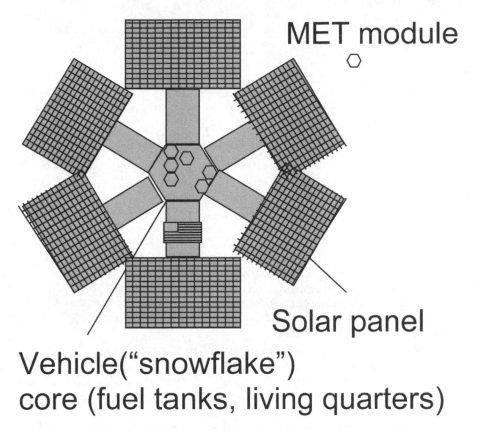

MET module

Solar panel

Vehicle("snowflake")
core (fuel tanks, living quarters)

Figure 1. SAMET interplanetary propulsion stage. Solar panels can be derived
from those used on the ISS and are much longer than shown here.

The Mars fuel plant and Mars ascent vehicle (Atlas + Apollo capsule) are sent up. The Apollo command module will be used because it is already a proven, man-rated design. Computers and other gear will obviously be updated. A pony version (shortened fuel tanks) of the Atlas V will be used as the MAV (Mars Ascent Vehicle) since it can be accommodated in the Shuttle bay. The vehicle will have approximately the same performance as the Atlas that lifted John Glenn into orbit, except that in the thin atmosphere and weaker gravity of Mars, it will lift a three-person Apollo-derived capsule from Mars surface into low-Mars orbit (LMO). The vehicle will be assembled and tested in LEO and then readied for its departure to Mars.

It is assumed that a precursor mission has already landed on Mars and has demonstrated in situ Mars fuel production and the launch of the MAV (Mars Ascent Vehicle) mission under Mars conditions, as well as precision Mars landing. Such a full-up test would be considered essential for safety and minimizes risks for the HMM by prepositioning supplies. The MAV carries ten MT of water into Mars orbit as a test payload that serves as an emergency fuel return package for the HMM. The full success of such mission would be considered essential to this HMM. This precursor mission also establishes a Mars habitat supply dump for the HMM and is landed as close to the desired HMM landing site as possible.

Mars Orbit Injection – Pioneer Mission
The Mars Pioneer Vehicle MPV will undock from the ISS and begin a three-month-long burn with 20 MET thrusters consuming 500 kW of power. Burns will be intermittent until sufficient orbital altitude is achieved to minimize the Earth's shadow. The first two months will be spiraling out to escape velocity. Because of gravity losses, 6 km/s will be required to achieve escape velocity. The Mars transfer orbit injection burn will begin when escape velocity has been achieved. This burn will occur continuously for one month to achieve the required three km/sec for a Hohmann-like transfer orbit to Mars. Ten months of free flight will follow until the MPV and its propulsion section reach Mars. It will aerobrake at Mars to assume LMO (Low-Mars Orbit).

In a realistic scenario, this craft will be the second or even third such MPV vehicle launched to Mars, with the initial ones serving to perform a full-up test of all systems on Mars surface – especially the crucial launch of the MAV into Mars orbit, a non-trivial task. Such a precursor mission would not only increase crew safety but could also establish the beginning of the Mars base for the human landing, and also emplace vital return fuel, water, in orbit around Mars to serve as emergency return fuel for any late abort of the human landing.

LMO (Low-Mars Orbit) Operations

Once in LMO, the MPL will detach and descend to the surface of Mars. At the designated landing site the Mars tractor will prepare a bed in a crater and deploy the nuclear reactor to it as in Mars Direct. It will also deploy the Mars crew habitat some distance from it in a similar prepared bed. The main landing body will serve as the launch pad of the MAV.

Mars In Situ Fuel Production

The Mars in situ fuel plant will use electric power from the nuclear plant to make liquid oxygen and kerosene for the MAV. It will use five tonnes of hydrogen to make approximately 100 tonnes of LOX-RP-1 and water propellants. It will do this by making use of stabilized plasma discharges to make a mixture of hydrogen and Martian atmosphere into "syn-gas" (synthesis gas), a mixture of CO_2, H_2 and H_2O that is the starting point for the Fischer-Tropp process to create kerosene. The Fischer-Tropp process has been employed by governments with abundant coal and water who are cut off from petroleum. It was employed successfully in large scale plants by the Germans during WWII to make diesel fuel when they were cut off from petroleum.

Kerosene in refined state is RP-1, the fuel for many existing rockets, and diesel oil for the tractor is easily produced by the same process. This will also be used to fuel the Mars tractor, together with oxygen, for constructing the Mars base and other tasks. The advantage of making RP-1 on Mars is that this allows production of propellant of comparable I_{SP} to methane (I_{SP} of LOX-RP-1 = 360 seconds versus I_{SP} for LOX-Methane = 370 seconds), but at twice the mass efficiency of hydrogen (1 MT H_2 yields 40 MT of propellant for LOX-RP-1 versus 20 MT for LOX-Methane). The use of RP-1 also allows the ready adaptation of an Atlas 5 or similar rocket designs for use on Mars and means that no new engine needs to be designed for a MAV, thus saving much R&D cost. The manufacture of RP-1 on Mars will also hasten the expanded settlement of Mars, since it is easier to store than methane in the thermal environment near human habitats and will allow the use of a broad range of legacy rocket and diesel technology. What we need is Detroit Diesel power on Mars.

Kerosene is a "paraffin," or straight line hydrocarbon, of approximate formula $C_{12}H_{14}$, that can be made by the Fischer-Tropp process. It is created with other hydrocarbons of greater or lesser molecular weight, so the product vapor stream from the Fisher-Tropp catalyst beds will have to be distilled to yield kerosene of the required purity. The light and heavy hydrocarbons can either be stored for other uses or simply burned and resupplied to the syn-gas generator to be sent through the catalytic bed again. The manufacture of kerosene from Martian atmosphere by the Fischer-Tropp process has already been demonstrated under an SBIR program by TDA research in Colorado in 1998 (Peter Geberstein, private communication).

Landing Site Selection

The Mars landing site should be strategically located for a Mars base of human operations, and will be the eventual capitol of the Mars Republic. Primary in its consideration should be that it is near water, and secondarily it must be centrally located on Mars near many points of interest in the planet's exploration.

The shoreline of what appears to have been a Paleo-ocean on Mars, named the Malacandrian Ocean,[2] runs around the northern hemisphere of Mars at the zero kilometer elevation line, and may contain a fossil water table. The coincidence of this paleo-shoreline with other areas of scientific interest may make an ideal region for a landing site.

The end of the Mariner Canyon is considered an area rich in scientific targets, and the landing site should be near it. This may rule out other sites better suited to robotic exploration. Gusev crater, for example, is apparently water rich but is far from the Mariner Canyon terminus. A site close to the Viking I and Mars Pathfinder site would probably be better, and near the Canyon terminus at Chryse Planum. Such a site would be near zero longitude, close to the equator to provide good sunlight for base auxiliary power, and near the zero kilometer elevation (possible paleo-ocean shore) line. Exploration of even the mouth of the Mariner Valley will be the great prize of any scientific expedition because its exposed strata and debris talus from past floods will reveal much of Mars geological history.

Launch of the Mars Exploration Team

The Mars CV vehicle undocks from the ISS where it has been assembled and tested and is launched to Mars. It spirals out as it gains speed. The Mars crew joins the vehicle after passage through Van Allen belt using the Apollo Command Service module for crew transfer. The commit to Mars transfer orbit injection is given, and the Mars transfer orbit injection burn is initiated. We, our own human flesh and blood, are off to Mars.

Once the Mars transfer burn is completed, we will stow the solar arrays and spin the vehicle for gravity. Spin induced gravity in small vehicles, requiring spin rates of perhaps once a second, have been criticized due to the Coriolis forces that this will induce on the crew in addition to simulated gravity. However, as has been pointed by Dr. Claes-Gustaf Nordquist, M.D. (Private Communication), the human body and nervous system are very adaptable and the crew will quickly develop "sea legs" to adjust to this rotating environment. Such a prediction could be easily tested in LEO with a special space station module, or even part of the Mars mission crew quarters. Following Doctor Nordquist's prescription, crew quarters should be toroidal and stretched around the ship to minimize spin for 0.4 g, and sectioned into small rooms to minimize rapid movement. These provisions will minimize Coriolis effects, aiding adaptation. We will also endeavor to keep crew busy during ten months by turning part of their water tank into living quarters.

We use aerocapture / aerobrake at Mars and assume a desired orbit. Mars decent for the three person Apollo capsule is accomplished by heat shield, parachute and landing rockets. The robot tractor comes out to pick up the crew and their capsule for transport to the habitat. The capsule will be added to the habitat to enlarge it. The tractor will be used to expand and construct new living quarters by entrenching and covering constructed tubular sections.

Mars surface activities will consist of enlarging the habitat with partially buried structures, and exploring for and exploiting water resources. Both of these activities will be the beginning of the Mars colony. Harvested water can be used directly on later missions for propellant and hydrogen production. It is believed at this time that Mars water use on initial missions is probably not advisable unless those resources are very well characterized. Finally, the crew must prepare for departure and board the MAV rocket by a rope ladder.

LMO Final Operations

The MAV will not only lift the crew from Mars but also bring up 10 MT of water for a departure burn to get out of Mars' gravity well. The trade-off involved is the desirability of lowering the amount of time spent spiraling out from Mars gravity as well as the desirability of minimizing the amount of mass lifted from the Mars surface with the crew. In one limit, the crew and large mass of water go up to LMO and rendezvous with the MITV for the trip home. The water is used to reach Mars escape and Earth orbit transfer injection. In the other limit, the MITV stays in high-Mars orbit and the MAV lifts the crew and a small amount of water to fuel the return home. This requires a larger launch from Mars surface, however, which is more difficult and certainly more dangerous. For the initial mission, then, we will go for MOR in LMO to minimize MAV launch mass. It is also possible that the initial return trip water will be lifted up into Mars orbit by a previous Mars precursor mission to demonstrate Mars fuel production and launch from its surface.

Once MOR is achieved the MITV spirals out of LMO; this will take approximately two months. The Earth transfer orbit burn is then initiated. This mission plan is time consuming, but it is simple, and for an initial mission simplicity is desirable. The ten month cruise to Earth is then done. There will then be an aerocapture at Earth and a stay on the ISS for debrief and quarantine.

Figure 3. The 50kW, 915 MHz MET thruster being run using air propellant.
Right-hand image shows its size relative to human being.

3. Summary

Mars X builds on the genius of Mars Direct and applies many other tricks, including using R&D strategies, to reduce costs. It must be remembered that R&D is expensive, and so it must not be used to reinvent capabilities that have already been demonstrated.

Accordingly, Mars X relies primarily on vehicles that either exist or once did. Even the MAV is simply a modified version of the Atlas V. The only truly new vehicle is the SAMET-propelled MTOIV, which consists of a ring-shaped, space station-derived crew habitat, a propulsion unit consisting of modular solar panels derived from the space station, and a cluster of twenty MET modules which have low technical risk and use present day technologies. Thus technical risk and R&D are concentrated on the crew habitat on Mars, which should be made as similar to the crew habitat of the ship as possible, and in the MAV and its in situ fuel plant, which by themselves are straight forward applications of presently understood technologies. The small nuclear reactor, necessary for the whole enterprise, is similar to demonstrated small nuclear power plants used by the US Army in the Cold War for portable Arctic bases. Given the low technical risk of this enterprise, it appears reasonable that it could be done in ten years for $20 billion.

Mars X is flexible; the rapid development and public acceptance of space nuclear power can be incorporated in the architecture to speed Mars X. It is suggested that a nuclear-electric propulsion flyby mission to Pluto, since it would be robotic and one way, would help build public confidence in such propulsion technology.

The use of electric propulsion produces dramatic reduction of mass in LEO. The use of water fuel reduces launch costs by allowing many small cheap launches to bring up fuel and accumulating it. The use of kerosene at Mars for MAV fuel allows utilization of legacy systems for Mars ascent vehicles. Cost can be reduced in a ten year program to under $20 billion and allow a rapidly mounted mission.

References

1. Robert Zubrin and Richard Wagner, *The Case for Mars: The Plan to Settle the Red Planet and Why We Must*, New York, Simon and Schuster, 1997.
2. J.E. Brandenburg, John Kline, Ron Cohen, and Kevin Diamante *Solarius: A Low Cost Mars Mission Architecture Using Solar Electric Propulsion*, Abstract, 2001 Meeting of the Mars Society.
3. Kevin Diamante, John E. Brandenburg, Ronald Cohen, and John F. Kline, (2002) *The Operation of a MET (Microwave Electro-Thermal) Thruster on Water Vapor*, AIAA / IAS Joint Propulsion Conference.

//

The MET (Microwave Electro-Thermal) Thruster Using Water Vapor Propellant†

John E. Brandenburg, Florida Space Institute and University of Central Florida, jbranden@mail.ucf.edu
and John Kline, Research Support Instruments, kline@researchsupport.com

Abstract

The research program to develop the MET (Microwave Electro-Thermal) thruster at Research Support Instruments, Inc. (RSI) using a variety of propellants is described. The MET has undergone dramatic evolution since its first inception, and it is now moving towards flight development. The MET uses an electrodeless, vortex-stabilized microwave discharge to superheat gas for propulsion. In its simplest design, the MET uses a directly driven resonant cavity empty of anything except gaseous propellant and the microwave fields that heat it. It is a robust, simple, inexpensive thruster with high efficiency, and has been scaled successfully to operate at 100 W, 1 kW, and 50 kW using 7.5, 2.45 and 0.915 GHz microwaves respectively. The 50 kW, 0.915 GHz test was perhaps the highest power demonstration of any steady state Electric thruster. The MET can use a variety of gases for fuel but the use of water vapor has been shown to give superior performance, with a measured specific impulse (I_{SP}) of greater than 800 seconds. When this added to the safety, ease of storage and transfer, and wide availability of water in space, the potential exists for using a water-fueled MET as the core propulsion system for refuelable space platforms.

Introduction: MET Thruster Research

The MET thruster was first conceived in the 1960's when it became apparent that microwaves could be used to create electrodeless discharges in cylindrical resonant cavities to heat gas for space propulsion. The electrodeless discharge, it was thought, would allow high efficiency heating of gases with a large reduction in erosion of thruster surfaces (Jahn 1968). However, early versions suffered from inefficiencies and problems of plasma stabilization – the tendency for the plasma to move to the antenna – that made simpler electric thrusters appear far more promising. Increases in microwave source efficiency led to a resurgence of interest in the 1980's, first at University of Michigan (Whitehair 1987), and then by Dr. Michael Micci at the Pennsylvania State University, supported by NASA and the Air Force Office of Scientific Research (e.g., Sullivan 1991). NASA efforts culminated in an 11 kW thruster being run using 0.915 GHz (915 MHz) at NASA Glenn Research Center (Power 1993).

Evolution of the MET

Early models of the MET were hindered by the requirement of a quartz tube to stabilize the plasma in the resonant cavity (Whitehair 1987). The heat loss to the tube walls limited efficiency and high power operation. To overcome this problem, Dr. Micci found that the use of bluff bodies to create a turbulent eddy in the fuel flow and axisymmetric drive of the cavity could stabilize the discharge (Balaam 1995). This was successful due to the importance of the Paschen parameter E/P for gaseous discharges, so that the plasma would tend to be stabilized by the low pressure region of the turbulent eddy, and the coaxial antenna would tend to create a field maximum at the preferred location of the plasma. It was found the that the TM_{011} was the superior cavity mode to heat the gaseous fuel near the nozzle. Inspired by the work at NASA Glenn involving use of a vortex to achieve similar effects in a large quartz tube, and with the encouragement of workers at RSI, Dr. Micci and graduate student Daniel Sullivan were able to create plasma discharges stabilized by vortex gas injection and coaxial drive alone, so that no quartz tube or bluff body was necessary (Sullivan 1994). The MET had thus evolved into a much simpler and more efficient device, with a thrust chamber containing nothing but gas dynamic flow and microwave fields. However, the device still remained a bulky and complicated device due to problems of microwave delivery to the thrust chamber, which utilized a waveguide, three stub tuner, and ferrite isolator between the cavity and the magnetron microwave source.

MET in Compact Form

In the new device, it was found by workers at both Penn State and RSI that the cavity electric properties changed dramatically upon plasma ignition and assumption of steady state operation. Before plasma ignition, the cavity behaved as a very high Q resonant cavity with narrow resonance and high reflected power; however, after ignition, the cavity Q collapsed to near unity due to the power absorption in the plasma, with almost no reflected power and a broad resonance. This can be easily seen from the equation for frequency spread in the cavity:

$$\Delta f = f / Q \qquad\qquad (1)$$

And with power absorption per wave cycle,

$$dW = W/Q \qquad\qquad (2)$$

Obviously, for $Q \approx 1$ the resonance conditions of the cavity are almost non-existent, and the energy per cycle absorbed is almost total; both conditions lead to almost no reflected power from the cavity.

Therefore, for a cavity loaded with a very strongly absorbing plasma, the three stub tuner was not needed, and neither was the ferrite isolator. Thus, the physics of a low Q cavity allowed the design of the MET to be greatly simplified by inserting the output antenna of a magnetron directly into the base of the resonant cavity. The resonant cavity then functioned like a holrahm or optical furnace with two focal points: the microwave source sat at one focal point, and the target plasma sat at the other. Once the plasma ignited in the cavity, the absorption of power from the magnetron was so complete that the magnetron needed no reflection protection. This resulted finally in 1994 in a very light, simple design for a MET thruster that could be mounted on a vacuum thrust stand or a spacecraft with only a fuel line and high voltage connections being necessary. The design of the MET has evolved quickly and is now a compact unit suitable for space flight development (see Figure 1).

The MET thruster was found to scale successfully to both small power and higher power by designing the devices so that the parameter Γ = Power / (wavelength)^3 was held constant. This meant that a higher power device needed to go up in wavelength, and smaller powers went to shorter wavelengths. Successful MET operation at 50 kW at 915 MHz and 150 W at 7.5 GHz was achieved easily. The operation of the 50 kW and 150 kW was very similar to that of the 1 kW version. Operation and approximate cavity dimensions of all three models are shown in Figure 2.

Operation of the 2.45 GHz MET in the 1 Kw Range

The MET was developed using 2.45 GHz microwave sources because of their availability and low cost. It was possible to modify microwave oven magnetrons for use in MET thrusters for experiments that placed the magnetron at risk, such as the first direct drive experiments. Because commercial magnetrons use air as a high voltage insulator, in order to run the MET on a vacuum thrust stand, the magnetron at the MET base were enclosed in an airtight sealed container with ports for cooling water and high voltage cables. A picture of the MET in this configuration is shown in Figure 3. The magnetrons were modified to accept water cooling by removing air cooling fins and wrapping a cooling water pipe around the core.

MET research was historically limited in budget, so the cost of materials was a constant concern. For the RSI design, aluminum was used in the MET structure because of low cost and oxidation resistance; the diaphragm separating the discharge chamber from the aft magnetron output antenna was made of boron nitride because of its high temperature limit and low loss tangent; and the nozzle was made of graphite because of its high temperature limit and low cost.

Once the MET was constructed in this way, the thruster could be operated on a vacuum thrust stand using nitrogen and helium as propellants. In general, the MET performed with impressive I_{SP}, efficiency and reliability, particularly considering that no optimization on nozzle design had been done. Operation in vacuum was particularly simple: A photodiode-triggered solenoid valve controlled propellant flow. With the valve closed, the magnetron was energized, and residual low pressure gas in the thrust chamber formed a diffuse glow discharge; when this occurred, the photodiode triggered the solenoid and, as the flow initiated, the plasma collapsed rapidly from the wall into an ellipsoidal discharge, with the plasma forming a virtual plug in the nozzle (explanation in Sullivan 1994) and causing a chamber pressure rise from ~ 1 Torr to greater than 500 Torr.

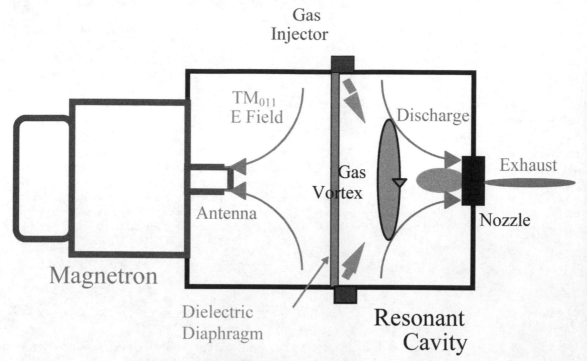

Figure 1. A diagram of the MET thruster in its present form.
The magnetron generates microwaves from high voltage DC power.

Figure 2. Photos of MET operation and approximate cavity dimensions at 100 W, 1 kW, and 50 kW.

High Pressure Operation

It was discovered that high pressure in the discharge chamber (P > 500 Torr) was essential to stable and efficient operation at low Q. This fact has been recently supported by work at Aerospace Corporation, with the assistance of RSI (Diamant 2001). The exact dynamics of the gas, plasma and microwaves are quite complex and so could only be explored experimentally in MET research programs. However, it is apparent that a key parameter is the rate of electron-neutral collisions, which is controlled by gas density or pressure in the chamber. If pressure is high, microwave energy is converted to heat in a highly localized region, and a sharp plasma-gas interface is formed. If the pressure (and thus collision rate) is low, then the interface between plasma and gas becomes diffuse and enlarged, and heating is less effective. High pressure means high gas density and a highly resistive plasma, because of the high rate of electron-neutral and neutral-neutral collisions transfer energy rapidly from microwave fields to heat. The process of energy transfer is apparently a multi-step process, with electrons being accelerated by the microwave electric fields, the electrons transferring energy to molecular or atomic vibrational modes through collisions, and then vibrational modes equipartitioning energy into thermal kinetic energy (primarily via neutral-neutral collisions). During MET development at RSI, pressures below 500 Torr often resulted in unstable operation, with pressure slowly dropping, leading to a "glow mode" similar to that seen by DC arc jet experimenters. In this mode, the electrons decouple from the ions and neutrals, becoming essentially collisionless. When this occurs, a highly conducting, very low pressure plasma forms, and electrons reflect large amounts of power back into the microwave source, leading to direct drive magnetron shut down or burn out. At the same time, the electrons couple very little energy into neutral or ion heating. Electrons are thus hot, but neutrals and ions remain cold. In such cases, the MET would give an I_{SP} only like a cold gas thruster (indicated neutral gas temperatures of room temperature), while the electrons had reached 10,000° K or above. Thus, high pressure operation was essential to good MET operation and high I_{SP}.

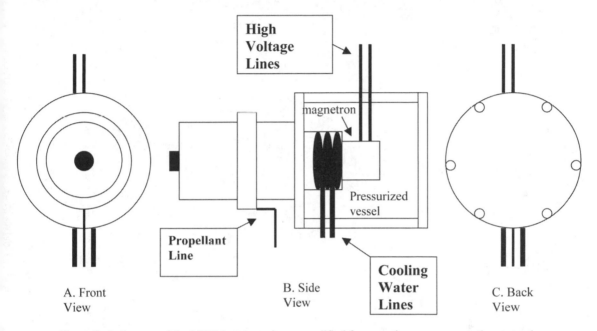

Figure 3. A diagram of the MET thruster as it was modified for operation on a vacuum thrust stand. The commercial magnetron was enclosed in a pressure vessel because it uses air as an insulator.

It became apparent that the MET discharge behaves in a very generic manner for a variety of gases when run at high pressure (low chamber Q) manner. In this mode, both helium and nitrogen formed similar discharges and ran at similar thrust efficiencies.

The performance of the MET was measured for nitrogen and helium gases at 1 kW by taking the MET thruster to NASA Glenn and running on NASA's inverted pendulum vacuum thrust stand. In these tests, an I_{SP} in the range of 250 seconds was found for nitrogen at thrust efficiencies of 35% in terms of power to the magnetron. These results were duplicated at RSI using a different thrust stand design. One method for reducing the engineering difficulties of MET propellant delivery (particularly useful for water operation) was to use a "Chinese Fish Trap" thrust measurement device (Cheng 1984). This device absorbed the momentum of the plume rather than the force applied to the thruster. The momentum trap resembled a muffler (cylinder with perforations to relieve pressure build-up). The nitrogen propellant results agreed within 10% to those of the NASA Glenn data obtained on a conventional inverted pendulum thrust stand. On the RSI stand, helium gas achieved 400 seconds I_{SP} at 35% efficiency. The results obtained by RSI are shown in Figure 4. Recent results at The Aerospace Corporation have validated these values; Aerospace researchers have achieved 418 seconds I_{SP} with helium and 243 seconds I_{SP} with nitrogen. It should be noted that the efficiencies in Figure 4 include the factor of the magnetron efficiency, which for commercial magnetrons is approximately 70%; thus the thermodynamic efficiency of the thruster is close to 50% in terms of conversion of microwave energy into directed kinetic energy.

Water Vapor as Propellant

At the request of NASA Goddard personnel, RSI attempted to use water vapor as a fuel in the MET. This request was motivated by a desire to launch small satellites from the space shuttle that could be boosted to higher orbits. Operation on water vapor at RSI was surprisingly successful and resulted in stable compact discharges at high pressure. Operation of the MET on an inverted pendulum thrust stand in vacuum would have required either the piping of hot steam through flexible hose to the thruster, or a self contained vaporizer to place on the MET. The transmission of steam to the MET thrust chamber by flexible hose would have created the opportunity for condensation of the steam in the lines, leading to errors in mass flow measurement. The most straight forward solution to this problem would have been to create a self-contained vaporizer to mount on the MET having short steam connection to the thrust chamber. This vaporizer would either continuously vaporize liquid water supplied by a flexible hose or else have its own metered water reservoir. The construction of a self-contained water vaporizer for the MET posed serious problems, not the least of which was the possibility of venting boiling water into the vacuum system and shutting down the experiment. For these reasons, the momentum trap thrust stand was used. This method of thrust measurement enabled the MET with its water vaporizer to be mounted outside the vacuum thrust chamber, where its operation was much simpler.

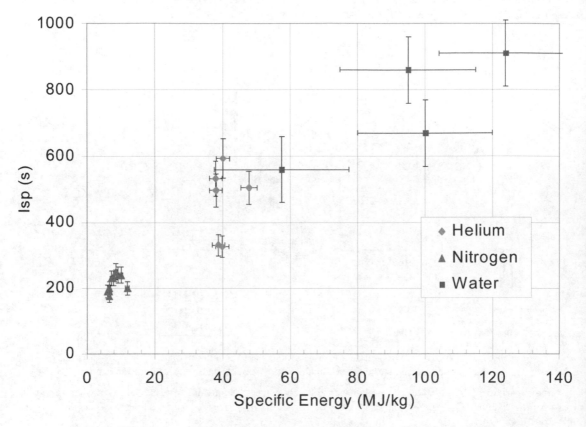

Figure 4. RSI data for specific impulse with various propellants as a function of specific power
(input high voltage DC power divided by flow rate).

Preliminary data, taken by the authors at RSI using these methods, showed that water gave excellent performance, with $I_{SP} > 800$ seconds (see Figure 4), apparently due to the lightness of the water molecule and possible recombination of the hot atomic oxygen and hydrogen created by the thermally disassociated water molecules streaming through the exhaust nozzle. This preliminary data has been approached by the performance measured at The Aerospace Corporation, which found an I_{SP} of 428 seconds in high pressure discharges, with some indications that this could be improved (Diamant 2002). This high measured performance, and the lack of visible emission from the water fueled MET exhaust plume (suggestive of thorough recombination in the nozzle), indicates that water may be a nearly ideal fuel for many space propulsion applications. However, water, because of its two-phase nature and ability to quench electric discharges (possibly through electron-oxygen attachment) if vapor pressure is not well regulated, presents many engineering difficulties to the experimenter.

The Water Vaporizer
Experience at RSI showed that operation of the MET on water vapor appeared to require that the ratio of electric field to chamber pressure stay within a tighter range for water vapor than for other gases, presumably because of water vapor's tendency to attach electrons and thus quench discharges. Because of this, water vapor pressure had to be well regulated. Running water vapor in the MET is not for the easily discouraged experimenter. Operation of the MET at high chamber pressure, above 500 Torr, was essential to high I_{SP}. Indeed, it was not straight forward to achieve this high pressure; at very low pressures (below 500 Torr), the tendency of water to quench the discharge prevented increasing the pressure to the easier-operate high range. As a result, it was found at RSI that best way to

run the MET thruster on water at high chamber pressure was also the simplest: a boiler with a constant pressure, toggled by a solenoid valve, quickly increased the chamber pressure of the MET into the desirable > 500 Torr range, avoiding a slow transition through the low pressure regime. In the boiler, liquid and gaseous water were kept in contact at an interface. When the boiler was then connected to the thrust chamber, the liquid gas interface allowed the thrust chamber pressure to be the vapor pressure of the hot liquid water in the boiler. If a fluctuation in the thrust chamber plasma caused the pressure to drop, the liquid water simply boiled more vigorously to raise the pressure back to its proper point of operation. If pressure went up, the liquid boiled less vigorously and the pressure relaxed back to steady state. A boiler vaporizer with a liquid-vapor interface thus allowed the liquid water to do pressure buffering on the system. However, it must be stressed that workers at The Aerospace Corporation recently succeeded in running the MET on water vapor at high I_{SP} using a continuous flow vaporizer (Diamant 2002) by simply increasing the microwave power to overcome water quenching during the slow transition through low pressure.

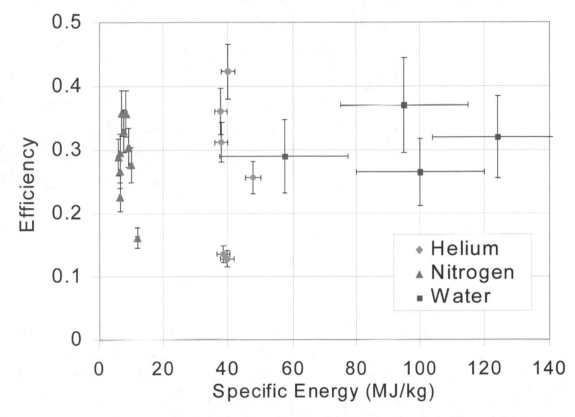

Figure 5. RSI data for efficiency with various propellants as a function of specific power.

It can be said while the nitrogen and helium MET thruster is in the process of space hardware development, the water MET is still a physics experiment, and has not yet been reduced to engineering practice. The difficulties of propellant delivery must be overcome if the MET running on water is to gain acceptance as a space propulsion option. In particular, engineering problems of vaporizing and delivering water at regulated pressure into the MET discharge chamber, in a compact unit that can be made integral to the MET and thus function on a vacuum thrust stand, are considerable, and require further research and development.

Summary and Discussion

The MET thruster has now been run in a compact 100 W form at 7.5 GHz, 1 kW at 2.45 GHz, and at 50 kW using 915 MHz microwaves. Microwaves can be generated efficiently (> 90% at 915 MHz) and beamed for long distances coherently. On a spacecraft, the microwaves can be received on an antenna and fed directly into the thrust chamber without any intermediate conditioning. This suggests the MET thruster is highly compatible with a space transportation infrastructure based on large scale nuclear or solar power stations and antennas beaming power for orbital transfer, or even for powered landings in reduced gravity environments such as the lunar surface.

The MET is simple, efficient, can be scaled efficiently to high powers and high I_{SP}, and it can be run on water vapor, perhaps the most common chemical compound in the Cosmos. Water is found on the poles of the Moon, Mercury, and Mars, and is abundant in the outer solar system. Water is easy to transfer between spacecraft and to store and accumulate in space. Water is also an element of all human space flight, where it is byproduct of fuel cell operation, a source of oxygen, and essential for human life and health functions. This means that a water handling infrastructure exists on all human flight missions in space and can thus be expanded to accommodate propellant water. All of this means that the MET, using water as a propellant, has enormous potential for human space flight missions, most principally those to the Moon and Mars.

References

1. Jahn, R.G. 1968. *Physics of Electric Propulsion*. McGraw-Hill. New York.
2. Whitehair, S., J. Asmussen, and S. Nakanishi. 1987. *Microwave Electrothermal Thruster Performance in Helium Gas*. J. Propulsion. 3 (2): 136-144.
3. Sullivan, D.J. and M.M. Micci. 1991. *The Effect of Molecular Propellants on the Performance of a Resonant Cavity Electrothermal Thruster*. IEPC 91-034.
4. Power, J.L. and D.J. Sullivan. 1993. *Preliminary Investigation of High Power Microwave Plasmas for Electrothermal Thruster Use*. AIAA 93-2106.
5. Balaam, P. and M.M. Micci. 1995. *Investigation of Stabilized Resonant Cavity Microwave Plasmas for Propulsion*. Journal of Propulsion and Power. 11(5): 1021-1027.
6. Sullivan, D.J. and M.M. Micci. 1994. *Performance Testings and Exhaust Plume Characterization of the Microwave Arcjet Thruster*. 94-3127.
7. Diamant, K.D., J.E. Brandenburg, R.B. Cohen, and J.F. Kline. 2001. *Performance Measurements of a Water Fed Microwave Electrothermal Thruster*. AIAA 2001-3900.
8. Cheng, D.Y. and Chang, C.N. 1984. *Deflagration Plasma Thruster*, pp. 371-384 in *Orbit-Raising and Maneuvering Propulsion: Research Status and Needs*. AIAA. New York.
9. Diamant, K.D., R.B. Cohen, and J.E. Brandenburg. 2002. *High Power Microwave Electrothermal Thruster Performance on Water*. AIAA 2002-3662.
10. Brandenburg, J.E. and J.F. Kline. 2001. *Solaris: A Low Cost Human Mars Mission Architecture Based on Solar Electric Propulsion*. Proceedings of the Mars Society Symposium, Stanford University.
11. Brandenburg, J.E. and M.M. Micci. 1995. *The Microwave Electro-Thermal (MET) Thruster: A New Technology for Satellite Propulsion and Attitude Control*. AIAA 96-16827. 9th Annual Conference on Small Satellites.

//

Spacecraft Computer System Design Considerations for a Piloted Mission to Mars†

Edward N. Brown

brownec@surfside.net

– Part 1 –

Introduction

Piloted deep space missions, such as a human mission to Mars, present new challenges in the design of spacecraft computer systems. Because of the human element involved, the demands of the operational mission, and the constraints imposed by weight, electrical power, reliability, environment (especially radiation), and data processing performance considerations, general purpose consumer or business oriented computer solutions are not acceptable.[1] That does not imply that every circuit and every microchip for the computer system needs to be designed and constructed from scratch. Some may indeed be, but most will be developed by utilizing, modifying or tweaking existing designs and products. But it's not as simple as just selecting top-rated top-value components from a computer catalog or store shelf, hooking them up, and then fine-tuning the result. That system won't do the job. The solution for this application must be engineered. What this really means is that it is not sufficient to simply state that the system must be "user friendly, long life, low weight, low power, high reliability, robust quality, and high performance." Like many other complex designs, "the devil is in the details"[2] and that is especially true in this application. While these design characteristics are valid, exactly how best to achieve them is a very difficult task. There are many questions to answer. How reliable does it need to be? How much power should it consume? How small does it need to be? How much processing throughput should it have? These are just some of the major and more obvious questions. A myriad questions will present themselves as each area is investigated. The engineering challenge is to determine the optimal or appropriate levels and trade-offs (quantitative where possible) among these characteristics such that all the operational, system design, and programmatic requirements (cost and schedule being just another two parametric constraints) are satisfied and the system contribution to overall mission success can be assured to a level that is acceptable to all the stakeholders. The best way to systematically go about doing this is to utilize standard Systems Engineering procedures and techniques to flush out the driving design requirements and eventually synthesize a working design.[3] This two-part paper represents a combination of a presentation made at the 2002 Mars Society Convention (Part 1) and one made at the 2003 Mars Society Convention (Part 2).

Approach

To determine the design drivers for a system definition, four different perspectives can be combined in an analytical framework. First, we must look backward to the experience gained and lessons learned from earlier human space flight endeavors, such as Apollo, Skylab, Space Shuttle, and International Space Station (ISS) [although Space Stations have a different design perspective because they are more akin to habitats than to transportation vehicles], as well as from unmanned deep space missions. This can provide valuable insight into the best way to move forward by not making the same mistakes that were previously made over again, and by keeping certain design approaches that previously were found to be successful as precepts. Second, we must understand the vision, goals, objectives and operational concepts of reputable entities associated with Mars exploration, such as the Mars Direct Mission, NASA Mars Reference Mission, or ISTC Project 1172. This is required since the design must be tailored to fit the mission and not be a general purpose or one-size-fits-all type of design. Thirdly, we must understand the current commercial technological state-of-the-art by examining the computer products of industry leaders in both the manufacturing (hardware and software) and system integration areas (especially in the telecommunications industry). This will likely strongly affect the actual design synthesis because it is usually more expeditious to utilize or adapt an existing design than to initiate a design from scratch. Finally, we must review active computer technology R&D programs currently being pursued by government, university and industrial R&D Labs that have applications to spacecraft design, in order to ascertain the directions, thrust and focus of these endeavors. This will insure that our design perspective is not out of kilter with other efforts and that we have a project level information and technology base in which to leverage off.

Therefore, the process can be simplified to the answering of four questions:

1. What are the design requirement drivers based on the raison d'être and "lessons learned" from both past and present human space flight and robotic space flight program designs?
2. What are the design requirement drivers as dictated by the most generally accepted mission concept?
3. What are the design requirement drivers that result from an understanding of today's commercial state-of-the-art industrial computer product technology and projected future trends?
4. What are the design requirement drivers resulting from analyses of active ongoing R&D programs concerned with computer technology development applicable to space vehicle design?

In the process of investigating questions 1, 3 and 4 above, certain guiding principles of system design will emerge. These principles may be driven by performance, cost, or programmatic concerns, but they will represent the technological bounds within which the system will have to be designed.

Perspective

In human space flight projects to date, custom designs, military designs, and carefully modified commercial designs have predominated. However, since 1994 the DOD has led an effort to utilize open system architectures and ruggedized industrial grade commercial off-the-shelf components to the greatest extent possible in the design of vehicular and shipborne computing platforms.[4] NASA JPL has embraced this philosophy in concert with robotic space projects developed under the "faster, better, cheaper" mantra.[5] Although there has been mixed success in the application of COTS / Open Systems to date, this has now become a mature guiding principle of system design that is meant to counter the high expense, long lifetime inalterability, and undesired supplier dependencies associated with earlier custom Mil-Spec system designs.[6] Observance of this principle will allow future DOD and NASA programs to leverage off rapid growth of computer and information technologies now occurring in the commercial marketplace. There is no reason to believe that this overarching precept will not also be applicable to the design of the fleet of spacecraft that will carry humans to Mars (although some specific exclusions are probably inevitable). The benefits and payoffs of using COTS / Open Systems are just too great to ignore and it has become the modus operandi for implementing all but the most critical or the most sensitive functional requirements.

In addition, certain design integration approaches that have evolved over the past decade, such as distributed, integrated, modular, and miniaturized avionics,[7] will be the cornerstones upon which the computer design of the Mars vehicle will be based. Fundamental design concepts, such as the decentralization of functionality, the utility of embedded computer control of almost all spacecraft systems, the need for reliable and timely communications between all computing elements, and the ability to detect, isolate and reconfigure around faulty elements, will dictate that certain design directions be followed. An example of such a direction is the use of network-centric system topologies with protocol driven information transactions instead of unique point-to-point user defined data transactions. This broad understanding of design directions will provide the high level guidelines from which the more detailed drivers to the final spacecraft computer system design will emerge.

Therefore, it is necessary to add a fifth question to the above four in order to complete the overall investigative arena.

5. How do the answers to the above four questions synthesize together with the high level guidelines that have evolved from the COTS / Open Systems paradigm and distributed / integrated / modular / miniaturized avionics concepts?

When answers to these five questions begin to materialize, we will start to accumulate a knowledge base of information regarding the desired attributes of the computer and data processing systems that would likely be on board a piloted spacecraft on a mission to Mars. As the knowledge base matures, a set of preliminary system design requirements (including identified design drivers) will eventually be derived and formulated in appropriate specification type language. From these high-level system requirements, a preliminary architecture and notional system design, with performance metrics, will eventually emerge.

Manned vs. Unmanned Spacecraft

However, prior to delving into the analyses proper, it is necessary to understand and appreciate the key differences and similarities between manned and unmanned spacecraft, and in particular, between computer systems designed for manned spacecraft and computer systems designed for unmanned spacecraft. Although a human mission to Mars will obviously utilize a manned spacecraft, the operational mission design and logistics will be similar to those which have already driven the design of deep space unmanned robotic spacecraft. So there is a synergism that can be obtained by examining the attributes of each (note that unmanned Earth orbiting satellites are intentionally omitted from this comparison although powerful computer systems on board large communication satellites are worthy of review and should be evaluated in connection with the analysis involved in answering questions 3 and 4).

Spacecraft computers designed for both manned and unmanned missions share a number of key attributes:
- May need to operate over long mission times
- Need to minimize power, weight, size requirements
- Need for autonomous operation under certain conditions
- Need for robust fault tolerance and reliability
- Need for automatic FDIR (fault detection, isolation, recovery)
- Need for efficient and reliable telemetry
- Need for robust environmental qualification
- Need for resource management
- Need for high computational and processing performance

Both designs must accommodate these concerns although the degree of concern and the level of specification is really determined on a case-by-case basis and is often driven by the greater vehicle design that provides the platform for the computer system or by the mission requirements. For example, because of the small physical size of unmanned probes, the need to minimize power, weight and size is extremely important and is usually a fundamental driving requirement, whereas for manned spacecraft it may be a goal or negotiable requirement. The actual specified values in each case will be driven by the vehicle design constraints. Similarly, the need for autonomous operation and automatic FDIR is very important for unmanned probes because they are driven by mission requirements (e.g., long time out of communications, latency of command signal), but may be equally important for manned spacecraft because of different mission requirements (e.g., crew sleep cycle, small time window, hazardous mission phases).

However, because of these differences in spacecraft vehicle design and in mission requirements, there are some fundamental differences in the associated computer system designs that have been installed and operated to date. Table 1 presents a comparison of vehicle and mission characteristics between manned and unmanned spacecraft related to computer system design and operation. A computer system designed for a manned vehicle will likely differ from a computer system designed for an unmanned vehicle because of the differences shown in this table. It can be seen that some characteristics are unique to the manned system, some characteristics differ in relative value only (due to the vehicle design), and some characteristics differ because of mission requirements or perspective. The top five items form the crux of the comparison. The bottom line is that the primary objective of the computer system on a manned spacecraft is to operate the vehicle, including all the associated subsystems, and service the occupants in a safe and efficient manner, whereas the primary objective of the computer system on an unmanned spacecraft is to operate the vehicle and the sensor instruments in order to obtain the science data and transmit it back to Earth. As can be seen from Table 1, the presence of a human on the vehicle drives the design direction.

Computer related attributes that are unique to manned space flight can be summarized by the following:
- Overarching requirement for Human Safety
  - Safety Driven Design and Development Process
  - Safety Driven Design Architecture
  - Importance of FDIR to operational safety
- Need for User Friendly Human Interface for Situational Awareness
  - Visual Display outputs
  - Tactile Keyboard, Edge key and Switch inputs
  - Voice Recognition input
  - Auditory Alarm outputs
- Greater allowable power, size and weight constraints due to greater available resources driven by human needs
- High throughput and logical processing requirements to accommodate real-time automatic control and monitoring of spacecraft systems and to accommodate flight crew needs for manual monitoring and commanding of spacecraft systems

- Allowance for onboard in-flight maintenance and repair
- Incorporates capability for manual pilot-in-the-loop control of certain flight maneuver and control functions (rendezvous, docking, landing)

Manned	Unmanned
Vehicle centric functionality	Payload centric functionality
High logical processing requirements needed for vehicle and crew applications	High computational processing requirements needed for sensor and instrument data
Safety directed design - destination driven	Performance directed design - science driven
Need for onboard user interface for command and display	N/A
Need for accessibility and maintainability	N/A
Importance of auto and manual FDIR (including manually switched backup)	Importance of auto FD and correction
Onboard power generators necessary for life support and thermal control allow for higher power consumption	Low onboard power generators dictate need for low power consumption
Optional reversion to manual operation when comm to MCC is unavailable	Need for autonomous operation when comm to MCC is unavailable
Need to work over short and long missions	Need to work over longer mission times
Greater memory and I/O requirements	Lesser memory and I/O requirements
Basic reliance on tried-and-true technology	More tendency toward innovation and less mature technology

Table 1. Comparison of Spacecraft Computer System Characteristics.

Historically, when these attributes were considered during the development process, the resulting design usually contained certain characteristics that differentiated it from a computer system design for an unmanned robotic probe, such as:
- Greater throughput and processing power requirements
- Greater memory requirements
- Greater I/O requirements
- A greater number of larger and more complex programs
  - generally utilizes an Operating System and System Services software
  - more diverse application programs
- A sophisticated human interface
- More complex Redundancy Management and synchronization requirements
- Incorporate Central Processing Units (CPUs) and Microprocessors (µPs) instead of Digital Signal Processors (DSPs) and Microcontrollers (µCs)
- Greater power dissipation and cooling needs
- Greater concern for radiation-induced Single Event Upsets (SEUs) and Single Event Effects (SEEs)
- Larger footprint
- Longer development time and higher cost

In fact, because of these demanding requirements, multiple computers or processors were often used for different functions. For example, the flight control computer, the command and control computer, the systems management computer, the display driver computer, the engine control computer, and the I/O computer may all utilize separate processors and be located on separate cards or even in separate Line Replaceable Units (LRUs). Mass memory, timing controllers, and backup systems may also be in separate cards or boxes. That is not to say that future manned spacecraft computer designs will follow this tendency. That's just how it happens to be today. Although the attributes, characteristics and requirements will be similar, the actual design implementation will change according to the high level guidelines and directions that are in effect at the time.

The key to the design of the computer system for the manned Mars mission will be to determine what combination of characteristics and attributes, conventionally associated with either manned or unmanned spacecraft computer systems, represents the best characteristics and attributes for the manned Mars mission.

Summary

The analysis required to answer each of the five questions presented herein is non-trivial and will require careful research and investigation into areas normally affiliated with both manned and unmanned spacecraft design. But this will assure that only the best and most applicable aspects of commercial computer product technology and space vehicle related computer technology R&D will be properly integrated with mission requirements and experiential "lessons learned" type knowledge, in order to valuably contribute to the overall design of the spacecraft computer system for the human mission to Mars.

– Part 2 –

Introduction

In Part I the overall perspective and design approach was presented along with an analysis of the differences between computer systems for manned and unmanned spacecraft. It was shown that the computer system for the piloted Mars vehicle will leverage off certain key design technologies found on Space Shuttle and ISS, such as redundancy and fault tolerance, user friendly human interface, real-time processing and control, onboard maintenance and repair, and manual pilot-in-the-loop control; but will also benefit from evolving technologies found on the latest genre of unmanned robotic spacecraft, such as automated reasoning for autonomous system control, model-based health monitoring, diagnosis and recovery, and automated science data evaluation and observation planning. However, the overarching guiding design principle will be the requirement for human safety and the safety directed design process that necessarily follows. In Part 2 it is shown that this will drive a system-of-systems engineering approach that will employ network-based wide band communications between distributed embedded processing units and laptop applications program / user interface units for all but the most safety critical flight control or life support functions. These safety critical functions will be architected similar to those of the Space Shuttle, but with a drive towards low power, small size, high performance hardware devices, and advanced software techniques for improving fault detection, isolation, and identification, and for improving the situational awareness resulting from the user interface.

Approach

An understanding of the factors that influence the design of spaceborne computer systems illustrates how the overarching architectural and integration ground rules, that have evolved from contemporary NASA and DOD programs and studies, reveal the high-level design drivers. The following factors that influence the design of spaceborne computer systems include many of the same factors that influence the design of terrestrial computer systems, but must be optimized in a predictable manner for the space application:

- System Hardware Architecture
  - LRUs, Boards, Backplanes, Buses
- System Software Architecture
  - Operating System, System Services, Applications Programs
- Computational Capability
  - Precision, Speed, Throughput
  - Memory Capacity, Instruction Repertoire
- Software Language, Programming, and Code Size
- Input / Output Capability
- Communications Bus and Network Characteristics
  - Bandwidth
  - Availability and Determinism
- Reliability and Safety Related Items
  - Hardware Failure Rate
  - Software Integrity
  - Fault Tolerance and Redundancy
  - Data Error Correction and Compensation
  - Self Test and Health Management
- Mechanical
  - Weight – Volume – Packaging
- Electrical Power Requirements
  - Normal Voltage, Current, Duty Cycle
  - Tolerance to Abnormal and Transient Power Conditions
- Maintainability
  - In-flight Maintenance Capability
- Environmental
  - Temperature, Vibration, Electromagnetic Effects, Radiation

Central to the overall design process is the role of fault tolerance. Although the importance of fault tolerance has not really changed over the years, the methodology by which it is obtained most certainly has. History can be projected forward to identify the key requirements in the design of the computer systems for a piloted Mars vehicle. However, in recognizing the importance of fault tolerance, it is necessary to analyze the man-machine system as a whole and not just the computer components in isolation. This was true for Space Shuttle, much more true for ISS, and will be of even greater significance for the Mars vehicle because of the very large number of complex subsystems involved, the criticality of certain operational phases, and the overall mission duration. Fault tolerance is not just about redundancy of parts but also about efficient and errorless human involvement in the mitigation or correction of abnormal situations. To achieve this human performance requires incorporation of key design elements and careful attention to the user interface.

Evolution of Design Philosophy in Human Space Flight

Early space projects relied on reliability of components (Mil-Specs), process quality (QA Inspections), extensive component, assembly, unit and system testing, and minimal use of software. Apollo was a turning point. A dual redundant computer system was rejected in the early stages as being too complex and too uncertain in reliability. But designing around component reliability constraints and failure analysis revelations (FMEA, Fault Tree) turned out to be even more expensive. It was hard to get good reliability data on components, and testing revealed disconnects between theoretical reliability and real-world reliability. Furthermore, FMEA / Fault Tree results came too late to meaningfully influence system design. Even without hardware faults, small operator errors often caused computer restarts, with accompanying service interruptions and initialization problems. To overcome these shortcomings required a movement toward more redundancy and fault tolerance in design. Nowadays, the primary approach used in achieving safety of human space flight is the adherence to the "design to eliminate hazards" mantra and its resulting use of redundancy management and fault tolerance in design.

For unmanned robotic spacecraft computers, the highest driving fault tolerance requirement is that no credible single point failure shall prevent attainment of the mission objectives or result in a significantly degraded mission. This is not good enough for manned missions. For computers on crewed spacecraft, the highest driving fault tolerance requirement is twofold:

> No single failure or operator error can result in a non-disabling injury, severe illness, loss of life sustaining functionality, or loss of an emergency protection system; and

> No combination of two failures or operator errors can result in a disabling injury, loss of life, or loss of the spacecraft (excludes independent simultaneous failures).

This is known as the Fail Operational / Fail Safe (FO/FS) criteria. The human element adds complexity and presents significant design challenges. There is still today much controversy and intellectual analysis that surrounds this criteria. It was the result of an evolutionary design process on Space Shuttle. It is still true for ISS although slightly modified because of the longer time availability for human intervention and the lack of dynamic flight phases.

The original Shuttle design concept was FO/FO/FS utilizing seven parallel processors in seven strings with independent input data. This was rejected because of weight, power and cost constraints. Every alternative between this and three parallel processors in three strings was evaluated in detail. The final design chosen was a FO/FS strategy utilizing five parallel processors in five strings, where four processors in four strings comprised the primary system and one processor in one string provided a hot-switched backup system. This design incorporates replicated hardware and dissimilar software. A fault detection and isolation scheme is used to eliminate first and second faults, and it also protects against generic software faults. Redundancy management and computer synchronization are integral and necessary to the design but these functions come with a very high overhead cost in design analysis and verification testing. Voting is performed at the computer inputs and at the actuators / end effectors.

A significant limiting and complicating factor to a four string computer design is the fact that most spacecraft subsystems have evolved to a three string design. This is true for the Electrical Power, Engine Control, Navigation Sensors and Display Generators that interface with and are closely coupled to the computer system.

Shuttle Lessons Learned

The design goal started as two-fault-tolerance with failure coverage of 100%. In reality, many factors made achieving this nearly impossible. In striving to meet FO/FS requirements with a less than pure approach, every combination of failure, however improbable, had to be analyzed and verified as acceptable – magnified by reconfiguration actions which had to be verified – resulting in a design, testing, and training program of staggering proportions (> 255 exceptions to FO/FS requirements, > 700 pages of crew malfunction procedures, > 1,000 pages of off-nominal crew procedures).

One alternate approach would be to draw a probability "line in the sand." Unfortunately, there are significant drawbacks to this approach, such as the uncertainty in determining an acceptable P (probability), the uncertainty in determining a method for calculating P, and the uncertainty in the actual calculations of P.

A different alternate approach, which is the path followed today and likely into the near future, is to use a robust software-driven fault detection, isolation and recovery scheme that incorporates advanced software error detecting and correcting algorithms, along with advanced hardware design of sensors, monitors, memory and I/O (data bus) drivers. Improved accuracy of diagnostics and reduced false alarm rate are critical to an improved design for future applications.

High Availability Technology – Telecommunications Industry

Features of High Availability (HA) technology that are attractive in spacecraft design include the following:

- Hot Swap
- Plug and Play
- I/O Fail-over, CPU Fail-over
- Robust Error Detection and Correction (EDAC)

- Dynamic Reconfiguration
- Hot Restart

The question is whether these design features can be utilized as is, with modifications, or at all. Mission critical HA computing usually follows the "open systems" approach and demands continuous service with no loss of application state. It utilizes redundant resources of clustered processors and fault management software that typically recovers faults in tens of seconds and requires application program restart with loss of data (state) not already saved to mass memory. Can we utilize this technology for the computing needs on a piloted spacecraft to Mars? Unfortunately, in most cases the answer is "no" for a number of reasons.

Telecom HA systems are not in general "real-time" systems needed for critical flight phases such as take-off, landing, docking and rendezvous. Furthermore, they suffer from the following undesirable attributes:

- Susceptible to undetected faults – bit flips, polarity reversals, level shifts, etc.
  - non-detection of failure in active unit
  - non-detected latent failure in standby unit
  - standby failure can affect active unit causing additional failure (pollution)
- Total loss of availability for failure to switch-over or fail-over
- Overall outage time too long
  - even "6-Nines" allows an outage of 31.5 seconds/year
  - not good enough for life and death situations
- Active-Standby Mode switchover and fail-over times not fast enough
  - requires power-off of bad unit and initialization of good unit (20 seconds-1 minute)
- Fault coverage estimating methodology not that good
  - likelihood of successful switchover / fail-over not precise
- Service group autonomous recovery techniques not good enough
  - added complexity and failure modes
- "Role swap" accommodation not good enough
  - overlapping operation doesn't help for many faults

Space Qualified Microprocessors

Computer technology has come a long way in 30 years and will undoubtedly continue to advance during the next decade and beyond. The Shuttle computer runs about 500K software lines of code (SLOC) written in the HAL/S language. It is expensive to maintain, not easily expandable, and requires unique software tools. Figure 1 compares the Shuttle computer architecture to a modern space computer architecture.

Computer Technology Has Come a Long Way in 30 Years

Current: ~500 KSLOC HAL/S

GPC
25 Hz Frames

- Unique Language
- Unique Tools
- Expensive to Maintain
- H/W Dependent
- Not easily Expandable

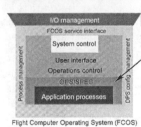
Flight Computer Operating System (FCOS)
PASS Architecture

Application Software Overlay (Choice of GNC, Sys Mgmt, or Payload)

- Minimal Change Containment
- Differing Load Configurations (doesn't all fit!)

New: ~10 MSLOC C++

VMC MC PLC
80 Hz Frames

- Commercial High Order Language & Tools
- Robust Partitioning – Changes Isolated
- Portable and Maintainable
- Scalable/Expandable

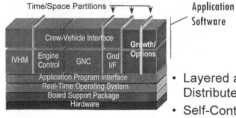
Open Systems Architecture

Application Software

- Layered and Distributed
- Self-Contained – No Overlays

Figure 1

Moore's Law is often used to model the development of information technology. It states that the density of devices on a single chip grows exponentially over time. By implication, a corollary to the law states that processor performance will grow at the same rate. Mathematically, this can be expressed as:

$$\text{Perf}(N) = 2 \exp{(N-N_0 / 1.5)} \times \text{Perf}(N_0)$$

Thus, a 10x improvement every five years is predicted, and has in fact occurred (more or less). However, while this may be good for the consumer electronics customer, it doesn't directly apply to space qualified processors. Because of the design and test effort involved, the small market, and the low profit margins, space qualified processors lag conventional processors by about three semiconductor life cycle generations (approximately 12 years). This is a fact of life, so it would not be wise to lock in the processor type today for the manned Mars mission if final design of the rest of the spacecraft is not going to occur until 2015, 2020, or later. Modularity of design is key. Being able to plug in the latest technology processor at the latest time possible, with minimal programmatic effects, is the goal. So a review of yesterday's or today's processors is not that meaningful. It is more instructive to analyze the future trends.

Figure 2 compares the performance of some popular space qualified processors (Note that this does not include Digital Signal Processors [DSPs] which are tailored for scientific data crunching and not really efficient at meeting the logical and interactive needs of a control / display oriented computer). A throughput of 300 million instructions per second (MIPS) is now state-of-the-art for the PowerPC 750, but already efforts are well under way to qualify the PowerPC 7455 to 800 MIPS. Projections of growth to 1,800 MIPS (the SCS750) are well founded.

Space Qualified Computers

Unfortunately, a space qualified processor does not a space qualified computer make. The space qualified computer must incorporate all the hardness (against environments and radiation) and reliability of the space qualified processor in all of the components that make up the computer. That applies to memory (Flash and DRAM), controllers, timers, and I/O. Plus, it applies to both chips and boards. A space qualified Single Board Computer (SBC) may be just one board in and among memory boards, controller boards, and I/O boards.

Near Term flight Processor Performance

Processor	closest commercial	clock speed	mips	mflops	SPECint95	SPECfp95
Rad750	PPC 750	150	300		7	4.7
Pentium	Pentium-I	166	178.4		4.5	3.73
RHPPC	PPC 603e	166	210		3.94	2.71
Rad6000	PPC	33	35		1.68	4.9
StrongArm	StrongARM	88	90	n/a		n/a
Thor	n/a	50	50	16		
MongooseV	R3000	12	10.6			
1750A	n/a	1	3			

Dhrystone measures integer performance.
Whetstone measures floating-point performance.

Performance of Selected Flight Architectures

Figure 2

In addition, the computer must be architected for fault tolerance. A flexible, low cost, state-of-the-art, fault tolerant computer would have the following characteristics:

- Fault Tolerant
  - masks all failures automatically
  - discriminates between transient, intermittent, permanent faults
  - prevents fault propagation between channels / strings
  - assesses remaining degree of fault tolerance
  - reintegrates "lost" fault tolerance when possible
- Low Cost
  - mostly COTS architecture to leverage upgrading with newly developed products
  - redundancy management transparent to applications developers
- Flexible design
  - processors operate independently or as members of a fault containment region
  - all processors have access to all data regardless of connectivity
- State-of-the-art technology
  - small, low power, low weight
  - fast
  - graceful degradation with four channels / strings

The canceled X-38, originally intended as a Crew Rescue Vehicle for the ISS, incorporated a fault tolerant computer design that advertised a reliability > 0.99999 and fault coverage of 100%. However, recent advances intended for introduction on the Orbital Space Plane (OSP) have already improved upon this design. The new design has a throughput that is 5x faster, while having weight, volume and power specifications that are 50% less than the X-38 design. Figure 3 illustrates a Single Board Computer using advanced processor and fault mitigation technologies.

Computer/Bus Architecture Segregates Functions

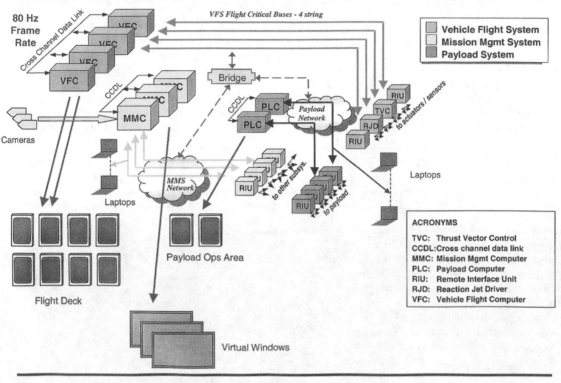

Figure 3

Space Qualified Computer and Avionics System

The overall computer system must accommodate a mix of safety critical, mission critical, and non-critical functions. To minimize the probability of a fault in a non-critical function from affecting the performance of a critical function (and also to minimize development and redesign work), functions of differing criticality can be segregated by the computer and data bus architecture.

Within the computer unit itself, advanced technologies such as Triple Modular Redundant (TMR) processing, single-event-upset (SEU) immune controllers and "glue" chips residing on radiation tolerant Field Programmable Gate Arrays (FPGA), and error-protected memory accessing, will be utilized (see Figure 3). In the software arena, functions of varying criticality will be isolated from one another in time and space using advanced partitioning techniques, as shown in Figure 4. Faults occurring in one partition will be contained and not propagate to functions in another partition.

The overall avionics system will be architected such that processors and buses are partitioned in order to segregate functions. For flight critical functions, highly reliable serial data buses (e.g., 1553, 1394) with deterministic data bus software protocols, will remain the design approach of choice. For non-critical mission and payload functions, high-speed switched fabric network technologies will transfer data between diverse and distributed processors, nodes and workstations throughout the spacecraft. This will include display processing for cockpit Multifunction Display Unit (MDU) screens, camera-driven virtual windows, and laptop / notebook computers. The network medium will include wireless, wired and optical communications components. Figure 5 illustrates a notional avionics system architecture.

The characteristics of modularity, integration and distribution will be strongly emphasized and supported as follows:

- Modular
  - building block approach
  - standardized interfaces and components
- Integrated
  - some components shared by different functions
  - processor hardware runs software functions of differing criticality levels – partitioning

- Distributed
  - processing embedded within various subsystems
  - need for communication networks, bridges, and small remote interface units
  - need for coordination and management of distributed components and processes
    - replication in mass and time
    - pollution protection

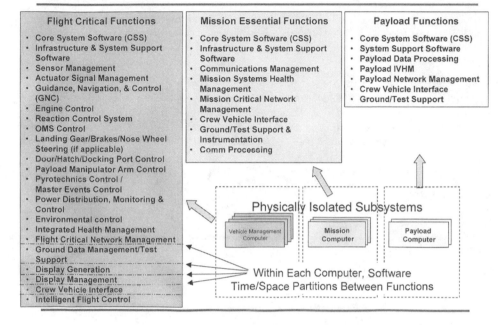

Figure 4

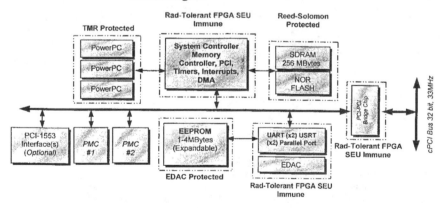

Figure 5

Future View of Computer System for a Crew Carrying Spacecraft

Using the design principles, methodical design approach, and analysis of technology previously outlined, certain design features and concepts become apparent. These will be the guiding principles for future designs. Each of these features can be elaborated upon in more detail in separate and more specific design documents, but a summary listing is included here:

- Networks of diverse, programmable information handling components / units, from chip level upwards
- Information may be digital or analog
- Components may be electronic, photonic, or electromechanical
- Communications and Availability are the key concepts

- Wireless, fiber-optic, and copper cable connectivity used together where appropriate
- Most user command and control from portable devices
- Capability to adapt to different operating modes and environments
- Capability to concurrently execute multiple applications
- Embedded autonomous vehicle health monitoring – Integrated Vehicle Health Management (IVHM)
- "State" centered approach to data and redundancy management
- Real-time system design accommodation
- Goal oriented automation
- Reconfigurable computing (use of FPGAs)
- System-On-Chip design (CPU, DRAM, SRAM, NVRAM, mass memory, power management, communication control, built-in self test [BIST], watchdog timer, error correcting)
- Plug-and-Play Common Modular Avionics
- Fault Tolerant Software

To better understand how general hardware and software design characteristics and features will become important in future Mars missions, it is useful to identify current conditions and project them into the future. In Figure 6, a selected set of hardware and software design characteristics are statused today and extrapolated into the future. These future characteristics will be integral to the design (and in many cases will become design drivers) of the computer system for the flagship spacecraft of the human mission to Mars.

Conclusion

There are no show stoppers and no technological leaps are required. The basic concepts, trade-offs, and approaches have been presented in this paper and the design process is well established in industry. We are better prepared to go to Mars today than we were to go to the Moon in 1961. Only sound engineering judgment and practical engineering planning are needed (plus the requisite funding and management direction). That, together with some old fashioned hard work, and the spacecraft computer system will be ready to support the first human mission to explore Mars.

Looking to the Future

Current   *Future*

Hardware

Current	Future
• High power, heavy, actively cooled	• Low power, light-weight, reduced cooling
• Non-standard interfaces	• Standard interfaces
• Custom or VME architecture	• Modular architectures
• Single sensor packaging	• Multiple/multi-parametric sensor packaging
• Point solution sensors - copper wiring	• Fiber sensor networks
• Low data rate interfaces	• High data rate optical communications
• Centralized main computer	• Distributed computing
• Low density memory	• High density memory

Software

Current	Future
• Limited Qualitative Model Based Reasoners	• Systemic Qualitative & Quantitative Reasoners
• Statistical Analysis	• Simple On-Line & Off-Line Prognostics
• Expert Systems	• Adaptive / Self Learning software
• Ground Systems: Real-Time and Post-Flight	• Real-Time On-Board Flight Systems
• Limited Integration Across Vehicle Subsystem and Phases of Operation	• Completely Integrated Across Vehicle Subsystems and Phases of Operation

Figure 6

References

1. Tomayko, James E., 1988, *Computers in Spaceflight – The NASA Experience*, NASA Contractor Report 182505, www.hq.nasa.gov/office/pao/History/computers/Compspace.html
2. A common quotation often applied to computer programming and design meaning that if you do not pay enough attention to the details, you can get burned. See Martin, W. Mike, 1999, *The Devil is in the Details – The Difference Between Programmatic Concepts and Design Concepts*, Lecture, http://bucharest.stanford.edu/KitTemp/1999/Lecture-2-12-99/MikeMartin/Sld001.htm
3 Schoening, William W., 2000, *Exploring Concepts During Pre-System Definition*, Proceedings of INCOSE 2000, Paper ID 30, Minneapolis
4. Morgan, Rick, Mark Pacelle, 2001, *Designing Deployed Systems with COTS*, SKY Computers White Paper, www.skycomputers.com
5. Katz, D.S., P.L. Springer, R. Granat, M. Turmon, 1999, *Applications Development for a Parallel COTS Spaceborne Computer*, Third Annual Workshop on High-Performance Embedded Computing, Lexington, Massachusetts, http://pat.jpl.nasa.gov/public/dsk/papers/hpec99.pdf
6. Foreman, John, 1998, *On the Front Lines of COTS – Lessons Learned, Speculation for the Future*, Tutorial, Software Engineering Institute, Carnegie Mellon University, www.stc-online.org/cd-rom/1998/slides/tut2jforeman.pdf
7. Conmy, Philippa, John McDermid, 2001, *High Level Failure Analysis for Integrated Modular Avionics*, 6th Australian Workshop on Safety Critical Systems and Software, Brisbane, www.jrpit.flinders.edu.au/confpapers/CRPITV3Conmy.pdf

Passive Magnetic Support System for High-Speed Launch Sleds Assisting Martian Surface Departures†

John B. Barber, President, Modern Transport Systems Corp., barber.john@worldnet.att.net,
and Duane Barber, Modern Transport Systems Corp., menagerie@hevanet.com

Abstract

If humanity is to develop a significant and sustained presence beyond Earth's surface, the problems associated with extraterrestrial transport need to be addressed. Mars is often considered as a prominent site for human activity. Martian activities of serious magnitude would benefit from improved means of access for transport from the Martian surface to Earth, and to other destinations in the solar system. A technique being considered to assist launches from Earth's surface may be applicable to departures from Mars: high speed launch sleds, carrying a winged, Mars departure vehicle to a take-off velocity. Magnetic levitation offers an attractive support system for such sleds. Several types of magnetic suspension are available. Each has differing characteristics that influence their respective applicability to this function. Of these, the passive magnetic system possesses characteristics that appear to make it especially desirable for application to a Martian launch sled. Propulsion could be via reaction devices, making use of Martian resources. The low atmospheric pressure on Mars allows for high sled speeds with only moderate aerodynamic interactions or resistance, compared to Earth. At the same time, the presence of the Martian atmosphere allows for a lifting body ascent, reducing the thrust needed from the departure vehicle's propulsion system. Assisting the launch of Mars departure vehicles with such a high speed launch sled reduces the fuel the departure vehicles themselves must carry, and may help to enable a sustainable, intensive-use, extraterrestrial transport system.

Introduction

The limited degree to which humanity has to date established itself throughout the solar system is due in large part to the enormous difficulties faced in simply leaving Earth and getting around out there. Prospects for sustained human activities on bodies such as Mars are dim unless means can be developed to provide for access to and from there without the stupendous expenditure of resources our current methods entail. Although many feel that there will inevitably be a major human presence Mars, there is now little understanding about just what that might entail. Possibilities range from geological research bases to extraction of Martian resources for use elsewhere, to serving as a marshaling and logistics base for more widespread travel, commerce and exploration throughout the solar system. For this to occur in a meaningful way, though, transportation to and from the Martian surface must be facilitated.

Initially, our presence on Mars will be small and tentative, and considerations of large scale travel will probably not be an issue relative to other concerns that can be expected at that stage. If we are to progress to a more permanent and substantial presence, though, transport will emerge as a major consideration. Whether to ship Martian-based resources elsewhere, or to simply return Martian-bound cargo craft to their points of origin, as well as to facilitate passenger transport to and from the planet, it will be necessary to establish a Martian departure system that is practical and not excessively demanding of either Martian or off-Martian resources. The use of high speed launch sleds to assist vehicles in departing from the Martian surface may support that objective.

Potential Benefits of High-Speed Launch Sled

Considerable attention is being focused on alleviating the difficulties in leaving Earth's surface, with consideration being given to horizontally-launched lifting body vehicles as a possible means of improving the transport situation. If such a path is adopted for Earth departure, high speed launch sleds might be utilized to assist their take-off. A similar strategy could be beneficial for enhancing Martian transport.

There are, to be sure, significant differences between the Martian and terrestrial environments. The Earth has a large gravity field, leading to a high orbital speed, in the neighborhood of 26,000 fps, and a significant atmosphere, allowing for a lifting body type of ascent for a considerable portion of the necessary altitude gain. This latter factor alleviates the need for a vertical thrust significantly in excess of the sheer weight of a vehicle to simply achieve the desired orbital altitude. Mars, conversely, has a gravitational attraction only about $1/3$ that of the Earth, leading to a much more modest orbital velocity, about 11,000 fps, and a limited atmosphere to assist in ascent. Much less thrust is required to overcome the vehicle's weight and lift it from the Martian surface. It might be argued that these two factors render anything other than a conventional, vertical lift-off rocket technique as unnecessary for departing Mars's surface.

An assisted take-off from Mars, however, can potentially provide significant advantages to a Martian transport system. Shifting a sizable portion of the responsibility for velocity gain from the departure vehicle to a ground-based, captive launch sled, may be beneficial. And Mars' atmosphere, while considerably less than that of Earth's, is nonetheless able to provide lift for an aerodynamic body. Further, the lessened atmosphere means that high take-off velocities could be achieved on Mars' surface without the extent of aerodynamic effects that complicate such an action on Earth: dynamic pressure, shock wave blast and heating. A horizontally-launched vehicle, carried to a significant fraction of the required orbital speed on a powered sled, can utilize a much lower mass ratio of gross lift-off weight to burnout weight. This reduces the fuel that must carried by the launch vehicle.

Oxygen is in abundance on Mars in various forms, and fuel substances are available. Carbon monoxide, while not a terribly exciting rocket fuel, will work and can be obtained in abundance from the carbon dioxide atmosphere. It appears that water is likewise abundant, meaning that substantial quantities of hydrogen are also available. Thus, fuels for departure from the Martian surface are locally available. Even so, though, the extent to which their use can be minimized will ease the burdens on the local Martian economy.

Making use of an assisted launch can serve to reduce the demands on local fuel. A launch sled may be considered as a reusable, leave-behind first stage. By reducing the delta velocity that the departure vehicle itself needs to generate, it can serve to reduce the mass fraction of the departure vehicle significantly.

This has an additional beneficial effect. The advantage can be claimed by reducing the size of the launch vehicle. The launch vehicles themselves will either be produced on Mars, or produced elsewhere and shipped to Mars. In either case, being able to move the same amount of payload with a smaller vehicle will be of advantage. Conversely, if vehicle size is retained, the vehicle will be able to transport a larger payload from Mars. Either way, advantages will accrue.

Desirable characteristics for such a sled include:

• the ability to support the heavy loads associated with a fueled launch system.
• functionality at take-off speeds for the launch system.
• readiness for operation in minimal time with minimal labor.
• rapid deployment to support launches within short turn-around periods.
• minimal impact of launch process on track way.
• provide a smooth ride for the launch system during the take-off process.
• provide for graceful abort accommodation.

Several options exist for supporting a launch sled. Simple skids have been used in the past, but experience has shown that this technique is less than desirable. Problems of friction and vibration occur.[1] Steel wheels, such as railroad wheels, are possible, but steel wheels operating on steel rails require a very precise alignment if the ride quality is to be satisfactory. This in turn imposes significant maintenance demands on those responsible for the infrastructure. Further, wheels in the Martian environment, with an abundance of dust, may prove problematic.

Magnetically levitated systems (maglev) are frequently considered for this purpose, and have a number of potential advantages to offer over alternatives. Having limited or no physical contact with their guideway, they offer the opportunity for a convenient way to bring launch vehicles up to lift-off speed. The weight of the launch system would be spread over the length of the magnetic lifting elements, resulting in a far less severe load environment for both the lifters and their track way. Maglev systems can offer minimal running resistance, allowing more of the propulsive energy supplied to the launch sled to be used for actually accelerating the mass of the sled and the launch system, rather than for overcoming suspension system friction. Maglev systems can also offer a smooth and cushioned ride for the launch system, reducing the likelihood of damage to itself and its cargo.

Magnetic Levitation Technology
Several distinct maglev technology options are available for consideration for application to this purpose. The three most likely candidates include the traditional electromagnetic and electrodynamic approaches, and the more recently developed passive magnetic alternative. Each has a variety of characteristics and features that bear on their individual suitabilities for application to the process.

Electromagnetic
One of these three maglev technologies, the electromagnetic, works on the basis of magnetic attraction. It uses direct current electricity flowing in coils mounted on the vehicle to create magnetic fields that interact with steel levitation elements in its track to create a lifting force. The magnet units on the vehicle attempt to "snuggle up" to the underside of the track's steel levitation elements. This mode of lift is unstable, and must be constantly monitored and controlled to prevent either the vehicle from becoming overlifted and clamping up onto the steel of the trackway, or underlifted and falling free. Its configuration is somewhat limiting with regard to being able to resist upward lift of the launch vehicle's aerodynamic surfaces as the system increases in speed. It is essentially configured to provide lift upward, and as the system develops aerodynamic lift and perhaps goes into a negative weight condition, the electromagnetic suspension system may suddenly find itself totally unloaded and clamping upward onto its track surfaces from underneath. Thus, this technology would be difficult to adapt to the launch sled function, unless downward acting "reverse" lifters were also installed – at a significant penalty in additional weight and complexity.

Electrodynamic

The second category of maglev, electrodynamic, is a repulsion technology. With this concept, discrete magnetic fields are generated on board the vehicle and emanate from under its bottom. These fields interact with electrically conducting components embedded in the trackway underneath as the vehicle passes by, inducing electrical currents in these track way components. The net effect of this is a repulsive force that tends to lift the vehicle and suspend it above the track way. The greater the speed of the vehicle, the greater the lift force. There is an associated drag force that must be overcome by the propulsion system. Depending on the configuration, this drag force may become less of an issue at higher speeds. There are often instabilities associated with it that can lead to uncontrolled vibrations and rough riding characteristics. In particular, electrodynamic systems are not naturally well damped, and can be subject to excessive vibration.[2,3] Similar to the electromagnetic, it does not lend itself well to resisting upward forces due to the lift generated by the aerodynamic surfaces of the launch system as the speed increases towards that of take-off. A downward-oriented force system would need to be added to prevent premature lift-off of the system from its tracks, again with considerable additional expense and complexity.

Passive Magnetic

The third maglev alternative, passive magnetic, is an emerging technology. It has been developed by Modern Transport Systems Corp. (MTSC) in prototype form, in response to the perceived shortcomings of the electromagnetic and electrodynamic varieties. Although originally developed with more industrial uses in mind, such as the transport and handling of materials in factories and warehouses, ports and harbors, construction and mining sites, and similar applications, it possesses features which appear to make it highly suitable for application to a high speed launch sled for winged launch systems. The passive magnetic system works on the attraction principle, and is inherently stable in lift. It uses high strength permanent magnets mounted on the vehicle, interacting with steel elements placed in the trackway in one or more rails, in a sandwich configuration. Figure 1 illustrates the rail and magnet array configurations.

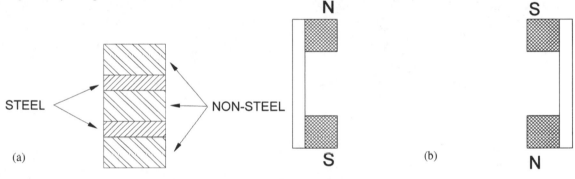

Figure 1. MTSC Magnetic Support System Components.
a. MTSC Magnetic Support Rail – end view.
b. MTSC Magnet Array – end view.

The magnet units hang to the sides of the trackway rails, and the magnetic fields are oriented such that as a greater weight is applied, the vehicle drops slightly until the corresponding magnetic restoring force balances it. It does not require an input of power for levitation, does not require the vehicle to be in motion for levitation to occur, and does not require gap sensing or other control systems in order to maintain stable levitation. Because the magnets hang to the sides of the rails, they seek to center vertically on the steel layers of the rails. They resist motion in both directions vertically, and thus provide equal resistance to both downward and upward forces. This causes it to restrain the lifting tendencies of an aerodynamic launch vehicle as it approaches take-off speed. Figures 2 and 3 illustrate these features of this technology.

The passive magnetic design provides a smooth and continuous levitation, minimizing vibration and shock to the launch vehicle during the take-off run. It can incorporate an inherent damping function, to minimize the effect of turbulence or other irregularities acting on the launch vehicle, and its trackway is of simple construction. It provides a simple, rugged and reliable support system.

The passive magnetic system does require lateral control. In MTSC's current prototype development, this is provided by steering wheels. Properly configured, these steering wheels bear very little load, except when the vehicle is passing through a curve, at which time they provide the necessary turning forces. This arrangement is suitable for low speeds on Earth. In the Martian environment, for the speeds likely to be of interest, an alternative steering principle is called for, such as a non-contact steering by means of opposing repulsive magnetic fields created by eddy currents.

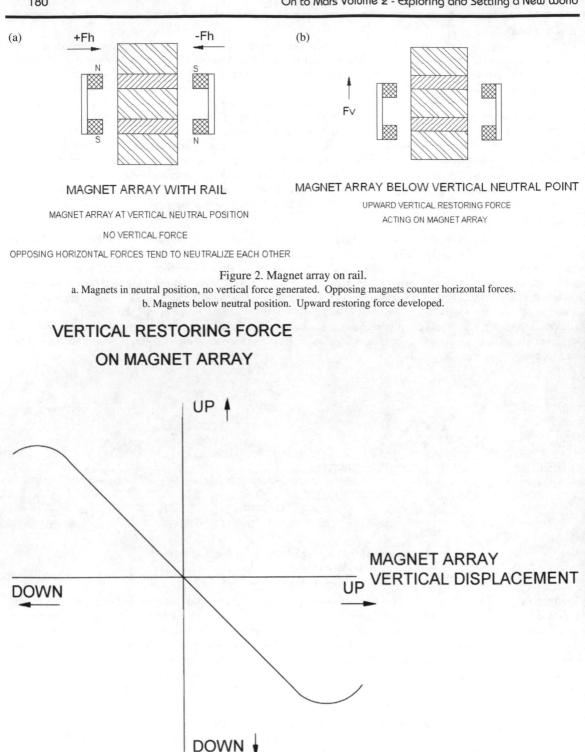

(a)

MAGNET ARRAY WITH RAIL

MAGNET ARRAY AT VERTICAL NEUTRAL POSITION

NO VERTICAL FORCE

OPPOSING HORIZONTAL FORCES TEND TO NEUTRALIZE EACH OTHER

(b)

MAGNET ARRAY BELOW VERTICAL NEUTRAL POINT

UPWARD VERTICAL RESTORING FORCE
ACTING ON MAGNET ARRAY

Figure 2. Magnet array on rail.
a. Magnets in neutral position, no vertical force generated. Opposing magnets counter horizontal forces.
b. Magnets below neutral position. Upward restoring force developed.

Figure 3. Relationship between vertical restoring force and vertical displacement. Vertical displacement in one direction causes an opposing restoring force, increasing with displacement.

Although the passive magnetic technology development to date has been limited to low-speed, modest-load prototype units, there do not appear to be significant obstacles to its application as a high-speed launch sled support technology. The levels of lift achieved to this point are on the order of 400 pounds per foot of rail. It is anticipated that this can be increased to a level of about 1,000 pounds per foot of rail, by both improvements to the efficiency of the magnetic lift circuit and by increasing the magnetic energy per unit length (i.e., bulking up the cross sectional area of the magnetic source units). Multiple rails result in a commensurate multiple of lift per unit length, and increasing the length of the magnets additionally increases the total lift. The low Martian gravity, approximately 1/3 that of Earth, helps in this regard.

Figure 4 shows photos of some of the prototype units that MTSC has constructed to prove and demonstrate the technology.

Launch Sled Characteristics

There are a number of configuration issues to address with a high-speed launch sled and track system:

- length of track
- geometry of track
- propulsion of the sled
- braking of the sled

Length of Track

The length of the launch sled track is determined by the necessary speed for take-off of the launch system, and the acceleration applied to it to propel it down the track.

Figure 4. Photos of MTSC Passive Magnetic system prototypes.

Although the characteristics of such a launch system itself are not established at this point, it is likely that useful take-off speeds will be in the range 2,000 to 4,000 fps. These provide a significant fraction of the kinetic energy necessary to achieve Martian orbit.

The acceleration of the launch system determines the necessary track length. Lower acceleration levels impose lesser burdens on the payload, an important consideration if items of some delicacy are being carried (e.g., people, sensitive equipment, etc.). They also place lesser demands on the propulsion system. However, lower acceleration levels result in longer track runs. For discussion purposes, a minimum acceleration of 1 g is assumed. For an upper limit, a maximum acceleration of 3 g is assumed. A 3 g horizontal acceleration combined with $1/3$ g of normal vertical gravity results in only slightly over a 3.2 total acceleration load. This value can be tolerated by most persons of reasonable health, and can generally be accommodated by most objects likely to be transported. These values result in a range of acceleration distances of about 4 miles to nearly 12 miles for a speed of 2,000 fps, and about 16 miles to 47 miles for a speed of 4,000 fps, in rough numbers.

It is apparent that the higher the speed, the more desirable it is to be able to utilize a higher rate of acceleration. Presumably, once at the desired take-off speed, the launch sled will cruise for some period of time while the take-off process takes place. The empty sled will then need to decelerate to a stop. These factors add to the necessary track lengths. For discussion purposes, it is assumed that the cruise time at take-off speed is 5 seconds, and the deceleration rate is the same as that for acceleration.

With acceleration, cruise and deceleration, the total track length varies from about 10 miles to 25 miles for a velocity of 2,000 fps, and from about 35 miles to nearly 98 miles for a 4,000 fps velocity.

Geometry of Track

For launch sled tracks, a popular conception is of a long acceleration stretch, followed by an upward curve so that the lift-off vehicle is catapulted into the air. In actuality, this is probably not a realistic scenario. As speeds increase, the radius of such a vertical curve becomes excessively large. For example, at a speed of

only 1,000 fps, in order to not exceed a 2 g centripetal acceleration, the curve radius must be no less than about 3 miles, and grows to nearly 12 miles at a speed of 2,000 fps. Achieving geometries of this sort requires extremely extensive site work. Hoping to find a suitable patch of ground, with something close to the desired slope geometry, in an acceptable setting, is most likely not a reasonable expectation. It is more likely that simply creating a straight and level track is the desired choice. What would then be desired is a site where several miles of relatively flat terrain exist, and where the necessary grading, cutting and filling operations would be minimal. Additionally, a location near the Martian equator would likely be desirable, to facilitate easy access to equatorial orbits.

Propulsion of Sled
The basic purpose of the sled concept is to provide a significant level of velocity for the launch vehicle. Thus, the sled needs a propulsion system of its own.

With a maglev system, non-contact propulsive techniques are typically used. Generally, these are linear electric motors of one sort or another. If significant masses are to be accelerated to the velocities of interest, a large amount of electrical energy is required. Energy requirements in the area of a gigawatt and more might be expected. A recent paper by Caprio, et al.[4] discusses aspects of this issue. Equipment must be installed to handle, control and switch it extremely rapidly, down the entire length of the track. Electrical current levels on the order of mega-amps may well be involved, with voltages in the range of kilovolts and higher. Massive cabling would be required. This has significant implications for system infrastructure cost and practicality. At speeds of several thousand feet per second, the launch sled is passing over the track extremely quickly and the required switching times are diminishing rapidly. Simply switching power flows of this magnitude on and off at the required frequencies, considering reactive effects, may be unrealistic.

An alternative approach to sled propulsion is the use of rocket engines, using locally available fuel materials. Oxygen is abundant on Mars, in the carbon dioxide atmosphere and in surface materials. The carbon dioxide atmosphere can yield carbon monoxide, a combustible gas. While not terribly exciting as a rocket fuel, it will work, and is available in abundance. Also, there are increasing indications that hydrogen may be present in sizable quantities, in the form of subsurface water / ice. Should this prove to be the case, this may represent a high quality fuel source. Combining hydrogen and carbon into methane for a fuel is an attractive option, also, as it achieves a higher performance than CO and is much easier to handle than liquid hydrogen.

Pursuing this avenue a step further, it might be feasible for the launch vehicle's own propulsion system to run during the take-off process, with propellant supplied live from the sled. Fuel and oxidizer for this portion of the vehicle's operation could conceivably be stored on the sled, and pumped to it during the acceleration run. Prior to the vehicle's actual separation from the sled and take-off, the propellant lines would be disconnected and the vehicle would revert to its on-board stores of propellant. This would be somewhat akin to aerial refueling, albeit taken to a more demanding level. It would reduce the space required on the vehicle for propellant storage, but simultaneously allow the use of the vehicle's own propulsion during the take-off phase. It might be noted that consideration has been given to aerial refueling of LOX, by Pioneer Rocketplane[5] in their development of a launch vehicle concept.

Figure 5 shows a conceptual illustration of the launch sled.

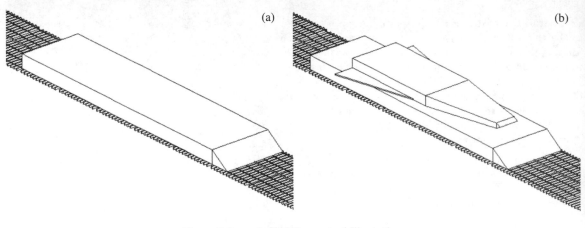

(a)   (b)

Figure 5. Launch Sled Conceptual Illustration.
a. Launch sled with track.
b. Sled with Mars departure vehicle.

Braking of Sled
Once the vehicle has taken off from the launch sled, the sled itself must be decelerated and brought to a stop, and then returned to its starting point. This requires a deceleration technique. Several methods are available, including

dynamic braking using specially placed magnets on the sled which interact with conducting elements in the trackway to create electrical currents which are dissipated resistively (i.e., eddy current braking). Another possibility is to use active linear electric motors in the trackway, which would drive the launch sled's speed down to zero. Retrorockets constitute an additional possibility.

Deceleration on Earth could make use of aerodynamic resistance, but the atmosphere of Mars is probably too thin to be of significant use for this purpose. However, Mars does have one material in abundance that perhaps could be used for energy dissipation: dirt. On many highways in mountainous areas on Earth, runaway truck ramps are provided. These consist of a long bed of sand or fine gravel, several feet deep. An out-of-control truck is simply driven into this, and plowing through the material it slows rapidly. Perhaps something similar could be used on Mars, using Martian sand.

A factor that must be considered is the possibility of an aborted launch, where the vehicle for some reason does not take off when the sled has reached the proper velocity. Now, the sled must bring both itself and the fully fueled launch vehicle gracefully down to zero speed, a process requiring much more braking effort than for merely an empty sled. This could have the effect of lengthening the launch track, to account for the added energy dissipation necessary.

To account for this possibility, it may be desirable to utilize a baseline braking system for normal deceleration of the empty sled, such as a combination of retrorockets and eddy current braking, with an additional option in the case of an abort, such as sand-based energy dissipation. For normal braking operations, only the baseline system would be used. In the event of an emergency, such as an aborted lift-off, actuators on the sled would extend into a sand trough to provide the additional deceleration necessary to achieve a stop in the available track length.

In addition to the need for a means of decelerating the sled to a stop, the sled must be returned to its starting point. Thus, a reverse propulsion system is necessary. This might be accomplished with a modest capacity linear electric motor.

Installation

At first glance the distances involved for a launch sled may seem extreme for consideration for implementation on the Martian surface. Going to the highest feasible acceleration rates helps in this regard. Even so, at an acceleration of 3 g, considering lift-off cruise time and deceleration, track lengths of perhaps 35 miles might be required. Although this may seem a daunting challenge for the Martian environment, with automated construction practices and equipment, and making use of Martian materials, they may well be quite achievable.

The track bed itself is simply Martian regolith, suitably leveled. There has been considerable thought given to use of lunar materials for construction of bases and facilities on the Moon, including lunar concretes. It would seem that similar use can be made of Martian materials. With the presence of Martian water, a Martian concrete becomes more feasible, perhaps, than lunar concretes on the Moon. The track itself for the system consists primarily of steel and aluminum, both of which are likely available in considerable abundance on Mars (the combination of plentiful Martian iron and carbon provide the base materials for steel, and the presence of additional metallic elements there allows for alloying to produce a variety of steels). Once a human presence of any size is established on Mars, it is conceivable that processing plants would have been established there to produce materials for habitats and other purposes. One such purpose could be the construction of a Martian launch sled track.

Much of the actual construction could perhaps be performed by robotic equipment. Most of the heavy work involved would likely be in the realm of dirt moving, to achieve a level path for the trackway. This is an activity well suited to robotic machinery. Development of this robotic dirt moving capability could well result in useful applications on Earth, for that matter, in such areas as large scale surface and subterranean mining. And in fact, efforts are currently under way in this regard. A team at the Commonwealth Scientific and Industrial Research Organization, in Kenmore, Australia, is developing autonomous equipment for handling the "load / haul / dump" procedures in mining operations.[6]

There is some advantage to be gained from even a relatively low-speed launch sled, with a short track, although the extent of advantage increases as the speed and track length increase. Perhaps a short section could be built first, and used, while construction continued with additional segments. As the track lengthened its utility would grow. A "build-as-you-go" philosophy could be employed.

A launch sled of the type discussed here must be capable of supporting a sizable load, and transporting it down the acceleration trackway without causing undo damage to the trackway or the surroundings. Sled-borne launch systems considered for Earth launch are likely to weigh in the neighborhood of as much as 1-2 million pounds. On Mars, due to the lesser orbital speed required, the launch vehicle weights could be substantially less. Combined with the reduced gravity of Mars, the weight burden to be borne by the sled and its track could be quite small in comparison to that experienced on Earth. Lunar soil has been found to have substantial bearing strength, and initial indications are that Martian soil is similar in this respect. So its ability to support the trackway should be satisfactory.

Thus, the construction of a track on the Martian surface, some 30-40 miles in length, might not be unreasonable at all.

Other Uses

The discussion here has focused on using a magnetic levitation technology to assist in Martian departures. The same technology could be harnessed to assist in providing a means for traveling on the Martian surface. On Earth, maglev trains are envisioned as a way of traveling between cities, as well as serving as urban transit systems. When human activities on Mars reach a stage where there are permanent, widely separated settlements, with substantial interchange and travel between them, it may be desirable to have a relatively speedy, surface-based transport system to facilitate these activities. Advantages that maglev brings for these purposes include the ability to travel at speeds considerably higher than likely feasible by wheeled systems, and a relative immunity to Martian dust. In these cases, propulsion would likely be electrical, using linear motors.

Conclusion

A high-speed sled for assisting the departure of horizontally-launched, lifting body vehicles from the Martian surface appears to be an attractive way to reduce the burden that Martian transportation imposes on the available resources. Magnetic support for such a sled seems to be a preferable technique, and the passive magnetic support system, such as that developed by Modern Transport Systems Corp., seems especially applicable to this purpose. Initial consideration of the concept suggests that, while the form and characteristics of such launch vehicles are yet to be determined, there are design options for a magnetically supported sled capable of accommodating a sizable envelope of likely designs. By facilitating the transport of people and materials to and from Mars, and by reducing the logistical burdens that such transport imposes, a passive magnetic-based launch sled may be able to play a role in helping open the door for a more significant and permanent presence in space for humanity.

References

1. Minto, D.W. and Bosmajian, N.; *Hypersonic Test Capabilities at the Holloman High-Speed Test Track*, published in *Advanced Hypersonic Test Facilities*, Vol. 198, Progress in Astronautics and Aeronautics, edited by Lu, F.K. and Marren, D.E., AIAA, 2002.
2. Kim, I.K.; Kratz, R.; and Doll, D., *Technology Development for US Urban Maglev*, presented @ Maglev2002 conference, Lusanne, Switzerland, August, 2002.
3. Rote, D., *Passive Damping in EDS Maglev Systems*, presented @ Maglev2002 conference, Lusanne, Switzerland, August, 2002.
4. Caprio, M.T.; Pratap, S.; Walls, S.A.; and Zowarka, R.C.; Center for Electromechanics, University of Texas, *Linear Electric Motors for Aerospace Launch Assist*, presented @ Maglev2002 conference, Lusanne, Switzerland, August, 2002.
5. Wolf, R.S., *Pioneer Rocketplane Reusable Launch System*, AIAA-2000-1611.
6. *Caves of Steel, The Economist* magazine, November 23, 2002, p.73.

//

Tools and Equipment for Mars Missions

//

Design of an Autonomous Harvest Robotic System and a Biomass Chamber on Mars†

C. Ham, R. Johnson, L. Retamozo, R. Patil, H. Choi and J. Brandenburg
Florida Space Institute / University of Central Florida

Abstract

In situ food production will become essential in the expansion of human exploration on the Red Planet. When colonizing Mars, consumables resupply from the Earth will become not only very costly due to constraints on mass and volume, but also may cause potential psychological problems to astronauts having a diet mostly of packaged foods. The diverse technologies required for the biomass production on Mars, providing food and advanced life support operations have not yet been adequately integrated and demonstrated. In this paper, we develop an autonomous harvest robotic system and a biomass chamber on Mars to supply fresh vegetables and fruits as a dietary supplement for the crew. The unique feature of our approach is the use of robotic systems to replace human labor and provide optimal environmental conditions for plant growth. The system will provide an optimal and autonomous biomass production capability to maximize the ability to grow plants with minimal human input. First, this paper presents development of a biomass chamber and selection of vegetables and fruits that can be suitably grown in controlled environments. Then, an advanced robotic system is proposed in order to maintain the biomass chamber autonomously. It harvests, transports and stores the fruits via remote operator commands. Sensors mounted to the robotic system will monitor the biomass chamber environments such as temperature, humidity, light intensity, etc. An integrated control system is also presented that provides stable operation of the robotic system assuming optimal growth and health of the plants under production. It provides autonomy, monitoring, diagnosis, fault-recovery and self-learning execution. Finally, we introduce our prototype system that has been operated on a near-continuous basis in the Controlled Ecological Life Support System facility at Kennedy Space Center.

1. Introduction

Ways to reduce the risk and cost of traveling to Mars have been well examined throughout the last two decades. One cost reduction method is to reduce the mass requirements and to provide the food and life support systems through bioregenerative power utilizing a biomass production chamber (BPC). The BPC is an essential structure. It would aid spacefaring crews in the generation of oxygen and food. The BPC also provides critical first steps in any terraforming venture. For any such efforts, crew self-sustainability is critical.

Reliable technology for biomass production in space that will provide food has not yet been developed. This system, in order to be effective, will need to use robotics to augment mechanization or replace human labor. The proposed research will evaluate current technologies and identify critical improvements necessary for the creation of an Autonomous Biomass Production System (ABPS) that uses robotic systems. In space or on other planets a controlled environment using ABPS is required to successfully cultivate food plants. Without an autonomous system, crops may not be uniform and could not be grown in non-human-rated environments, and may preclude human interaction. The ABPS can solve many problems associated with variability and accessibility of plant growth chambers, moreover such system can reduce the psychological burden of astronauts for harvesting and management. Such a system can maximize effective time utilization of astronauts. Consequently, there is a timely need to build and test ABPS in order to develop algorithms that will maximize the ability to grow plants autonomously. The Florida Space Institute and the University of Central Florida have actively participated in the BPC research activities at NASA performed at KSC on a regenerative life support system using hydroponics plant growth in a closed environment.

2. Mars Environmental Constraints

Requirements for the landing of initial NASA missions to Mars include a position within ±15 degrees of the Martian equator. Factors to consider in this region include temperature, pressure, solar irradiance and dust storms. Nevertheless, assumptions for the BPC will require slightly higher temperatures at the lower altitudes and that seasonal variation will be insignificant. Consequently, the expected daily temperature range selected for design purposes is 150° K to 300° K. However, depending on surface elevation, the pressure can range from 2 to 10 mbar. For the purposes of the BPC, atmospheric pressure will be assumed to be near vacuum. Daily winds have been recorded to be only 0.5-0.9 m/sec during the day and 4.5 m/sec at night. However, both localized and global (originating in the Martian Southern Hemisphere) dust storms can reach up to 25 m/sec. In the aphelion solar irradiance will be 473 W/m^2, while the perihelion irradiance will be 718 W/m^2. Thus, the BPC will be designed to withstand the worst case environmental scenario.[1]

2.1 Physical Limitation Concepts

2.1.1 External Structure Requirements

The external structure must be deployed from a manned or unmanned vehicle and must have an internal frame structure that can be stowed in a tight payload compartment, from which it can be deployed automatically. Due to the mass and power savings, transparent materials are considered as a good option, but should allow maximum light (400-700 nm) transmittance and be able to withstand ambient (Martian) UV radiation and other environmental parameters. Structural geometry should consider factors such as lighting, harvesting efficiency, orientation with regard to solar movement across the sky, and capability to withstand pressure differentials (from inside to outside the BPC could range from 10 to ~50 kPa).[2]

2.1.2 Plant Growth Area Structure Requirements

Plant growth will be implemented providing adequate volume to accommodate crop growth and any projected materials handling, and be capable of withstanding the rigors of the Martian surface environment. Structural components should have minimal mass and be able to be stowed and deployed effectively.

2.1.3 Water and Nutrient System Delivery System Requirements

A solid media or fluid system will be implemented for providing water and nutrients to the plant root zone; management of mineral nutrients and root zone aeration should be considered. Similarly, a system that manages the mineral nutrients recollection and root zone aeration should be considered.

2.1.4 Lighting System Requirements

In general, plants need a light range transmittance in the 400-700 nm range in order to sustain life. The incident light at the surface of Mars is a half of the incident light on Earth. Hence, a number of lighting schemes may be considered, that include artificial lighting systems. For the BPC plants to perform their functions properly, a mid-day intensity of 125 W/m^2 (400-700 nm), or about 500 mmol/m²•sec photosynthetic photon flux (PPF) is sought, with minimal intensities of at least 50 W/m^2 (~200 mmol/m²•sec) maintained at least for a 12 hour cycle each Martian day (minimum of ~2 MJ/m²•day or 8 mol/m²•day).[3]

2.1.5 Atmospheric Composition / Conditioning Requirements

Ventilation, temperature control, gas composition and relative humidity are important parameters in providing the appropriate environment for plant growth. The design concept must address the controls for providing a satisfactory environment. Atmospheric management will require separation and storage of photosynthetically generated oxygen (O_2), and systems or concepts for restoring carbon dioxide (CO_2) consumed by the plants.

2.1.6 Materials Handling Requirements

Depending on the management concept proposed, automation of plant harvesting and replanting might be required. Harvesting of crops would require removing plant materials for possible dehydration and storage. Edible materials might be separated and stored for crew arrival, or systems might be designed for human tending to reduce the need for automation. Similarly, replanting could be automated or operated as a human-assisted operation.

2.2 Temperature Control

For a Mars habitat, the cooling and heating requirements are major factors. During extremely cold night temperatures the BPC may require supplemental insulation schemes, depending on structural characteristics (e.g., nighttime covers). There may be a need to dissipate heat from the lighting system, while heat input may be required during the dark cycles. Thus systems that can collect, store and distribute waste heat should be included in the design concept.

2.2.1 Environment Temperature Range Discussion

According to the 1996 Pathfinder mission data, temperature near a 40° North latitude during a Martian day ranges between 200° K and 259° K. Nevertheless, assumptions for the BPC will account for slightly higher temperatures at the lower altitudes. Seasonal variation will be insignificant. Consequently, the expected daily temperature range selected for design purposes is 150° K to 300° K.[3]

2.2.2 Temperature Control Devices Utilized (see Table 1)

Polyurethane foam walls will serve as heat exchangers. Air circulation requires two 125 W and two 150 W blowers. The blower capacity is based on 246 m^2 of plant shelving. The air handling system provides from three to four air exchanges per minute, with air velocities ranging from 0.1 to 1.0 m/sec. Heat rejection and humidity control will be accomplished by chilled water coils located at the outlets of blowers. The condensation that occurs on the coils will help to monitor the evaporation rates. Atomized streams of water implemented directly in the air stream provide supplemental humidification if needed.

Subsystem	Range of Operation
Air Revitalization System	
Oxygen	18.5-23.45%
Carbon Dioxide	300-5,000 µL/L
Chamber Pressure	101Kpa
Ventilation and thermal control	
Air Temperature	15-35 °C
Relative Humidity	70-85%
Air Velocity	0.1-1.0 m/sec
Leak Detection and Control Leakage Rate	1% of the chamber volume/day

Table 1. Atmosphere Supply and Control Requirements.

2.3 Requirement of Bio Production (Crop Selection) (see Table 2)

Beans	Soybean	Carrots	Grasses	Italian Calbrese
Dwarf Rice	Tomato	Algae	De Cicco Green Broccoli	Spinach
Lettuce	Arabidosis Thaliana	Snowball Cauliflower	Asparagus	Potato
Amaranthus	Cabbage	Beets	Wheat	Ivy
Strawberries	Onion	Oats	Barley	

Table 2. Biomass Production Candidates.

The temperature range and the other physical conditions of Mars play a vital role in plant selection. A sample list of twenty-five acceptable plants has been collected which favor temperature range between 288.15° and 293.15° K. These plants have been selected for final review due to the fact that they are hardy species with high resistance to frost.

Among these candidates, sixteen will be brought to Mars in the bedding material as seeds. The growth of these plants will occur after the deployment of the BPC vehicle. Oxygen content needed inside the BPC is 10%-14% of the total air volume. The plants were chosen on the following basis:

1. The plants are high yielding and fast growing.
2. The nutritious biomass of carbohydrates and proteins is relatively high.
3. The ratio of edible to inedible portion of the plant is large.

4. The food processing requirements are minimum.
5. The likelihood of astronaut psychological and aesthetic acceptance is greater than for other selected plants.
6. They are not tall, so multiple levels can be grown.
7. Horticultural requirements are easily met through robotics.
8. Their environmental requirements are easily met with minimum power usage.
9. A large amount of information is available about each plant, including genotypes.

The plants finally selected after detailed study were: wheat, dwarf rice, soybeans, potatoes, dry beans, strawberries, lettuce, tomatoes, onions, broccoli, spinach, beets, baby carrots, amaranthus and arabidopsis thaliana.

3. Robotic Harvesting System

3.1 Control requirement
Processes and components in the system must be controlled continuously. Plant growth and health must be monitored in the presence of dynamic variations. Typically, control and monitoring are done conventionally using standard, off-the-shelve control / sensor modules. The problems arising in monitoring dynamic states need to be resolved.

Advanced controls (such as nonlinear robust control, adaptive control, etc.) are ideal candidates in the design of an autonomous system operating in an unknown and changing environment. However, space-bound computers have very limited computational power for analyzing all data in real time and synthesizing all control signals.

3.2 Dexterous End Effector (ALSARM) (see Figure 1)
The ALSARM is composed of a three degrees-of-freedom robotic manipulator with automated control. The ALSARM is to be equipped with an end effector that is capable of retrieving samples from the BPC. the end effector (EE) system is required to grip, cut and move plant material. To achieve these goals, the EE will utilize four motors to control the pitch, yaw, and roll motion along with gripping. The cutting of the vegetation sample is achieved through passive means.[5,6]

One of the requirements of the end effector is to harvest different types of fruit. This has led FSI to adopt the changeable end effector; a cone and adapter method is proposed for easy exchange of the end effector. The cone of each end effector will be compatible with a single adapter in the manipulator.

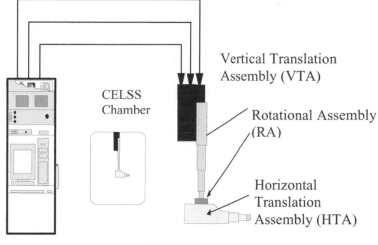

Figure 1. ALSARM design and control.

3.3 Manipulator
The robot manipulator has two telescoping arms, vertical and horizontal translation assemblies, and one rotational joint. The ALSARM end effector is an extension of the robot manipulator's horizontal telescoping arm. The end effector is supposed to be mounted at the end of the telemag (telescoping arm mechanism). The end effector that goes on the telemag of the ALSARM is capable of retrieving samples from the BPC.

3.4 Biomass Production Chamber Monitoring System
This mechanism is to operate in a Biomass Production Chamber (BPC). The BPC is an enclosed environment used for plant growth and oxygen regeneration. It is composed of two plant chambers. The BPC is to be sealed so that water vapor and air do not escape and the water can be recycled. The scientists in this program want to eliminate personnel entry, reduce the leak rate of air and water vapor, and obtain very consistent measurements inside the chamber. Sensors mounted to the end effector will be capable of measuring temperature, infrared temperature, relative humidity, air speed and light intensity. A computer is already programmed to take these measurements at hundreds of different points in the BPC. Another addition to the ALSARM is the placement of a viewing apparatus for manual direction. There have been for different designs for the end effector.[8,9]

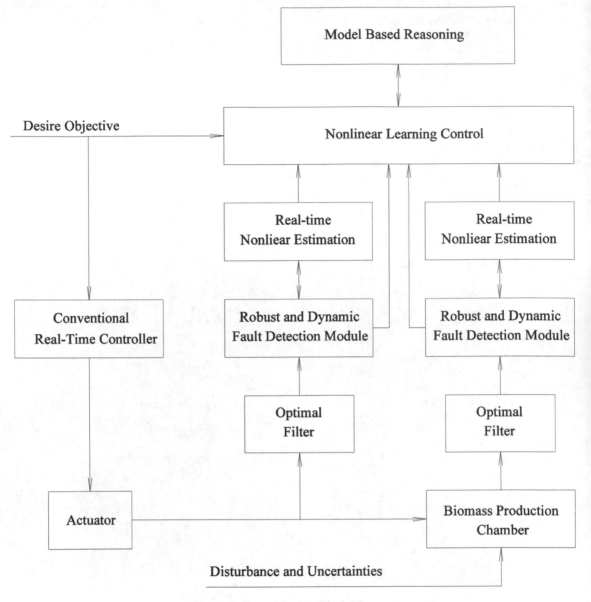

Figure 2. Control System Block Diagram.

3.5 Control System for Autonomous Harvesting System (see Figure 2)

By overcoming two major obstacles, the proposed intelligent control framework can achieve the following technical objectives:

- Robustness in a changing, uncertain environment: The system is capable of identifying external disturbances and operating conditions using the standard sensors and computational power on board, which can be done using advanced nonlinear robust control algorithms. The control system must also be robust so that all uncertainties within actuator capability can be compensated for.
- Fault tolerance: Upon automatic detection of a failure of any component, the control system can maintain system functionality by switching to the redundant backup. More importantly, in the case that no more redundant part is available, the control system is capable of automatically adapting and achieving the best possible performance.
- Autonomous operation and self-reconfiguration: Robust identification and control are integrated so that the system is capable of self-activating through environmental diagnosis, self-calibrating, self-deploying, and self-adjusting by intelligent reasoning. Upon detecting a fault, the robust estimation module will provide sufficient information for transient control after excluding feedback from faulty sensors.
- Intelligence: Model-based reasoning capability will be an integrated part of monitoring, diagnosis and recovery.

In summary, the proposed intelligent control system will reduce the cost, enhance reliability, and increase safety, so that the BPC can be operated in an environment that could change significantly. The block diagram of the control system considered in this research project is shown in Figure 2. The system basically consists of a set of actuators that drive the robotic system of the BPC. The system variables (or the states) are monitored by sensors. The measured signals will pass through optimal filters (such as a typical Kalman filter). The proposed real-time management configuration combines state-of-the-art tools. Specifically, optimal filtering, nonlinear and robust estimation, nonlinear fault detection, nonlinear learning control, and the expert module are integrated in such a way that they complement one another. Analysis and design will be done using nonlinear system theory (Lyapunov direct method), estimation and learning, and artificial intelligence techniques based on heuristic knowledge.

3.6 Unique Features

The proposed intelligent control framework has a hierarchical structure and consists of the following layers / components:

1. Local control at the process / component level (bottom and local level).
2. Discrete monitoring and fault recovery control (intermediate and local level).
3. Individual estimation and monitoring devices (intermediate and local level).
4. Nonlinear learning control (intermediate and regional level).
5. Model-based reasoning for intelligent control (top and global level).

The Biomass Production Chamber (BPC) at NASA's Kennedy Space Center has been operated on a near-continuous basis for six years, providing baseline data for using plants in closed life support systems in space. Total biomass yields of the different crops (wheat, soybean, lettuce, and potato) were strongly dependent on total lighting provided to the plants and generally close to anticipated values based on university research and preliminary growth chamber trials; however, edible yields and harvest index have sometimes fallen short of anticipated values.

Based on this analysis of closed system plant growth, it is shown that a well-developed ABPS robotic system can effectively reduce crew time and increase yields by having a consistent process to improve crop production.

4. Transportation and Storage

4.1 Transportation (see Table 3)

Transferring harvested fruits to the storage area is the final step and also an important task for the autonomous harvesting system. The transportation system should consist of an actuation system that drives the conveyor belt. An electrical actuation system has advantages over pneumatic and hydraulic systems because of lower pressure and gravity conditions in the chamber. The ALSARAM arm will harvest the fruit and will place the harvested fruit on the transportation system, which will carry the fruit to the storage system.

Function	Description
Safety of fruit	No damage during transportation
Adaptable	It should adjust it self according to the type of fruit or vegetable to be harvested
Exit	Gentle drop to storage box
	Harmonize with storage box movement
Reliability	Fruit should not be stuck
Sorting	Fruit Sorting and storing should be automated online
Inspection	Damaged or fruit with diseases should be detected automatically
Cleaning	Cleaning inside the passage in case of emergency

Table 3. Functional Description of Transportation system.

4.2 Inspection

An inspection system is required to detect the damaged fruit or fruit with diseases, which can cause health problems to the astronaut. The inspection system should be placed in-line on the transportation system to detect flaws before the fruit is stored.

4.3 Sorting and Packing

A sorting mechanism should be implemented that will sort varieties of fruit harvested. This might be useful for astronauts to easily identify the fruit they are looking for. A packing mechanism should be implemented before the storage section that will pack the harvested fruit. The packing system will preserve the fruit for longer periods of time.

4.4 Storage and Inventory

Normally, an artificial refrigeration system is required to preserve the fruit, but we can utilize the natural Mars environment for the refrigeration system. By using the natural environment of Mars and implementing a temperature control unit, it would be possible to maintain the temperature of the storage system.

A storage inventory system will be implemented in the storage system that will maintain the fruit inventory. This system can be automated; if fruit available is less than the target inventory level then a signal for next harvesting cycle will be given.

5. Overall Control Strategy

5.1 Temperature

Polyurethane foam walls will serve as heat exchangers. Air circulation requires two 125 W and two 150 W blowers. The blower capacity is based on 246 m^2 of plant shelving. The air handling system provides from three to four air exchanges per minute, with air velocities ranging from 0.1 to 1.0 m/sec. Heat rejection and humidity control will be accomplished by chilled water coils located at the blower's exits. The condensation that occurs on the coils will help to monitor the evaporation rates. Atomized streams of water implemented directly in the air stream provide supplemental humidification. The pathogen filtering system will be composed of electrostatic precipitators, which remove small debris, and coarse filters, which remove large debris, to prevent contamination of air ducts.

5.2 Humidity

Humidity in the Mars BPC needs to be controlled and will be monitored by means of a relative humidity (RH) sensor, which, when connected to a circuit, provides on-chip signal conditioning. These sensors contain thermoset polymers that interact with platinum electrodes and allow an interchangeability of +5% RH, with stable performance. The sensor will be located in the BPC ceiling and it will operate over the temperature range of 233.15° K to 358.15° K.

5.3 Pressure

Pressure will be measured and controlled by using a sputtered thin-film pressure sensor, which is stable even in extreme operating conditions. An Entran EXPT high temperature pressure sensor would be recommended. It has an operating range of 328.15° K to 523.15° K and a measuring range varying from 233.15° K to 423.15° K. The pressure sensors will be placed at the end of the air vents detecting air speed. If the air speed matches the desired speed, then no cleaning of the ducts is required.

5.4 Light

In order for the plants to receive natural sunlight, an Aerogel top round window has been chosen as the best option, because of the mass and power savings, which will allow maximum light of 400-700 nm transmittance and withstand the Martian ambient UV radiation, as well as the other environmental parameters addressed above. Additionally, to ensure the appropriate functionality of the plants, artificial illumination will be feasible by using 400 W high-pressure sodium lamps (six per plant area), which yield an average photosynthetic photon flux of 1,500 µmol/m^2•sec when operating at full power. The lamps are powered by dimming ballasts, which allow variable light levels for each crop area, which are controlled by the chambers data acquisition recording and control system. The crop will be separated from the lamp bank by means of a polycarbonate plastic sheet barrier. Because the sodium lamps are pressurized, concern arises when transporting them in a cargo hold exposed to vacuum. This is why light emitting diode (LED) banks will be implemented, since they are unpressurized and can produce the necessary light for plant growth. LEDs generate less heat and they are useful for longer missions, unlike incandescent and fluorescent bulbs.[5]

5.5 Atmospheric Gas

In order to maintain effective plant production as well as plant waste processing, the levels of atmospheric gases (oxygen and carbon dioxide) should be controlled. Therefore a system that can generate oxygen on demand, filter out carbon dioxide, and replace or remove the nitrogen is required. Replacement oxygen can be from two sources. First oxygen is a byproduct of photosynthesis, the plants produce a certain amount of oxygen and the crew uses some of that, then the excess gas can be separated using a commercial separator and stored for future use. The second method of oxygen generation involves separating elemental oxygen from gases in the Martian atmosphere. This process is called the Oxygen Generator System (OGS), which receives, compressed CO_2 and extracts oxygen. CO_2 coming into the OGS from the Mars Atmospheric Acquisition and Compression system will be electrolyzed at a very high temperature (750° C), causing oxygen ions to be stripped away from the carbon dioxide and will be filtered through crystal zirconia. The O_2 would then be pumped to a cylindrical tank used as storage unit. The storage unit would have pressure monitors to inform the computer whether or not it is full. If it is full then the computer overlooks the storage unit and O_2 would be transferred to another unit. The plants will use CO_2, which will be need to be replenished occasionally. A simple sensor implementation based on CO_2 detection can be introduced. Once enough CO_2 has been pumped in, two air vents located laterally to the CO_2 "pump tube" would turn on as result of the computer's instructions and spread the CO_2 uniformly throughout the BPC.[11]

5.6 Ventilation

Air velocity will dissipate the heat generated by the lighting systems. The surface temperature was assumed to be 400° K and temperature of the air stream to be 293° K. Nusselt number was evaluated for the velocities ranging from 0.10 m/sec to 5.00 m/sec. After calculations, a direct correlation between average coefficient of convection and velocity was found. Knowing the heat generated by the bulbs, an average convection coefficient can be calculated, yielding to an approximate velocity of 3.3 m/sec.

Four vents located on four sections of the top edge of the BPC will be essential. Between two of the vents will reside the CO_2 intake system. Air ducts will run along the ceiling of the BPC starting from a main pump on the outside of the BPC and ending at each vent. At some point the vent will have to split into four ducts, which will run along the inside of the dome in circular fashion. The pressure systems will detect how much air travels through the ducts.

5.7 Plant Growth Monitoring

The addition of the concept of Health Monitoring (HM) into existing automated monitoring and control systems will provide a real-time intelligent command and control system, which has the capability to monitor and observe transient behavior along with the dynamic parameters of the systems being tested. Current test capability cannot measure the dynamic behavior of "system under test" (SUT) in real time. Abnormal dynamic properties are indicators of an out-of-tolerance performance of the SUT; they can be a predictor of impending failures in those systems. This feature adds a new dimension to existing test control mechanizations that will greatly enhance the accuracy of the system state which, in turn, increases the reliability of test and evaluation processes over those currently in use. This attribute will speed up diagnostic analysis to seconds rather than minutes or hours, thus reducing significantly fault detection and diagnosis times. The system will periodically capture images of growing vegetation. By applying some image processing algorithms, we can approximate physical size, root length, and color histograms of the vegetation. With periodic size monitoring, the growth rate can be calculated in both root and stem. We can also detect some plant irregularities, such as disease, by comparing the color histogram with the standard chart or model.[12]

The target set point and the overall top-level control scheme is as shown in Figure 3.

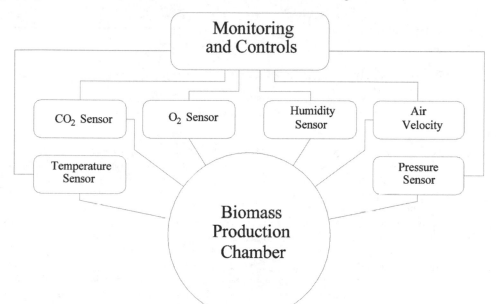

Figure 3. Overall top-level control scheme.

Parameter	Set Points
Temperature	10 to 30°C
Relative Humidity	40 to 90%
CO_2 Partial Pressure	0.1 to ~3 kPa
O_2 Partial Pressure	> 5 kPa
Light	400 to 700nm
Inert Gas Composition	Optional
Ethylene Gas	< 50 ppb equivalent at 100 kPa total pressure

6. Conclusions

The innovative control framework provides autonomous operation of the robotic system in the BPC. The autonomous harvesting system and control strategy provides synchronization, which is suitable for a compromise between the device's complexity and its manipulative ability. The autonomous harvest robotic system and biomass chamber will be a multi-tasking robotic system unit which will harvest fruit and measure temperature, relative humidity, air speed and light intensity with overall management and maintenance. This mechanism eliminates human intervention in BPC, avoiding contaminations and leakage of foreign elements, which allow consistent experimental procedures inside the chamber. These mechanisms can reduce the psychological burden on the astronaut associated with harvesting and management. The study of such a mechanism enhances our understanding of requirements for life support manipulation in the close test beds and it provides the baseline for space applications (i.e., International Space Station, Mars, etc.).

References

1. Hoffman, S.J., and Kaplan, D.L. (1997), *Human Exploration of Mars: The Reference Mission of the NASA Mars Exploration Study Team*, Exploration Office, National Aeronautics and Space Administration, Lyndon B. Johnson Space Center, Houston, Texas, July, 1997. This is version 1.0. This document may be found on the Internet at: http://exploration.jsc.nasa.gov/explore/explore.htm

2. Drysdale, A.E., Maxwell, S., Hanford, A.J., Ewert, M.K., (1999), *Advanced Life Support Systems Modeling and Analysis Reference Missions Document*. CTSD-ADV-383, JSC-39502

3. J. Appelbaum, G.A. Landis and I. Sherman. *Solar Energy on Mars: Stationary Collectors, Journal of Propulsion and Power*, Vol. 11 No. 3, pp. 554-561. Available as *NASA TM-106321.1993*.

4. Boynton William, *Mars Mercator Map of Neutron Levels*. Mars Odyssey Probe. University of Arizona. 2002

5. Building Technologies Program. *Lighting Systems. Advanced Lighting Systems*. 1994 Annual Report. http://eande.lbl.gov/BTP/pub/annrep94/Lsals.html

6. Drysdale, A.E., Thomas, M.M., Fresa, M.C., and Wheeler, R.M. (1992), *OCAM – A CELSS Modeling Tool: Description and Results*. SAE Technical Paper 921241. Presented at the 22nd ICES

7. Drysdale, A.E. and Grysikiewicz, M. (1996), MDSS-KSC Life Sciences Project Annual Report for 1995.

8. Drysdale, A.E. and Hanford, A.J. (1999), *ALS SMAP Baseline Values and Assumptions Document*. JSC 39317. 6/18/99

9. Dave Desrochers, Zhihua Qu, and Apiwat Saengdeejing, *OCR Readability Study and Algorithms for Testing Partially Damaged Characters*, 2001 International Symposium on Intelligent Multimedia, Video and Speech Processing, Hong Kong, May 2 – 4, 2001.

10. Apiwat Saengdeejing, Zhihua Qu and David Huibregtse, *Robust OCR System for Reading Optically Partially Damaged Characters*, The 13th International Conference on Computer Applications in Industry and Engineering (CAINE-2000), pp.128-131, Honolulu, Hawaii, November 2000.

11. Z. Qu, *Robust Control of Nonlinear Uncertain Systems*, John Wiley and Sons, Interscience Division, 1998.

12. B.C. Williams, *Model-Based Autonomous Systems in the New Millennium*, Proceedings of AIPS-96, 1996

//

A Modular Multi-Function Rover and Control System for EVA†

Stephen McGuire, Joel Richter and Kevin Sloan
The Penn State Mars Society, www.clubs.psu.edu/mars

Introduction

As technology and computing power increase at extraordinary rates, our ability to effectively explore our solar system increases to new levels. The immediate future will see the continual development of robotic exploration as our primary means of exploring other planets, and more specifically, Mars. Once the time does arise for mankind to again push his frontiers to new limits, our definition of space exploration will be completely redefined. However, human exploration of our solar system cannot happen without the assistance of our robotic counterparts, which helped blaze the trail into space. This transition to human exploration will see astronauts virtually isolated by lengthy communication delays, as long as forty minutes, and consequently being required to maintain an extremely high level of self sufficiency. Expectations of such a high profile mission will also mandate that copious amounts of field work and related studies be conducted over its duration. Due to their circumstances, early Martian explorers will be forced to work with equipment that was at one time controlled by large teams of scientists and engineers with immediate access to significant computing resources. In order for these astronauts to be able to maximize their time, and produce extraordinary quantities of data, new methods for human-rover interaction in planetary exploration must be developed.

One scenario that shows a strong need for more research into new methods is a small team of astronauts (or possibly even a single astronaut) on the surface of Mars conducting field work out of the immediate reach of their base. This seemingly commonplace situation will find team members out facing the elements while conducting research. It seems only natural for research rovers to accompany the team into the field. However, having to deal with robotic equipment while in a space suit presents several issues, the primary one being control. By incorporating virtual reality glove technology into the astronaut's gloves, his hands become a quick, easy and effective input device. A simple hand command can activate the gloves, and the rover begins to respond to hand gestures, which are interpreted as commands. Our isolated astronaut now has complete flexibility in the control over all of the different robotic equipment and machines that will be in the field with him.

All plausible mission outlines for the first manned missions to Mars entail the crew collecting extraordinary amounts of data in a large number of different areas. Most of the field work would be conducted with research rovers, such as those described above, each specializing in a different area. The number of rovers to be sent will quickly add up. A much smaller fleet of modular rovers allows for mission flexibility while significantly cutting back on overall mission cost and weight. These modular rover bases, which will accept a wide variety of scientific units, will operate on a very standard platform. Because all of the equipment will have interchangeable components, the astronauts will be able to troubleshoot most basic problems that could arise in the field. This modularity, coupled with the simplicity of the glove input, tackles many of the difficulties that an astronaut would face in the field that would otherwise severely hinder his productivity.

Gesture Based Control System
Motivation

The primary objective in developing a control system suited for the aforementioned situation is to compensate for the drawbacks that would inherently arise with the use of standard input devices in rather unconventional environments. Limitations will be imposed by gloves that will naturally be somewhat bulky and cumbersome, thus limiting finger dexterity as well as the ability to accurately press a small button. Compared to other commercially available input devices, we feel that a virtual reality glove is the best candidate to be adapted to use during EVA in a pressure suit. In order to use a keyboard, the individual keys would need to be large enough to be reached without trouble from pressure gloves. If the key size were to be scaled appropriately, the overall size of the keyboard would be ungainly. A traditional joystick is limited by the number of input dimensions; generally, only two degrees of freedom are present through the manipulation of the stalk, with additional degrees provided by trigger buttons or other small actuators. Even if the additional buttons were to be scaled to be usable from a pressure glove, there simply are insufficient degrees of freedom available for general purpose tasks.

Glove and Gesture Recognition Software
Current Work

We begin with a description of the input devices with which we are working. For this project, we have obtained two 5DT Technologies, Inc. virtual reality gloves, each with a serial interface to the host machine. These gloves have five finger sensors, as well as an auxiliary roll and pitch sensor. Each finger sensor consists of an optical fiber that wraps around the length of the digit, such that the flexion of a finger bends the fiber. In an electronics package attached to the body of the glove, a light source emits light, and photosensors observe the transmission through the fiber. The glove is calibrated based on the principle that, as an optical fiber flexes, the intensity of a transmitted light will vary as a linear function of the flexion. This flexion is represented as a single 8-bit byte value to the host. The roll and pitch sensors also produce 8-bit accurate results. After opening each device, we are ready to sample data.

In order to submit a sample to the processing pipeline, we must first read the data stream coming from the glove(s), and then assign the sample to the processing pipeline. To interpret the raw glove data, we utilize a vendor supplied library function that returns the actual value measured by the glove hardware. Since our project was designed to have the option of using multiple gloves, we keep track of the mapping between samples and gloves. From the input module's point of view, there is no more work to be done, and so the next sample is obtained.

Fundamentally, the system developed to interpret gestures has three major components: input filtering, gesture recognition, and device-specific output. Each of these components runs in a separate thread of execution, exchanging data through buffer queues. By multithreading the processing jobs, the overall program can process data without a dependence on the complexity of individual components.

Once a sample has been provided to the processing pipeline, the next stage is a simple exponential filter that serves to regulate noisy input data from the gloves. This stage was added after initial testing indicated that users have slightly shaky hands; after filtering, the data is much smoother and appropriate to use in a decision process.

After data filtering, logical gestures may be interpreted out of the physical data. We define a gesture to be a region of flexion for each digit. In practice, we have found that the output of the gloves can be divided into only three or four "zones," due to the fact that a human cannot repeat gestures with exact precision. Assuming that each finger was capable of producing each position independently, there are a maximum of 1,024 gestures; this number is entirely too optimistic. For example, as a limitation of the design of the individual gloves that we are using, the thumb measurement only has two zones; we have also found that users are not as comfortable with intermediate positions of the thumb as with positions of the fingers. Secondly, most people cannot move their pinky finger without incurring some movement in the ring finger; in the same vein, the ring finger generally implies a movement in the middle finger. Truly independent movement is only possible for the index and middle fingers. We divide the limitations into two categories: extrinsic for those limitations such as the thumb movement that are a result of the manufacture of the glove, and intrinsic for those limitations such as the non-independent movement of the pinky finger. Extrinsic constraints may be mitigated by investing in higher-quality gear; for our purposes, however, the performance of our gloves is adequate.

Since the virtual reality gloves are measuring one of the user's primary world interaction mechanisms, there may be situations where a user does not want his gestures to be interpreted. Similarly, a user may direct his gestures toward different targets. To accommodate these requirements, we represent the gesture recognition engine as a Mealy finite state machine, with glove data driving both transitions and outputs. Glove data is monitored for transition events, and then transformed into an output value appropriate for the device. We envision a future system in which there exists a hierarchy of states such that an initial gesture selects a device to control, and then subsequent gestures navigate the state space for a given device. In our current test bed, we have only one controllable device with one interpretation of glove data; thus, we have the two states illustrated in Figure 1. This single interpretation is a "direct-drive" state; the user's hand movements are directly interpreted into motion of the target. In more advanced control layouts, this direct drive state would be a child state of a general device selection state. For our purposes, this representation is appropriate for our prototype.

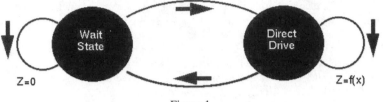

$$Z=0 \qquad\qquad\qquad\qquad Z=f(x)$$

Figure 1

If any device output is necessary, the gesture engine passes a request off to an appropriate output processing thread for the device at hand. Since both of our devices are locomotion devices, we have only one output thread. Depending on the capabilities of the controlled devices, different formats are supported. For our prototype device, we chose a basic serial format that can accommodate translational and rotational movement commands.

To facilitate the independent development of software from the underlying hardware, we opted to use a network robot hardware simulator. Called "Player / Stage,"[1] this simulator is designed to allow a controller to be developed under simulation, and then use the same binary code on the real hardware. One of our group members is employed at a mobile robotics laboratory on campus, and his managers have graciously allowed us to test the gesture control system on real robots. These robots are ActivMedia Pioneer 2-AT class devices; one under glove control is pictured in Figure 2. This image is a clip from a movie that shows the range of motion of the glove control system; the rover is put through a series of maneuvers combining forward and reverse rotational and translational movement.

Future Work and Extensions

In our current prototype, there are many more degrees of freedom in the controller than the device. Since one of the future goals of the project is to have glove control over several devices, we expect to take advantage of the excess freedom. However, a complicated control system will lead to human confusion and error. To make the gesture language easier to learn and apply, we will be adding a second glove to the system. In this extension, a gesture state transition can use one or both gloves for data input.

The use of two hands will make a system easier to interact with, but also raises the concern that a user will not be able to carry or hold anything while performing a two-handed gesture. In response to this issue, a complete backup gesture system could be implemented

Figure 2

such that no command is impossible to perform without two fully functional gloves. The two-handed state transition could serve merely as a convenient shortcut to the same destination state as a series of single-handed transitions.

To experiment with two-handed control, a rudimentary case in which the second glove controls an independent device has been implemented. This independent device is the gripper / lift combination on the front of the Pioneer robots; this device is only used to test out the capabilities of the gesture recognition system while our final project is under construction.

In the discussion regarding the hardware system of the virtual reality glove hardware, it was noted that each glove requires a serial port for communications. With our available computer resources, we were unable to run both gloves using the same host; the solution to this problem came through the previously mentioned Player robot server. Since both glove-host machines were on the same network as the robot, each could control its device independently. The point of this extension was to show that the same binary code could interpret gestures independently, as opposed to having a left-hand controller and a right-hand controller.

Although this scheme works as a proof of concept, we expect to use a host that is capable of driving both gloves simultaneously. The use of a robot server does point out a possible source of redundancy; if the gesture-based primary control system were to fail in the field, less efficient but functional alternatives could be engaged to complete the mission.

At this point it would seem prudent to discuss the field practicality of our glove control system. As noted previously, the gloves are hard wired to the serial input on a computer. This limitation is one that we are facing due to the model of gloves that we have been provided. Wireless gloves are commercially available; however they naturally are more expensive. Despite the need for the gloves to be physically connected to a computer, this does not limit their application to a laboratory setting. Our implementation of this control system will eventually entail a small, lightweight laptop that can be worn in a backpack. The glove-end laptop will communicate to the rover via wireless Ethernet. This modification will allow us to effectively simulate an actual glove control system.

Modular Rover
Motivation
The rover which our glove control system will be controlling highlights another unique feature of our project. One issue that mission planners will surely face with a manned mission (as well as they do for all missions, human or robotic) is the trade-off of reducing overall cost and weight, while still sending ample equipment. Most rovers and other robotic equipment sent will be optimized for one specific portion of the mission, and will consequently lay idle for lengthy periods of time. To solve this issue, we are designing a modular research rover, which will maximize the versatility of the available equipment.

Modularity
An issue being addressed in this rover design is long term usefulness and flexibility. Extremely specialized rovers are the most logical solution when sending single unmanned missions that will only last a period of weeks or months. When humans travel to Mars, they will most likely take numerous rovers with them, as well as spare parts to help ensure their longevity. With this being the case, it only seems natural to maximize each rover to its fullest potential. The balance between specialization for a specific goal and long term usefulness leads to one conclusion – rovers that are sent to Mars on a human expedition will almost certainly be modular. This will allow the rovers to serve several purposes throughout the duration of the mission. The idea is that a rover chassis can be built to accept payloads with standardized connectors and control implementations. The modular system will consist of multiple bays atop the rover into which scientific instruments will dock. There will be a connection for power, data transfer to the rover computer, and mechanical connections to secure it in place. By simply plugging in the module and connecting the latches, the new module will be ready to use. The rover chassis will provide every module with locomotion, communication back to the astronaut and habitat, and a computer to process the data collected by the module.

This modularity allows for more types of science packages than there are rovers, increasing the mission capability at a lower cost than with specialized rovers. Redundancy is also increased, because a failure will most likely take place on the rover chassis and not the module, due to the complexity of the drive system and the harsh Martian environment. If the chassis fails, the module can be placed in another rover with no loss of functionality.

The key to modularity is a set of standards with which all electronic equipment will operate. While this is not a new idea, it has not been widely employed. The need to adopt a standard is clear, considering the time required to rewrite drivers and software to transfer hardware from one rover to another. One member of the Penn State Mars Society experienced this hassle while working at NASA Ames. While this leads to wasted time on Earth, astronauts will not necessarily have the luxury of time to reprogram equipment, leading to equipment laying dormant if there is a problem on the rover carrying it unless a standard is adopted. The benefits will naturally carry over to Earth when time is spent transferring hardware from rover to rover.

Rover Overview
When designing the overall rover layout and major systems, one of the most heavily influencing factors is intended usage. A smaller rover proves ideal for investigating where a human cannot, namely small crevasses and caves. On the other hand, a larger rover provides more independence and flexibility, since it can traverse longer distances at higher speeds, and carry a rather high amount of scientific equipment. Obstacles would also prove less difficult to avoid with a larger rover. However, for the purposes of this project, we soon realized that a compromise would suit us best. A mid-sized rover can hold all of the major equipment that we need it to and accept multiple modules at once, but at the same time be small enough to necessitate modularity. Furthermore, a rover of this magnitude is the most feasible to construct, both from the point of view of cost and general ease of construction. Microrovers require working with tiny devices which are very expensive, in addition to requiring significantly more precision on the fabrication end. Large scale rovers begin to involve components from larger vehicles, such as all-terrain vehicles and even standard automobiles. Although these parts are easily accessible, they are somewhat expensive and also become very cumbersome when producing a research rover.

The rover we are constructing (sketches for which may be seen in Figure 3) is 24" long, 16" wide and the main compartment is 10" high. This compartment will sit about 8" off the ground, suspended by shock absorbers on a bottom plate containing the motors and axles. With the separate bottom plate, containing the motors and axles, we are intending to isolate minor vibrations that occur as a result of driving on mildly rugged terrain. This suspension system is designed to compensate only for small jolts, and will not allow the rover to become an aggressive off-road vehicle. The wheels chosen will be 10" in diameter. While the ultimate objective of our design will be the utilization

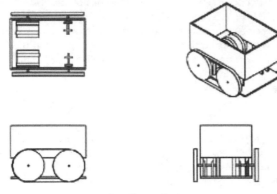

Figure 3

of treads, wheels may need to suffice for short-term demonstration purposes. The reason for this decision is one of availability. Extensive searching for treads has turned up only a limited number of options. Snowmobile treads are the most readily available, but their excessive width and robust construction make them impractical for our purposes. Snow blower treads are of a more reasonable scale in comparison, but tend to be largely unavailable on the internet and in large home improvement stores.

An on-board laptop will provide both computational resources and communications via wireless Ethernet. The primary function of the computer will be to receive input data from the glove-end computer and transmit vital information to microcontrollers which will directly interface with most of the hardware. Two firewire cameras mounted on the front of the rover will provide several opportunities for stereo vision applications. Perhaps the most obvious application would be for autonomous navigation. The rover would gain increased self-sufficiency with its ability to detect hazards and locate targets. The stereo vision would also provide useful in terrain mapping and related scouting missions. Once modules for the rover are developed and we begin to perform field tests, the laptop on board will serve to establish the rover as a mobile laboratory. Any samples that are collected could easily be analyzed in the field for quicker results.

Project Time Frame and Future Work
Project Time Frame and Possible Extensions
By the end of 2002 designs will be finalized for the rover base, as well as plans for extensions of the glove control system. Construction will begin early in 2003 and will focus primarily on the rover base. A fully functioning mobile base, complete with on-board computing systems, will be running by May of that year. In addition, this stage will see basic stereo vision applications implemented. This includes simpler tasks such as locating and driving to an astronaut in the field. Glove control will also be expanded to a two-glove input system that will begin to allow us increased flexibility in the control of the rover. Over the course of the 2003-2004 academic year the rover base will be completely finalized and fine-tuned, and we will begin to develop different modules that can be used in field tests.

Because this project has been built with modularity in mind, there are inevitably countless extensions which could easily be explored and implemented. The stereo vision system can be expanded to perform more advanced autonomous navigation tasks. This includes hazard detection and obstacle avoidance, two very crucial abilities. Beyond even the rover we will have built, we can begin to explore team robotics to tackle even more varied situations. One example would be a separate module which serves simply to deploy a microrover. Microrovers would be designed for the sole purpose of going where a larger rover simply cannot access. Since the glove control system is designed with a universal language, controlling an entire team of rovers would be a natural extension of the basic command language. While a multitude of possibilities exist, this outlines just a few of the possible extensions for this project.

Field Testing / Mars Desert Research Station
Once the mobile rover base has been completed, we will begin with very basic field tests. By analyzing its response to different situations and testing environments we may gain added insight that can be applied towards latter systems on the rover, including the individual modules. As we develop the rover and glove control system into a fully mature state, field testing will obviously become a very logical step in this project. Because the focus of this project isn't so much the actual technology as it is the implementation of that technology, most of the knowledge to be gained will come from these experiments. An ideal testing situation would be at the Mars Society's Mars Desert Research Station (MDRS) in southern Utah. At MDRS, Mars Society members conduct studies in manned Mars missions from an operations standpoint.[2] More specifically, they look at many of the human factors of a manned Mars mission, including how work will be performed. A project like ours, which deals with human-rover interaction, would be an ideal test subject at the station due to its investigation of research methods. A field test at MDRS would naturally be a beneficial partnership for both parties.

References
1. Brian P. Gerkey, Richard T. Vaughan, Kasper Stoy, Andrew Howard, Gaurav S. Sukhatme, and Maja J. Mataric; *Most Valuable Player: A Robot Device Server for Distributed Control*; Proceedings of the IEEE/RSJ International Conference on Intelligent Robots and Systems (www.icra-iros.com/iros2001/), pages 1226-1231, Wailea, Hawaii, October 29 – November 3, 2001.
2. The Mars Society; The Mars Society: Mars Desert Research Station; www.marssociety.org/mdrs/index.asp

Acknowledgments

We would like to thank our sponsors and supporting groups:

- The Pennsylvania Space Grant Consortium
- The Pennsylvania State University College of Engineering
- The Schreyer Honors College at the Pennsylvania State University
- The Department of Computer Science and Engineering at the Pennsylvania State University
- The Applied Research Laboratory at the Pennsylvania State University

//

Terraforming Mars with (Largely) Self-Reproducing Robots[†]

Robert Alan Mole
Ramole@aol.com

Abstract

The advent of self reproducing robots within fifty years will make possible the terraforming of Mars in a few decades. All schemes for terraforming should be judged against this likelihood, and we should think in terms of the next century, rather than the remote future. Such robots can also help terraform the Moon and many other bodies in the solar system.

Introduction

There are various schemes to provide Mars with a CO_2 atmosphere by causing the evaporation of the dry ice at the South Pole and in the regolith. Although no one knows how much is available from these sources, getting it to vaporize may not be too difficult and a reasonably thick atmosphere may be attainable in a hundred years or so.[1,2]

However the next step in terraforming, changing some or all of the carbon dioxide into breathable oxygen, is far more difficult. It inherently requires about twenty times as much energy to split a molecule into carbon and oxygen as to sublime and warm it. Some terraformers hope to sow plants or bacteria to do the job by photosynthesis. But as Zubrin and McKay point out, plants turn less than 0.01% of solar energy into this chemical reaction, and would take over 100,000 years to complete the task. They hope genetic engineering can improve this efficiency a hundred times, but even then 1,000 years would be needed.

Other terraformers suggest the use of nanobots, tiny self reproducing robots that will rip the CO_2 apart physically. But nanotechnology is so far in the future that no one can really guess its capabilities or limitations. Thus the plan is not provably wrong (or right) and simply amounts to saying that the human race will make vast improvements in technology sometime in the future, and make the new atmosphere using nanobots or some other form of "magic." Actually this is very likely, but it really says very little about how we'll do it.

The Plan

Here is a clearer plan. We do not need nanobots, only robots able to reproduce themselves. Presumably such machines will be roughly man-sized, not microscopic. Nor need they be able to reproduce every part of themselves, just the heavy parts. Light parts – e.g., computer chips – can easily be brought from Earth. Even today a chip weighs half a gram, and two million of them weigh only a ton, so we can transport the brains for a robot army in a small space probe.

Once there are many robots on Mars they can make solar panels or reflectors enough to cover the planet, and the power can be used to run cracking plants to split the CO_2. Solar cells are 20% efficient, much better than the 0.01% efficiency of bacteria. So 100,000 years shrinks to 50. Solar cells of 75% efficiency are in the works,[3] so the time could be shorter still.

Even more promising is the process of simply heating CO_2 until it dissociates, and then pulling the oxygen out through a zirconium membrane. This process and variants have been studied as a means of in situ fuel production, with CO as a byproduct or fuel component.[4] Carbon monoxide is poisonous and explosive though, and would not make a good component for an atmosphere. However, via a catalyst such as hydrogen-reduced magnetite, oxygen and pure carbon can be produced at just 300° C.[10] Using solar reflectors and "ovens" with heat exchangers, this method could approach 100% efficiency and produce a 120 millibar oxygen level in just ten years.

Of course we may wish to cover only a part of the planet and cannot achieve perfection, so these numbers are only approximate. Yet uncertainties should not obscure the critical, the absolutely salient point: even 20% efficiency is more than 0.01% – two thousand times more.

We do not have manlike robots today, but our path to them is straight forward and one would expect them to exist before a thick CO_2 atmosphere on Mars (which will take at least 60 years from today to achieve). To understand why rapid progress is likely, see the addendum below. A possible scenario for Mars follows.

The year is 2050. Robots have been designed and tested extensively on Earth. Ten robots, at 100 kg each, are sent to Mars, where a scientific station or small colony already exists. (Note that 10 x 100 kg is one metric ton, which we surely can afford to send.) The colony is nuclear powered and produces adequate hydrogen for the robots' fuel cells. A couple of engineers will direct and maintain the robots, while hundreds of planners on Earth send directions with a few minutes' time lag. Some robots are manlike and dexterous, while others are small earthmovers. At first humans help with much of the work.

The robots begin by building an aluminum smelter (basically vats to fuse Martian ore, plus electrodes), surface mining ore with earthmovers, and smelting it into big sheets of aluminum. These will be used to build the bodies of robotic scrapers and mining machines, to be completed with lighter parts from Earth. From here out aluminum is plentiful.

Next the robots make dies and simple presses, and thus a stamping plant. Now they can turn aluminum sheets into the heavy structural parts of robots (perhaps exoskeletons like plate armor.) These parts are the heaviest single component of a robot, perhaps 60% of its weight. The robots assemble these parts together with a ton of parts from Earth, but this time the ton from Earth is enough for twenty-five robots instead of ten.

And so it continues, with more robots building factories for assembly, for gears and tubing, for bearings and motors, until it takes only a kilogram of Earth materials to build a robot; until it takes only a gram . . . There is no need for a fanatical insistence that the robots completely reproduce themselves on Mars – only that, in a practical way, most of their mass be produced there.

Soon there will be thousands, then millions of robots, and then they will set to work making solar panels and setting them up over much of the surface. A good atmosphere will be produced in decades. Thereafter they will remove the apparatus and oxygen will be maintained by plant life or just a few panels.

The advantage of this idea is that no magic is needed, just continued progress of the kind we are making now – better chips, better programs, and improved robots. We already have dexterous robots on assembly lines, and automated factories where robots completely assemble calculators and almost-completely assemble VCR's and even robots like themselves. We have experimental bipedal robots, autonomous rovers, experimental excavating and ore hauling robots, and so on.[9] True, we must improve these devices before we'll have robots that can do everything we want, but such improvements appear fairly straight forward and require no nano-warp-drive breakthroughs.

Details and Doubling Times

Let us ask how many machines we need to cover half a planet with solar reflectors. Mars' surface is 1.4×10^8 km^2, so we must cover 0.7×10^{14} m^2. What do we have for a basis of comparison?

We can certainly carry to Mars a ton of chips, enough for two million machines, so let's use that as a baseline. Spreading the reflectors or cells, once they are produced, should be easy. We imagine a man pushing a lawnmower, moving at 1 m/sec, cutting a 1 meter swathe, covering 1 m^2/sec. Now we imagine a farmer on a big tractor, going 10 m/sec (about 20 mph), unrolling a 10 meter wide roll of cells, covering 100 m^2/sec, (x 3600 sec/hour x 24 hour/day x 365 days/year = 3×10^9 m^2/year). At this rate a machine would need 22,000 years to cover the area. Or we could do it in a year using 22,000 machines. This is just one percent of two million, leaving the other 99% to mine the materials, make the cells, etc. Furthermore, silicon, aluminum, and iron are among the most common elements on Earth and Mars – so common one might almost say they are Mars. If we must cover the surface of a large silicon body with a 1 mm thick layer of silicon solar cells, we are not likely to run out of silicon, which extends to 4,000 km deep (from the surface to the center of the planet.) Nor should even thinner aluminum foil reflectors deplete the aluminum.

But two million robots, including man-like tractor and mining models, is a lot. How long would it take to make them by present day standards? The answer is surprising. In America today, a basic 1,000 kg car can be bought new for about $10,000. An automotive engineer has said that just 43% of that is production (not just assembly, which takes only fifteen hours, but all production from mining the ore to polishing the nameplate.) At an industry average wage of around $43 per hour (counting fringe benefits), that's just 100 hours. Hence a 100 kg robot should take 10 man-hours. (A tenth the weight of a car, a tenth the time.) The Fanuc factory in Japan produced a thousand robots per month, using what appears from illustrations to be about 100 to 200 robot workstations plus some human workers, all in 1988. This would give a doubling time of three to six days.[11] Allowing for the humans, let us assume seven days for argument. This means robots could reproduce themselves, doubling their numbers every week. In ten weeks ten robots would be ten thousand; in twenty weeks, 10 million, and so on.

To be sure, we on Earth already have the infrastructure, the steel mills and power plants and roads that make this possible. On Mars, ten robots will take a while to build aluminum plants (for structure and wires), a steel mill (for magnets and motors), a stamping plant and so on. But with those, production can take off. I would expect the infrastructure to take several years, but after that the doubling time might approach days. And even if each doubling required a year, we'd have our two million robots in under twenty years. If we needed more, an initial lot of ten would grow, in forty years, to eleven thousand billion (10×2^{40}, 1.095×10^{13}). This is an absurd number, nearly two thousand times the human population of Earth, but it illustrates the point that we can quickly have as many as we could possibly need. Given self-reproduction and a reasonable doubling time, virtually anything is possible.

How Much Ore is Needed?

Two million 100 kg robots represent about 2×10^8 kg of aluminum or, allowing for 40% ore, 5×10^8 kg of ore (500,000 metric tones). The great Bingham open pit copper mine produces 320,000 tons/day, so this is under two days' production. Harder is covering half the surface of Mars with reflectors. Aluminum foil is "quarter mil," or 0.00025 inches thick. Covering 0.07×10^{14} m^2 requires 4.4×10^8 m^3 of aluminum (1.2×10^9 tons), or 3.1×10^9 tons of 40% ore, which is 26 years' production at Bingham. This is a large but reasonable task. The mine employs 800 men; there will be millions of robots.

Waste Disposal

Two hundred millibars of oxygen, the Earth standard, represents 3 lb of oxygen in a column above every square inch of the surface. On Mars, with its lower gravity, 7 lbs would be required for the same pressure. Seven lbs of oxygen is contained in ten lbs of CO_2, the other three lbs being carbon. So if we produce seven lbs of oxygen above every square inch we'll have three lbs of carbon above the same area. At a density of 0.063 pounds/inch^2 for solid carbon, this is a carbon column an inch square by 48 inches tall (1.2 m), or probably 2 meters counting voids. That is, if we produce an Earth-normal pressure of oxygen from CO_2, we'll cover the whole surface of the planet with two meters of carbon dust – i.e., lamp black, a vile dirty substance that coats everything and shorts out electronics. This will not be Utopia. To avoid this coal miner's nightmare we must pelletize the carbon as it is made and either bury it in ravines and low spots or pile it into hills. If these are 200 meters high or deep they will cover just one one-hundredth of the surface, which is reasonable. They will, of course, be sealed with soil to prevent the wind from eroding them and scattering lampblack, and to prevent their catching fire. Thus, just as so much carbon is stored on Earth beneath our feet as seams of coal, so it can be stored harmlessly and neatly on Mars.

This dirty problem exists, by the way, no matter how oxygen is produced from CO_2. McKay et al note that "The limiting step in the production of any O_2 . . . [may be] the necessity of sequestering the organic material in places (presumably deep sediments) where it is not reoxidized [which could be very difficult]."[1] On Earth, biology has made coal from CO_2. On Mars, biology would do the same, but it might be millions of years before coal formed and was buried, during which time carbon would continue to pollute everything. No matter what, we must dispose of the carbon.

Heat Balance and Good Global Warming

If we use a lot of the solar energy striking Mars to break up CO_2, there will be less left to warm the planet and it will grow very cold – other things being equal. Fortunately, we don't have to leave those other things equal – we can change them drastically for the better.

The temperature of an object in space is determined by three things: the amount of sunlight striking it, the percentage of visible light (sunlight) absorbed ("absorptance," a_s), and the emissivity in the infrared (the percentage energy emitted by a surface compared to the amount a perfect emitter at the same temperature would radiate, e). A perfect mirror ($a_s = 0$) could grow very cold. An object with low emissivity would grow warm.

Mars' soil surface absorbs visible light well. Its a_s varies from 0.75 to 0.85.[5] We could achieve a little by darkening the light areas by spreading dark soil over them, but the a_s is already so high that we cannot do great things here.

For emissivity the story is different. Soil emissivity is nearly 1.0 (100%), the maximum possible. But emissivities as low as 2% are achievable, meaning that at the same temperature, the surface could radiate away just one-fiftieth the energy it now does.

The situation is complicated somewhat by the atmosphere, which radiates weakly itself and also insulates the ground via the greenhouse effect. The author has discussed the matter with a planetary atmospheres scientist, who confirms that a low-e surface on Mars would raise the temperature of both the surface and (by contact and convection) the atmosphere. But the effective emissivity of the atmosphere is only a few percent over the entire IR band (high at the 15-micron wavelength of CO_2 but low everywhere else.) Thus the atmosphere would not radiate much. However, a dust storm would make the atmosphere look like the surface, with an e of 1.0. (So dust storms would have to be avoided by covering dusty areas with foil and rocks.) As the atmosphere becomes like Earth's, with many gasses radiating at many wavelengths, the emissivity will approach the Earth's level of 50%. (But this is still only half the present 100% level of the surface, and the greenhouse effect of such an atmosphere would keep the planet tolerably warm by itself.) In any event, under present conditions, the atmosphere can be ignored and we can base calculations on lowered surface emissivity.[12]

If, then, we covered the ground with low-e material and used most of the sunlight for making oxygen, then just a fiftieth of the solar energy could keep the ground and air at their present temperatures. Actually, various thermal losses in the CO_2 cracking process will exceed 2% and the excess will go to warm everything above present norms.

Thus one may imagine thin aluminum mirrors covering the surface and focusing on high temperature retorts in which the CO_2 is cracked. The absorptance (a_s) of the mirrors is 8%, so they reflect 92% to the retorts. The emissivity is just 2%, so they radiate almost nothing to space – just a fiftieth of the current loss. The backs of the mirrors, facing the ground, are treated for high emissivity, so the excess heat is coupled to the soil and warms it.

Of the incoming solar energy, 8% is absorbed and turned into heat by the mirrors, and perhaps 24% is lost from the O_2 process, so overall only 32% of solar energy goes into heat (vs. about 80% today). But only 2% as much energy is lost. We go from today's a_s/e of 0.8/1.0 = 0.8 to 0.32/0.02 = 16.0 – a factor of twenty improvement. We net about 30% of the incoming solar energy for warming soil and air, and still have 68% to break CO_2.

McKay et al.[1] calculate that to warm a 1.0 bar atmosphere from 150° K to 288° K requires four years of Martian solar insolation, while to warm the top ten meters of soil will need 0.3 years.[1] That is, a total of 4.3 years of sunlight is needed to warm Mars to Earth-like temperatures. If we use only 30% of the energy for warming, it will take 17 years. If we cover only enough surface to capture half the sunlight, 34 years is required. Conveniently, that is about the time needed to make the oxygen (at 68% efficiency and catching 50% of the total insolation, 29 years is needed for 120 millibars). So, in 29 years, we can have a warm wet Mars with breathable atmosphere.

Other Uses
Given as many robots as we want, we can do many things besides split CO_2. Nitrogen may be locked in chemical compounds on the Martian surface. We can split those compounds and produce a nitrogen atmosphere. CO_2 thousands of meters deep in the regolith may take thousands of years to warm and evaporate, because heat transfer through kilometers of soil is poor.[14] But we could sink heat pipes at small intervals and get it in decades. We can produce other greenhouse gasses (per Lovelock and Allaby[15]) and speed warming. We can build many fission or fusion power plants and obtain energy in addition to solar, perhaps far surpassing it. We can build the giant orbital reflectors also envisioned to heat the planet, and build mass drivers or slingatrons to launch them.

Terraforming Other Bodies
The Moon
Oberg[6] tells of various schemes to heat Moon rocks and liberate oxygen, producing a thick atmosphere that would last but a moment in geological time – but a geological moment is 10,000 years, or forever in terms of human history. Clearly, robots plus solar concentrators (reflectors or Fresnell lenses) could heat the rocks this way. The Moon would still be deficient in hydrogen and nitrogen, and still have too long a day, but providing even a partial atmosphere would be a good start towards terraforming.

The Moons Of Jupiter
Jupiter is so far from the Sun that light is weak and its moons are very cold. However, the temperature of an object in vacuum is proportional to both light intensity and a_s/e, and an increase in one can compensate a decrease in the other. If we could cover the moons with thin layers having a high a_s/e, we might warm them to Earth-like temperatures. Gasses would sublime, creating thicker atmospheres, and greenhouse effects would warm the bodies further. (Again, the gasses are only weak emitters, so although they will be warm they will not radiate away much energy.)

Sunlight at Jupiter is only 1/27th as strong as at Earth. For Earth, a_s/e is about 0.6/0.5 = 1.2. To increase that 27 times we would need an a_s/e of 32. The best artificial surface for high a_s/e is vapor deposited gold, a_s = 0.19, e = 0.02, a_s/e = 9.5. This falls short, yet close enough to suggest that an adequate material might be developed.[7] Or, reflecting sunlight from vapor deposited aluminum (a = 0.08, e = 0.02) onto a high absorptance material arranged to heat only atmosphere (and not radiate directly to space) will raise effective a_s to 1.0 while not changing e at all, so a_s/e goes to 50, more than enough.

Again we have a good start, though other problems remain: long days and mostly-ice surfaces that might melt to produce universal oceans.

(Saturn's moons cannot profit from this method – all are too small to hold an atmosphere, except for Titan, which already has a thick atmosphere whose emissivity sets the temperature, making surface emissivity irrelevant. An a/e of 108 would be needed at Saturn, above the maximum shown here, and far better ratios would be needed for more distant planets.)

Mercury
Mercury gets 6.7 times the sunlight of Earth, but if its a_s/e was much less than Earth's it could have the same temperature. Earth's 1.2 divided by 6.7 gives 0.18 required. The white coating of barium sulphate with polyvinyl alcohol has an a/e of 0.06/0.88 = 0.068 – much better than required and enough to make the planet icy. (Obviously we could use a different material or cover only parts of the surface.)

If robots then built slingatrons and threw off material to increase the spin rate and shorten the day, if volatiles could be found in the crust (perhaps methane brought by comets during the original accretion as proposed by Thomas Gold for Earth[8]), and if these could be released to form an atmosphere, then even Mercury, most daunting of all the planets, might be fully terraformed. (However this is meant more as a vision than a suggestion; speeding the rotation of a planet is a stupendous undertaking.)

Conclusion

We can, today, build robots for mining and assembly, and will soon be able to make largely self-replicating robots that can increase their numbers to any desired level at little cost to us. Cheap robot armies give us a powerful new tool for terraforming whose uses have barely been touched on here. Others should consider the possibilities that this idea opens up.

For Mars we should have sufficient capabilities in fifty years, though it could be more or less. But surely we will have macrobots before nanobots, and whether it takes fifty years or a hundred and fifty we will have them, and be able to turn Mars' CO_2 into O_2 in a few decades.

We should therefore abandon the idea that it will take a hundred millennia to terraform Mars, and turn our thoughts to a livable O_2 atmosphere in a century or less. In any event, we may find a better way, but this believable scenario should be taken as an outside limit.

Let us explore Mars soon, and then make it a garden.

Addendum: Advances in Technology and NASA Work

Is it somehow "cheating" to assume we will have robots in fifty years? Is it wrong to base estimates on anything besides today's technology?

Martin Fogg describes a Freeman Dyson idea to place a fully self-replicating robot factory on Enceladus and comments:

> ". . . Dyson's reason for indulging in [this thought experiment] was to show just how unbounded the future might be. However most terraforming researchers leave von Neuman machines alone, as one of those tools of the "arbitrarily advanced civilization" mentioned in Section 3.1. Once such capability is attained, all bets are off and almost anything not ruled out by physical law is possible."[14]

The author agrees as regards full replication, though the partial replication described seems quite likely in the near term. But in a broader sense, is it wrong to assume any capability that we don't have today?

No. Practically our whole civilization was invented in the last hundred years. Far more scientists and engineers will work in the next fifty years than in the last fifty, and it is widely remarked that knowledge is increasing exponentially, so our capabilities will increase more rapidly during this time than in the past.

There is opposition to terraforming Mars before it has been studied in its natural state, and since people will not even walk on Mars for twenty years it will probably be fifty before studies are complete and terraforming can begin. Fifty years ago we had no robots (nor even any computers); now we have Mars rovers, assembly line robots, and vehicles that can move off-road through obstacle fields at six miles an hour.[9] In a few years we should have robots for strip mining and smelting. (Robotics students say a major equipment company could introduce robot mining trucks within five years, provided union opposition can be overcome, though the company itself is reluctant to discuss the matter – in spite of having demonstrated the hauler at a major mining equipment show in 1996, and having run it 8,000 miles in a strip mine![9]) In a few decades, we could have the more complex models here assumed. Thus these prognostications are not unreasonable. By the time we're ready to start changing Mars, the technology will be there to let us.

Previous NASA Work

After this paper was completed it was found that in 1980 NASA had held a summer workshop to examine the idea of a self-replicating "seed" factory: a 100-ton factory that would reproduce itself in about a year.[13] They too considered use of some outside materials, or "vitamins" as they called them. Based on the average composition of the lunar soil (no ore bodies were postulated) and a reasonable ratio of material processed to material output, they found a "closure ratio" of 90%-95% was feasible – i.e., only 5-10% of mass would be imported.

Their 146-page report is comprehensive and thorough. They examined terraforming, finding 100 millibars of oxygen on Mars was possible, through reduction of SiO_2 in the soil, in about sixty years (with power from a giant solar satellite, existence assumed.) The Moon, Venus, and more could be achieved. They found it possible from a machine shop viewpoint (producing mainly molded parts finished with laser cutters.) They postulated front-end loaders to bring in soil. A 4,400 kg loader running a 40 km round trip could deliver 4×10^6 kg per year. (Incidentally, this is all that is required to make the robots in the current scheme, and if the smelter were located next to the ore deposit and the round trip were only 1 km, then just one 100 kg loader would be required. Of course, more and bigger loaders would be needed for the reflector material.)

They mapped the steps needed to develop such a factory. They estimated it would take eighteen years. Although nothing was done with the idea, the past twenty-two years have seen great progress in allied fields, as computers have shrunk from the refrigerator size shown in one of their sketches to the latest single-chip models of today.

It is odd that this work has been forgotten, but obviously it has – else why do we continue to talk in terms of a thousand centuries to terraform Mars?

In any event, it is not unreasonable to assume that we can do fifty years in the future what this serious NASA study concluded we could do twenty-two years in the past.

$$CO_2 + H_2O \longrightarrow CH_2O + O_2.$$

Zubrin and McKay show that, with this reaction, 200 millibars of O_2 requires 17 years of the total solar energy striking the planet. Sublimation and warming of 1,000 millibars of CO_2 require only four years' worth of sunlight. The reaction above is shown to require 8×10^6 joules to produce 540 g (both per cm^2 of Mars' surface), which works out to 4.74×10^5 joules per mole of O_2.[1] The reaction here proposed, $CO_2 \longrightarrow C + O_2$, uses slightly less energy, 3.94×10^5 joules/mole.[16] For this paper it has been assumed these numbers are the same and the numbers in Reference 1 have simply been ratioed to obtain required times, because the values are close and it is not known what reaction would ultimately be used.

References

1. McKay, C.P., Toon, O.B., Kasting, J.F., *Making Mars Habitable*, Nature, v352, pp 489-496, August 1991.
2. Mole, R.A., *Terraforming Mars with Four War-Surplus Bombs*, JBIS V 48, pp321-324, July 1995.
3. Anon, *Plastic Power*, Scientific American, December 1996.
4. McMillen, K, *Methanol, a Fuel for Earth and Mars*, The Case for Mars V, Univelt, San Diego, 1997. and: Zubrin, R., Clapp, M., and Meyer, T., Pioneer Astronautics, New Approaches to Mars In-situ Resource Utilization Based on the Reverse Water Gas Shift, AIAA 97 0895, American Institute of Astronautics and Aeronautics, Reston, VA, 1997
5. Averner, M.M., and MacElroy, R.D., *On the Habitability of Mars – an Approach to Planetary Ecosynthesis*, Scientific and Technical Information Office, NASA, Washington, DC, 1976.
6. Oberg, J.E., *New Earths, Terraforming Other Planets for Humanity*, Stackpole Books, Harrisburg, PA, 1981.
7. Henninger, J., *Solar Absorptance and Thermal Emittance of Some Common Spacecraft Thermal Control Coatings*, NASA Reference Publication 1121, NAS 1.61:1121, NASA, Washington DC, 1984.
8. Gold, T., *The Origin of Methane in the Crust of the Earth*, in the US Geological Survey Professional Paper 1570, The Future of Energy Gasses, approx. 1993.
9. Various World Wide Web (Internet) sites at Carnegie Mellon University and NASA:
   Autonomous Mining Truck: www.frc.ri.cmu.edu/~ssingh/fastnav.html
   Excavator: www.frc.ri.cmu.edu/~axs/mass_ex.html
   Autonomous Off-road Navigation: www.frc.ri.cmu.edu/~tugv/home.html
   Nomad Autonomous Rover: www.ri.cmu.edu/atacama-trek/
   Pathfinder Mars Rover: http://mpfwww.jpl.nasa.gov/default.html
10. Halman, M., *Chemical Fixation of Carbon Dioxide – Methods for Recycling CO₂ into Useful Products*, CRC Press, Boca Raton, FL, 1993 (p 37). This references: Ishihara,T., Fujita, T., Fixation of carbon dioxide to carbon by catalytic reduction over metal oxides, Chem. Lett., 1991, 2237-2240.
11. Schodt, F.L., *Inside The Robot Kingdom: Japan, Mechatronics, and the Coming Robotopia*, Kodansha International, Tokyo, 1988.
12. Personal conversation with Dr. Bruce Jakowsky, U. of Colorado Laboratory for Atmospheric and Space Physics, August 1997.
13. Freitas. R.A., and Gilbreath, W.P., Editors, *Advanced Automation for Space Systems*, NASA CP 2255, National Technical Information Service, Springfield, Virginia, 1982.
14. Fogg, M., *Terraforming: Engineering Planetary Environments*, Society of Automotive Engineers, Inc., Detroit, 1995
15. Lovelock, J., and Alleby, M., *The Greening of Mars*, Warner, New York, 1984.
16. Lippencott, W.T., *The Study of Matter*, 3rd Ed., Wiley and Sons, NY, 1997.

//

Human-Assisted vs. Human-Assisting Systems in Mars Missions†

Ned Chapin, InfoSci Inc.
NedChapin@acm.org

Abstract

Human-crewed space missions are often contrasted with robotic space missions. Actually, human beings are part of both types of missions; the distinction really is one of the roles or functions done by the people, of where the people are, and on the systems included. For many purposes, the human-crewed vs. robotic mission distinction is inadequate, as when the concern is getting the best value for and from the people involved in the mission. Then the focus shifts to be on the mission's systems. With current technology, a more useful distinction often is between human-assisted vs. human-assisting systems.

With older technology, the systems were primarily human-assisted, and often concerned with the interaction of people and the mission infrastructure. Two examples are the breathable atmosphere system, and the inflight navigation system. With current technology, the systems can be human-assisting, where the system effectively acts as an extension or enhancement of the human contribution to the mission. Some human-assisting systems may be concerned with the interaction of people and the mission infrastructure, such as a physical sampling system onboard a human-crewed Mars rover. Some may be concerned with acquiring data, such as about soil compaction. Some may be concerned with analysis procedures, such as ascertaining soil composition. Some may help assess the meaning of acquired data, such as in doing pattern recognition in data accumulated over a sol. Some may help integrate work progress to date with limiting and environmental factors, such as in on-site ongoing planning of specific activities on Mars. Using human-assisting systems in Mars missions can multiply the capability of the people, thus making their efforts more effective and productive.

Performance Characteristics

Introduction

Since the most expensive kind of object to send on a Mars mission is human beings (people), what should be the role of people in Mars missions? Both the robotic missions and the human-crewed missions have people serving in on-Earth support roles. But human-crewed missions also have people going to (and from) Mars, with the expectation that they will do something valuable while on Mars (and upon their return). How can the value contribution of the people serving on Mars missions be maximized?

This paper starts by looking at the performance characteristics of systems, both living and non-living. The distinctively human characteristics are especially relevant for Mars missions. Then this paper contrasts human-assisted systems with human-assisting systems as it summarizes key characteristics of both types. Next, this paper considers the relevance of both types of systems for Mars missions, and sketches some of the trade-offs involved in practice. A conclusion ends this paper.

System Characteristics

Viewed broadly, systems are means of getting something done by a combination of materials, equipment or structure, and method.[1] Viewed broadly, the things called "systems" include non-human-creations. They may be inanimate, such as the solar system, or animate such as plants (such as trees or seaweed), animals (such as dogs or bacteria), and people. In general, the animate systems have four characteristics that are relevant here: they themselves normally incorporate (are composed of and use) systems, they are sensitive to stimuli, they do something (take action), and they use methods to link stimuli to actions.

For animate objects, the stimuli may be anything, such as radiation (examples are heat and light), presence (such as of shelter or a predator), or contact (such as pressure or water). Different animate objects have different sensory capabilities (for example, cats do not have color vision and honey bees sense ultraviolet radiation). The action taken may be anything, such as movement (such as a tree bending in the wind, or a worm wriggling, or a woodpecker knocking), or ingestion (such as roots taking up water or a bear eating a fish). The animate object uses the method to select the action as a response to the sensed stimuli. For example, a duck paddles to move in water, a mouse runs away from a snake, and a plant positions its leaves to receive light. In short, the animate objects have stimuli-responses methods, but the linking methods are limited by what stimuli the object can sense, and by what actions the object can take – for example, grass does not and can not run away when a grazing cow approaches.

Distinctively Human Characteristics

Human beings (people) have a far richer repertoire of linking methods and responding actions possible than other animate objects. People have access to tools that can extend the types and ranges of stimuli that they can sense. For example, tools enable people to sense the presence of electromagnetic radiation from radio frequencies to X-rays, and pressures from micropascals to megapascals. Tools also can extend the reach and scope of responding action for people. Pens, space vehicles, backhoes, telephones, computers, shaped charges, and cranes are but a few of tens of thousands of examples. People also have far more capability than other animate objects to determine the timing in linking stimuli and responses, from fractions of a second to centuries.

People's language abilities in transmitting knowledge and culture also strongly affect the linking methods.[2] People can be trained to accept stimuli they otherwise would not notice, and to search a wealth of possible actions in selecting a response. Examples abound, such as physicians, plumbers, bioinformatics researchers, chefs, detectives, and bulldozer operators. Some of this wealth comes from the different classes of responses. For instances, responses may range in space as well as in time, from within arm's reach, to local (such as within unaided sight or hearing), to extended (such as by telephone or video conference), to selectively very distant (such as to a comet or Mars). All of the responses beyond local are tool-assisted and have some sort of power source. Some of the responses to the stimuli are mostly routine (such as in driving an automobile), whereas some are more open (such as in playing sports) or involve a lot of in-flight creativity in the linking method (such as in playing a first-time jazz solo, or doing research on prions).

People commonly cast themselves into systems of their own creation, where usually they provide the methods for the system to use, or sometimes provide the methods even as the system operates.[1] For example, people may design a robotic rover for use as one of the systems for exploring Mars, or in a crewed rover system exploring Mars, the in-rover crew may be selecting on the fly what to do and when and how to do it. In either case, the characteristics of the system become very close to the characteristics of an animate object, as sketched above. In most human-created systems, people include themselves either as direct components of the system, or as the original providers of the method of linking the stimuli and responses, or as both.

Human-Assisted System Characteristics

In a human-assisted system, the system uses the help of people to do what people want the system to do. An ordinary automobile is an example because it is a system that requires a driver. Or for example, in a communication system using a typewriter, a person has to activate the typewriter, letter by letter, by pressing on the typewriter keys. Such a system is a way of communicating from person to person, but it only works when the sender or someone on behalf of the sender spends some time and effort in serving as a typist.

The roles of the people in a human-assisted system typically are concentrated at the interface between the system and the people involved with the system. As for example in using a word processing system such as Word™, people not only direct the system, but also apply their skills, training, experience and decision making in getting Word™ to work satisfactorily. The skills needed are often in data preparation and data entry. To acquire and keep the skills, both training and experience in using the rest of the system (the equipment and materials including software) are required. In practice, in feature-rich systems the people operating the system usually are proficient only in a subset of the features that they use often, and have to look up or get help on features not commonly used. The method demands are relatively low-level usually and implemented often by selecting from lists of specific alternatives, as from pull-down menus.

The system performance of human-assisted systems often gives the impression that the people are behaving in a habitual manner. In the systems, the stimuli-responses linkages are mostly built-in and usually involve only selection by a person for their activation. Sometimes the linkages are more distant and less obvious, as in the use of control keys or macros in word processing systems. Typically the system's function and limits are set by the equipment, and augmented and instantiated by people's selecting among predetermined alternatives. To the people serving as participants in a human-assisted system (providing its "assisted" character), the system typically seems dependable, reliable, consistent, and familiar (but with its own personality).

Some major systems used or proposed for use in Mars missions provide some examples. Since the Mars atmosphere is not suitable for human beings to breath, a system for providing a breathable atmosphere within an enclosed volume on Mars will probably be implemented as a human-assisted system. The system's function is to provide a specified atmosphere. People will have to clean it, adjust it, tweak it and monitor it. To enable it to operate without continuous or frequent human direction, people will have to program it, probably have to revise the program many times, and may even have to reprogram it.

Another example is the in-flight navigation system for use on the space vehicles transporting people between Earth and Mars. The system's function is to provide navigation for the space vehicle it is in. People will want to have the ability to obtain data from the system and enter data as well as commands – i.e., people will tend it and may redirect it.

A third example is the on-Mars ground portion of a robotic sample return system. The system's function is to traverse an on-the-ground distance to a specified point, collect a sample of the geology there, and return to its starting point with the sample intact. The people enter data and commands to specify the point and the kind of sample to be acquired.

Human-Assisting System Characteristics

In a human-assisting system, people use the system to help them do what they want to or have to get done. The system serves the person like or as an assistant, an aid, a resource, or a tool. An example of a human-assisting system is an automated directory in a large retail store that can promptly answer such questions as "Where are plastic measuring cups?" Another example of a human-assisting system (this one computer-implemented) is one that calculates and reports the basic statistics (count, range, mean, median, variance, standard deviation, histogram, etc.) from data designated by the person.

The roles of the people in a human-assisting system usually center on the person's desired result or purpose in invoking the system's performance – i.e., on how the system can help the person achieve something or get it done more quickly or accurately or at a more favorable cost. The people do little in guiding or directing the system, but instead accept or reject what the system does for them. The skills needed are in selecting the system to use, in recognizing the strengths and limitations of the system, in discovering or enhancing the use of the system, and in planning to make the use of the system to be effective in helping the person. The training and experience are mostly on the facilities and features of the system, and on delegation and collaboration. That is, the emphasis is on how to marshal the system's capabilities and limitations in combination with those of other systems, and with those of other people. The decision and direction needed are mostly relatively high-level focusing on the recognition of goals and constraints.

The system performance of human-assisting systems often gives the impression that the people are being quite varied in their behavior. Partly this is because the people are varying their selection of the systems to use. The stimuli-responses linkages exhibited by the people are less cut-and-dried, even though the stimuli-responses linkages in the selected systems are mostly built-in and mostly operate in a self-directed manner. The selection of systems gives the opportunity for the people to extend their stimuli sensitivity and the scope of their responses. Communication may figure significantly in either the stimuli or the responses or both.[3] Dynamic access to databases is usually present.[4] The combination of people and systems sometimes adds effectively a learning capability to the combination for the people involved. For some situations on Mars missions, human-wearable computer and communication gear appears to have an attractive potential.[5]

People working on Mars missions will likely find good use for human-assisting systems. One example is devising a recovery strategy for a major shortfall in greenhouse productivity. The people will want to call upon a variety of systems to help in diagnosis (why the productivity has fallen off), additional systems to help in selecting the action to take, and a still different set of systems to help carry out the selected action. In devising responses to the situation and its attributes (the stimuli), the people will seek the assistance of an assortment of systems and want to draw upon records stored in databases to get data that may be analyzed to provide help.

A second example is assessing the presence of lithic organisms in collected Mars samples. This can be done manually but somewhat tediously with the aid of minor laboratory equipment, reagents, microscope, etc. An automated self-contained system could save time and effort for the people, and save on materials. A third example is classifying weather patterns on Mars. This can be done by a computer or by people doing a manual analysis of recorded Mars weather measurements. An enhancement of the weather recording systems could classify the weather pattern as measurement were made and take into account measurement trends. This enhancement could save some personnel time, and provide more current weather data for the people.

A fourth example is doing data reduction and data analysis on the data originating in Mars field work. The raw field data can be very varied because of the sometimes unanticipated situations encountered in field work, vagaries in the measurement and recording process, changes in field work objectives from analyzes, differences in the disciplines involved, etc. Getting the field data in order and duly made a matter of record (data reduction) can be done manually at a major cost of personnel time, even with currently available computer assistance, as has been reported from the Mars Society field stations.[6] Added to that is the personnel time and effort to do the analysis work that should be done as fully as possible to get the value from the data in guiding subsequent field work. An important area for human-assisting systems is in more capable systems for helping personnel get field data captured, edited, recorded and analyzed.

A fifth example is replanning a Mars rover reconnoiter. Environmental conditions, such as dust storms and unforeseen physical obstacles and/or constraints, and changed objectives from fresh data analyses, can be the stimuli leading to a recognition that previously planned and scheduled Mars rover activities have a low probability of being worth the time and resources. Replacement activities of higher value then are needed, and have to be planned and scheduled. Doing such planning and scheduling can be a manual process, but can be costly in terms of personnel effort and quality of the result. Human-assisting systems can help the people create faster and better quality replanning and scheduling responses at less cost of personnel time.

Relevance for Mars Missions
Systems occupy positions on a continuum from pure human-assisted to pure human-assisting. Nearly all systems for use on or in Mars missions have some human-assisted attributes and some human-assisting attributes. With current technology, the typical system is a mixed bag but with a bias or weighting toward being either human-assisted or human-assisting. When the concern is providing the most value to people or getting the most value from people, the systems that contribute most are the more strongly human-assisting ones. In other words, when the concern is "bang for the buck," the focus moves in the direction of the human-assisting systems.

Both types of systems are influenced in their capabilities by changes in the technology and in the infrastructure. Both technology and infrastructure change in a jumpy fashion from the irregular flow of usually incremental improvements. This and previous Mars Society conventions have included reports on technology and infrastructure improvements.[7]

For use on or in Mars missions, four concerns affect the relevance of the balance between human-assisted and human-assisting systems. One concern in reliability. People, even those intensively trained, are famous for their relative creativity in responding in new situations, and their relative unreliability in providing consistent unfaltering performance in routine situations. Software failures and shortcomings have been a continuing concern in space missions, including Mars missions. Hardware or equipment failures or malfunctions have also adversely affected space missions, including Mars missions. Given the current technology and infrastructure, neither human-assisted nor human-assisting systems are totally reliable. Hence, risks must be assessed for them as for any systems proposed for use on or in Mars missions.

A related concern is system robustness. To paraphrase an old advertising slogan, systems should be able to take a licking but keep on ticking anyway doing their functions. Since unanticipated circumstances are likely to arise during Mars missions, robustness is important in systems, and risks should be assessed.

A third concern is maintainability. On Mars missions, repair or enhancement services for systems are sometimes in very short supply, such as on Mars itself. Yet systems may fail or need to be altered to meet newly recognized needs, whether they be human-assisted or human-assisting systems. For Mars missions, systems should have maintainability facilitating features built-in.

A forth concern is substitutability. If a system fails or becomes unavailable, what is the work-around or the substitute source of functionality? And what is the cost of that work-around or substitute functionality in reduced capability, flexibility and versatility for the people involved? For human-assisted systems, a common but relatively costly approach is redundancy. For human-assisting systems, a common approach is multifunctionality in the systems provided. This is parallel to providing cross-training for the mission personnel.

Trade-Offs

In considering the trade-offs for Mars missions of human-assisted vs. human-assisting systems, one important fact has to be recognized – on Mars, the people's environment will be effectively completely paperless.[4] Hence, one of the major needs is for human-assisting systems that compensate for or take advantage of that paperless environment. This applies both for communication of data through space, as between people on Mars or between people on Mars and Earth, and through time, as from one sol to later sols from or to any person on Mars.

The difference on Mars missions between human-assisted and human-assisting systems on weight and space are of concern. The human-assisted systems will be justified for weight and space on the basis of what the system does, such as provide human-breathable atmosphere. The human-assisting systems will require justification for weight and space by the value that they are expected to contribute to the mission and the people associated with the mission. Since systems typically include both equipment and materials, some burden of weight or space can be expected for each human-assisting system taken for use on Mars. Different systems will have different features to trade-off against differences in weight or space.

For a system to be used on Mars, durability of the system and its components is not a trivial matter. Dust and radiation are two of the major known stresses on system durability. The gear or equipment in the human-assisting systems should have no small moving parts, no parts needing lubrication, and no hard to clean (for dust) parts or surfaces or spaces.[3] Compromises that may adversely affect system functionality may be needed to gain needed durability. While similar requirements apply to human-assisted systems, some parts of those systems are likely to be physically larger (such a crewed Mars rover) and hence can be made more durable more easily.

The hardware or equipment chosen for use in systems on Mars missions also involves trade-offs. These are not new, and apply to both human-assisted and human-assisting systems. Should the hardware be commercial off-the-shelf with a known track record on the attributes noted above and in relevance, or be custom made for the mission? Equipment custom made to fit the need may be easier or more "intuitive" for people to use, especially in human-assisting systems. But off-the-shelf hardware (and associated software) may be less expensive and more quickly available.

Conclusion

The trade-off and relevance considerations are mostly of concern when the issue is how to most empower the people serving in Mars missions, and how the Mars missions can get the most value from people's involvement in Mars missions. The people and systems involved in Mars missions receive stimuli, and make responses in carrying out their roles in the missions. Human-assisted systems use people to perform functions within the system as the system operates. People use human-assisting systems to help themselves perform their roles in the Mars mission better, faster, or at a lower cost. Historically systems have been primarily human-assisted. As crewed Mars missions are considered, planned and performed, human-assisting systems can help get more return from the Mars missions by making the people associated with the Mars missions more effective and efficient in their roles.

References

1. Ned Chapin. *Computers – A Systems Approach*. Van Nostrand Reinhold: New York NY, 1971.
2. Richard G. Kline. *The Dawn of Human Culture*. John Wiley and Sons, Inc.: New York NY, 2002. Justin Gillis. Gene mutations linked to language development. *Washington Post*, August 15, 2002:A13.
3. Ned Chapin. *Communications Infrastructure to Support Human Activities on Mars*. On To Mars. Apogee Books: Burlington Ontario, Canada, 2002.
4. Ned Chapin. *What About the Data?, On To Mars*. Apogee Books: Burlington Ontario, Canada, 2002; 154–159.
5. Christopher E. Carr. *Applications of Wearable Computing to Exploration in Extreme Environments. On To Mars CD-ROM*. Apogee Books: Burlington Ontario, Canada, 2002; 2000 paper # 5.
6. Heather Chluda, Jennifer Heldmann, Steve McDaniel, and Troy Wegman. *MDRS Crew Rotation 1: Doing Science Under Challenging Conditions*. See elsewhere in this proceedings volume. Pascal Lee. *Simulating Mars on Earth – A Report from FMARS Phase 2. On To Mars CD-ROM*. Apogee Books: Burlington Ontario, Canada, 2002; 2001 paper # 1.
7. Examples in this year 2002 5th Convention are presentations in tracks 1B, 5B, 1D and 2D.

Increased Cost Effectiveness of Mars Exploration Using Human-Robotic Synergy†

Richard L. Sylvan M.D.
rlsylvan@aol.com

Abstract

The most effective way to rapidly explore the entire surface of Mars may be through the use of intelligent robots controlled in part by scientists on Earth and in part by astronauts on the surface of Mars. As was presented last year, the use of local controllers on Mars would eliminate latency and allow robots to cover more ground and travel to more interesting areas, while still allowing scientists on Earth to control their instruments. This combined control has the potential to increase data return by several orders of magnitude.

A number of advances in robotics and changes in NASA's rover concept make this idea even more attractive. The overall concept is further developed in the context of these advances and changes.

Overview

The exploration of truly a virgin territory is an experience that has few modern parallels: the late 19th and early 20th century exploration of the polar regions, and the more recent exploration of the Moon. In both of these cases, the explorers entered unknown territory having incomplete information concerning the characteristics and dangers of the area being entered. They had to rely on equipment that could not be fully tested until actually used in the virgin environment. During their period of exploration they were cut off from direct assistance from the remainder of humanity.

Of the two, the Moon exploration appears to be the one closest in analogy to the exploration of Mars. In the case of both the Moon and Mars, direct aid may not be possible, but assistance via communication will still be available

The exploration of the Moon took place in a relatively orderly manner:

1. Simple robotic devices go to the Moon (Pioneer).
2. Somewhat more complicated robotic devices do more detailed exploration (Ranger).
3. Detailed robotic evaluation of the surface prior to human landing (Surveyor).
4. Humans land on Moon.
5. Humans explore Moon (humans take over).

The exploration of Mars presents similar problems when compared to the Moon, but the increased distance and the at least theoretical possibility of improvement in the capacity of the robotic devices allows a slight modification of the above model:

1. Simple robotic devices go to Mars (*Sojourner*, Lewis and Clark).
2. Somewhat more complicated robotic devices do more detailed exploration (2009 long duration rover to be followed by similar robots of a similar, but more advanced design.
3. Detailed robotic evaluation of surface prior to landing (still be designed)
   3.5 Robots carefully examine and possibly even prepare potential landing site(s) (still to be conceptualized).
4. Humans land on Mars.
5. Humans explore Mars (humans take over, possibly with limited local robotic help, perhaps with independent unmanned missions continuing).

In both cases, robotic devices pave the way, after which humans, perhaps assisted by robotic devices in the landing zone, are the dominant explorers of the Martian landscape. Once humans arrive, the era of robotic exploration ends. In the case of the Moon, with the exception of a limited Russian program, the model obviously held. NASA, although commencing by fits and starts, appears to be heading in the same direction in its plans for Mars. The Mars Society model also does not contain a robotic component once humans arrive, except for local assistance of the human exploration process. This model although attractive, may not represent the best way of exploring the entire Martian surface.

Concept

In the exploration of Mars, both humans and robots have well established advantages.

Robots are relatively inexpensive and their failure costs substantial loss of time, money and potential science, but not human life.

Human exploration carries much higher potential costs, both in risk and money, but allows the use of an observer unmatched in its capacity to evaluate and interpret information.

Mars is big and by no means homogeneous.

Problem

In the exploration of Mars, or of any body more than a few light seconds from Earth, one is faced with the problems (among many others) of latency (the delay in communication caused by distance), bandwidth and signal strength. This latency greatly increases the difficulty and reduces the efficiency of robotic movement. Movement of the robot has to be controlled very carefully to avoid damage to these most precious of machines, limiting the area that can be explored by robots controlled from Earth, a distance that involves substantial latency.

In addition, communication at a distance causes a loss of signal strength and a reduction of effective bandwidth, reducing the rate of information transfer. Instruments are often down-designed to match the availability of bandwidth and signal strength.

The latter two problems can be partially ameliorated. Increased artificial intelligence in the distant robot may allow the performance of complex tasks without the immediate intervention of the operator. Narrower transmission beams of greater energy, perhaps driven by nuclear power, better multiplexing and data compression, spaced based relay stations, and more sensitive receivers, etc., can improve the acquisition and transfer of data. Neither problem can truly be eliminated, and the problem of latency is unsolvable within the context of known physics.

Latency causes the movement of Mars rovers to be painfully slow. The *Sojourner* rover moved a total of some 90 meters during her stay on Mars. Some of the limitation of movement was caused by the need to do science, once the rover reached its scientific target. The majority of the limitation derived from latency. Each movement of the rover had to be programmed, sent to the rover, and then the rover had to be checked to make sure it was where it was supposed to be. Each of these processes had to wait for a good portion of an hour before the next portion of the process could continue.

The Lewis and Clark Expedition is considered to be one of the best and most complete of the 19th century scientific explorations. They traveled about 10,000 miles, in round figures, to explore about 1,000,000 square miles. They did an excellent job. If one were to use a robot traveling at *Sojourner*'s rate of velocity, it would take a billion years to explore Mars to the same degree of detail. Robots may improve. *Lewis* and *Clark*, those well named 2003 rovers, will be 100 times as fast. (5,000,000 years for the pair). Assuming that we have a Moore's-law-like improvement in the efficiency of robots, (which is probably a relatively liberal assumption), in 15 years robots alone would be efficient enough to perform the task in 5,000 years. This is obviously far beyond the time frame one would like to explore Mars.

Also, current landings, both manned and unmanned head towards what mission controllers call mattresses, relatively flat landing areas, to cut landing risk. Unfortunately many of the more interesting formations are in rocky or even vertical areas, which are very difficult to impossible to reach with current and foreseeable rover concepts. One has to traverse the distance between the mattress and the area of interest, which takes substantial time. Movement, which was slow in a relatively simple area, becomes even slower when the terrain becomes more difficult.

Human exploration has its own set of problems. Human exploration is actually a dual mission:

1. Human exploration
2. Human survival

Actually, if one delineates the actual importance of the two missions, the actual mission is more like:

1. Human exploration
2. HUMAN SURVIVAL

Human survival is actually the primary mission. Because of the importance of the first explorers surviving, the initial landing will almost definitely be on a particularly flat, soft, relatively rock free mattress. In addition, for safety's sake, it will be extremely well explored, evaluated, and perhaps even prepared robotically prior to the human landing. The initial human landing site will be the most scientifically and physically evaluated area of Mars. We will know more about that spot than any other spot on Mars. And having explored that area better than any other, we will land there. Because of this need for safety, the initial human landing on Mars under the current model of exploration may actually slow the process of gaining information about Mars' surface. Only after an initial shakedown period will we be capable of freely exploring Mars as a primary mission. That will last several weeks to months at a minimum, and may take several expeditions before our equipment is sufficiently reliable to undertake long distance explorations. In Apollo, broad based scientific exploration in the fullest sense didn't get into high gear until *Apollo 15*.

In addition, both Dr. Zubrin's model of Mars exploration and its NASA equivalent use the concept of a chain of linked bases as a safety factor in the exploration of Mars. Each new base will be close enough to the previous base or bases to allow each to be used as a backup for the others. Because of the danger of Mars exploration, this is a more than reasonable concept. It means, however, that much of Mars would be unexplored for an extended period of time, until either the chain reached that point, or a new chain was initiated.

As mentioned above, the area of Mars that will be the initial site of landing will be the carefully evaluated before landing, to maximize the safety of the astronauts. We will explore that area so that we know it far better than any area of Mars, and then we will land there, Therefore, ironically, the area we will initially explore with humans will be the area of Mars that we know the most about. Human exploration of that area will add to that knowledge, but it would add greatly to our knowledge base if that human oversight could be extended over greater areas of exploration.

Proposed Model

I propose that the control of Mars robots be divided, with the movement of the devices being controlled by explorers on Mars, while the primary control of the instrumentation be retained by the scientific group on Earth. Mars based control of robotic operation has been suggested for local exploration of the area near the landing site. I suggest that Mars based control of movement be extended to robots on the entire surface.

Control of movement by a Mars based operator has several advantages. Latency would be reduced from many minutes to a fraction of a second. The rate of robotic movement could be increased by several orders of magnitude. One still would need to use exquisite care with the precious robotic explorers; however, instructions can come every few seconds, rather than every 1/4-hour or more. That resulting increase in speed would allow for much more rapid data acquisition. It would also render practical the concept of landing in a safe area and progressing in a relatively short time to an area which is uneven or rocky but geologically more interesting. It will hence allow more complete access to a much broader group of geographic areas, including difficult, rougher and more complex portions of the Martian terrain. This will facilitate a more rapid and complete evaluation of the entire surface of Mars.

In addition, as both NASA and Mars Society experience (at Devon Island Base) has shown, human EVAs are a stressful, tiring and potentially dangerous experience. They have to be very carefully planned. The equipment that they depend on cannot be tested completely (at least on the first few expeditions), and may not be capable of being used for the full duration of the mission. The Mars explorers may find themselves in a situation where they are on Mars, but are limited in their ability to actually explore Mars. By using their ability to control exploring robots on the entire surface of Mars, they will be given a mission that is important, that only they can do, and is interesting, while giving them a respite from the very stressful, although magnificently exciting EVAs.

Requirements

Although the model is conceptually simple, it requires several modifications and additions to the standard concept of Mars exploration. These can be divided into five basic areas

1. Alteration of the robotic explorer.
2. Alterations and additions to the equipment in the habitat.
3. Changes in astronaut training.
4. Modification of the workings of the human team on Earth.
5. Additions to the communication system orbiting Mars.

Alteration of the Robotic Explorer

The robotic explorer has to be modified in several ways. The proposed model requires a dual set of controls, one for the Earth team, and one for the Mars team. There has to be a careful and distinct division between the two. The Earth team should have control of the entire spectrum of potential instructions to the rover. At times when there is no Mars controller present, either because there are no humans on Mars, or due to unavailability of a Mars operator, the Earth team would control the robot just as it always had. Certain instructions, particularly those that modify the software controlling the instruments, should be unavailable to the Mars team, except in very carefully defined circumstances. Release of this lockout should only be available to the Earth team, and only with very careful control of security

Although the shuttle (and to a limited extent the ISS) has given scientists the experience of having experiments done by astronauts, these were in a relatively limited context. The Earth scientists using Shuttle or ISS experiments are doing a defined experiment that was designed to be performed in microgravity. The were not performing the open ended exploration of a partially or completely unknown environment

The Shuttle and ISS experiments are either self-controlling or run by an astronaut in a defined way. In this case of a robot on Mars, however, the Earth based scientist is exchanging increased information for loss of control. The robots and their instruments are extremely important to their developers, almost to a personal level. A large chunk of the experimenter's professional lifetime is put into the Mars explorer. It is the most important thing in his professional live. It will take substantial effort to allow even limited control of the precious machine to be placed in the hands of someone not on the initial team. At least initially, control of movement only should be in the hands of the Mars group. As experience is gained, and confidence in the system (hopefully) grows, some of the scientific functions can be performed on Mars as well.

The design of the controls should be standardized, from the point of view of the controller. There is substantial precedent for this in the aerospace world. All of the Airbus cockpits have extreme communality. This makes retraining easy, pulling of the stick or use of controls has similar consequences in all of the Airbus planes, regardless of size or design. The fly by wire system translates the pilot input into orders specific for the model of plane being flown. Similarly all of the designers of robots that would use a Mars controller should have an identical toolset. This would not only simplify training of the Mars controller, but would also limit confusion and reduce the potential for subsequent damage to the robot. This would not limit the computers that the designer wished to use. It is very common in the computer world to have a common high-level language that is translated into a broad range of lower level languages at the operating level (Java is an example of such a translation scheme). In addition, one could extend the toolset in later models as long as the system is upwardly compatible.

In addition to alterations in the control system, some degree of modification of the robot would be useful, although not mandatory. Both human and robotic missions to Mars will be planned years before their expected arrival on Mars. Both may have alterations of launch schedule. Any robotic device that might be on Mars during a human period of exploration should have the equipment to be dual controlled. It should be capable of utilizing the higher rate of instruction that would come from the Mars operator. Its cameras should be capable of an output rate that allows the Mars observer to get a picture of a robot's location between several times a second to (at a minimum) every few seconds. Otherwise the advantage of reduced latency will be lost. To increase the probability that it will be functioning on Mars when human explorers are there it should be made for maximum duration of survival (as the Viking landers were, but *Sojourner* was not).

Instrument power and communication systems need to be made more robust, so as to accept the increased number of duty cycles per unit time. They should also be capable of responding to instructions every few seconds rather than many minutes apart. If a rover has artificial intelligence that enables it go to a particular location, it should be capable of being overridden by the Earth team to allow fine control by the Mars team. Any self-protection programs to protect the robot from loss by striking an object or falling into a hole should obviously be left intact even when under Martian control.

If the program actually is put into place and matures, instruments in both the rovers and in the habitats may be altered and improved to take advantage of the increased bandwidth available to the Mars observers.

Alterations and Additions to the Equipment in the Habitat
In its initial configuration, the system will need relatively little equipment. The habitat will probably have several computers, using whatever configuration (laptop, PDA, or something that is created over the next ten years or so) is current at the time of its final design and construction. Its software should fit into any current computer, let alone one several generations more advanced. It will require a visual screen that shows sufficient detail to navigate. It will require a dedicated controller. The experience of both pilots and gamers has shown that keystroke control is not as facile nor as accurate as a joystick like device.

Most of the planned rovers have fields of regard (what the controller sees) that are changeable without moving the rover. This allows one to scan the surrounding territory without turning the rover. This is substantially more flexible than a system that only looks straight ahead. If one wants to go into rough countryside, such flexibility is obligatory. Without the ability to look around without moving, planning paths through complicated territory would be almost impossible. Although the final design should be subject to actual testing, my preference would be to have the controller of the visual system be an independent device, separate and distinct from that of the movement control. One could save weight by having them combined into one controller, switched by a toggle. The risk of moving the machine by accident when one only wanted to change the field of regard, however, would be high in such a system. The gain of lighter weight would not be worth the increased risk of rover loss in rough terrain.

Changes in Astronaut Training
Changes in astronaut training should be limited. The control system would have similarities to joystick controls found both in aircraft and video games. Acclimation should be easy. The astronauts would have to learn the specific needs of each robot and its scientific team. They would also have to become familiar with the characteristics of each Earth based team from the point of view of their objectives and personalities. Once again, this is just an extension of what they already do.

As mentioned above, however, astronauts have been doing science for others since the inception of the space program. If the control systems were standardized, the astronauts would have to learn only a single set of control systems for each broad type of robot, and continue to practice with these controls during their flight to Mars. The only addition to training is that the astronauts have to be willing to consistently ask for confirmation if they have any doubt concerning an instruction set, and avoid the very tempting path of saying "I know what that means," just to save a little time. Latency is not well tolerated by humans. However, to misinterpret an instruction is to put a robot or instrument at risk.

Modification of the Workings of the Human Team on Earth
The Earth based engineering / scientific teams may require more work than the astronaut team. First of all, while there will be one astronaut team at any one time on Mars in the near term, there will (hopefully) be several robotic teams. Other than budgetary constrains and some degree of oversight from NASA, each of these teams will have become used to designing, building, and then operating their robots in relative isolation, at a pace defined by the team. Essentially all communication has been within the team, using jargon and tacit agreements common to the

team members. They now will have to give up some of their control to someone outside the team. While this program has the potential to increase scientific output substantially, some members of the robotic team may resent the introduction of outsiders.

In my first Mars Society paper on this subject, I discussed in some detail the potential models for communication with latency between experts and others, and I will not go over that ground here.

I would reiterate however, that communication in the presence of latency is not simple. It is to some extent an art. Whichever model of communication is utilized, the key ingredient is that the expert must realize that it is the expert's responsibility to communicate successfully, not the recipient's. Because of the expert's familiarity with his machine or problem, he has a tendency to act as if the less expert receiver of instruction has his expertise, database and knowledge of the jargon used in the field. Any time two individuals with differing levels of knowledge and experience communicate, errors in communication may occur due to misunderstanding. Latency will only add to this problem, since the ability to ask, "What did you mean by that?" and get an immediate response will be lost. A call back for clarification should be welcomed, not felt to be an intrusion. Not all communicators can act in this manner, and if they cannot, they should not be the individuals who communicate for an Earth based team

Additions to the Communication System Orbiting Mars
The minimum system usable in this model is a single equatorial satellite just above one Mars radius high. This would allow a base within 60 degrees of the equator to have contact with robots close to the Martian base that most of the time, and cover the 240 equatorial degrees of the Martian surface centered on the Mars base and bounded by 60 degrees North latitude and 60 degrees south at least part of the time.

Two equatorial satellites 120 degrees apart would allow complete but intermittent coverage of the entire equatorial band, but would still leave the northern and southern thirty degrees about the poles uncovered.

Three equatorial satellites would give continuous coverage of the entire equatorial zone. Adding a single polar satellite would give at least intermittent coverage to the entire planet.

Ideal coverage would be three polar and three equatorial satellites. This system would allow continuous contact throughout the planet with a maximum two-way latency of less than $1/6$ second and a minimum one-way latency of slightly more than $1/25$ second.

Advantages of the Model
This model has several advantages over the classic program of Mars exploration. It allows exploration of large areas of Mars efficiently, even those distant from the landing site. The amount of time a robotic explorer spends moving to a site or actually doing science rather than waiting for commands from Earth would be greatly increased perhaps by as much as 100 to 1,000. This would make placing humans on Mars expensive, but cost effective from the point of scientific production.

It would allow one to reach difficult locations in a timely manner. The rate of movement of rovers, even "smart" rovers, is relatively slow. To get to a geographically rough but interesting area would involve either substantial landing risk, or landing in a mattress, then slowly moving cross-country. This model would allow one to land in an easy area and rapidly progress to one of geological interest.

It would allow missions to be successful in the presence of limitation or failure of the ability to perform EVAs. Particularly early in the period of Mars exploration, suits, rovers and airlocks may not be as tolerant of the Martian environment as one would like. If Mars based rover control was available, the mission would still be extremely valuable even if EVA was unavailable. This portion of the mission could be performed even if the mission was unable to land. If would also allow the astronauts to look at locations distant from their landing site, breaking up the monotony of isolation. If robots were to land on Mars during the mission, it would allow a new extension of their mission, which would not only add new science, but would be psychologically useful in reduction of isolation monotony. If fast new propulsion systems are available but not man rated when humans arrive on Mars, they could be used to land short duration inexpensive rovers that could benefit from Mars control.

Finally this is the only portion of a planned mission that can be almost completely tested on Earth.

Problems in Initiating the Program
Many of the problems in the initiation of such a program were discussed in paper one. In brief, the three groups of major problems are the budgetary, the psychological and the logistical.

Budgets are controlled by one of two separate arms in NASA, human and robotic experimentation. The program overlaps the two and is at home in neither. Although the scientific reward for such a program would be great, at its onset it would be seen as taking time and resources from the programs in place, the sending of either humans or robots into space. It would require some degree of redesign of instruments and communication packages. It would require robotic design teams to accept relinquishing some degree of control to outside operators. It would require resolution of language and training barriers. Cross training is never easy.

Logistical barriers are of three types. Both robotic and human missions to Mars have very long lead times. To get the maximum benefit from even the simplest form of this program, any robotic probe that might be active during a human Mars mission needs to be modified to be capable of Mars based controls. With the unpredictability of launch of both robotic and human missions, some of these robots will not be on Mars during a period of human presence. This has to be accepted as a risk of doing business, a risk that can be reduced by increased robotic lifetime

The training for robotic control has to be added to an already tight astronaut training program, and the skill maintained while en route to Mars. I think this is the least of the difficulties.

As mentioned above, the human and robotic teams, with substantial differences in style, must compromise for the common good. While on Mars, communication is between groups speaking different technical languages, with the additional barrier of latency

What's New This Year?
Having made a preliminary presentation on this topic last year, I had not intended to return to it for some time. Several reasons have induced me to return to the topic sooner than planned.

There has been a substantial improvement in robotic intelligence and robotic speed. Despite this improvement in robotic speed via artificial intelligence, this improvement is slower than Moore's law by several orders of magnitude. In the absence of human input Mars will not be well defined for one to two human generations. In addition the realistic EVA simulations of the Mars Society on Devon Island have confirmed once again the value of the human observer, but have confirmed as well the potential dangers and difficulties of EVA.

On a positive note, NASA's acceptance of the concept of nuclear power allows easier design of robust long duration rovers (albeit expensive). These rovers would most benefit from human Mars based control. In addition, the ESA has activated the Aurora project, and the Russians have expressed a strong interest in a human Mars mission in the 2018-2022 range. The private sector is already interested in going to Mars. The only group not on board is NASA. Possibly because of their problems with ISS, they have pushed Mars further and away. It is now a fourth to sixth level mission. L1, a repeat Moon landing, L2, one or more asteroid explorations, and the moons of Mars are to be visited in some order. Only then is a Mars landing to be considered. By that time the NASA astronauts could check into a Russian hotel with a French chef, and then watch the British and Italian Mars soccer teams on TV.

Suggestion: I recommend that experiments be performed that compare robots subject to latency with those under dual control, preferably under the aegis of the Mars Society. I believe these will show the cost effectiveness of humans on Mars due to increased science production of rovers distant from the landing site This would add additional weight to the argument that humans are needed for the exploration of Mars.

Conclusions
The use of astronaut controllers located on Mars has the ability to vastly improve the rate of scientific return of Mars rovers. It would allow a single Mars base to aid in the exploration of the entire Martian surface. A Mars base could be productive even in the absence of the ability to perform EVAs. This low risk, high reward mission would make the placement of humans on Mars cost effective in the scientific exploration of the Martian surface. Although there are some potential difficulties in initiating this program, the difficulties are definable and surmountable.

//

Information Systems Gear in Mars Analog Research†

Ned Chapin

Abstract
The reports from the Mars Society's Arctic and desert research stations show the serious efforts the crews have made to create appropriate Mars analogs. Those efforts have resulted in findings that will be valuable in preparing for human beings to survive and work effectively on Mars. Increasing further the realism of the Mars analogs will continue to produce valuable findings. One area that so far has received relatively light attention in the analog efforts is information systems gear. What could be done to strengthen the analog work in that area has both "do" and "don't" items.

Among the seventeen major "don't" items, here are four: no pens, no pencils, no fans unless with seals on the rotating parts, and no gear with moving contacting parts exposed to the atmosphere (like many Earth keyboards, pushbuttons and antennas). Among the seventeen major "do" items, here are four: gear with rechargeable power packs, glove-compatible tethering for small hand-held gear, multi-function gear such as send-receive-capture of audio and images, and gear with no user-maintainable parts. To try to minimize the use of the "don't" items and to maximize the use of the "do" items, planning will be required in preparing the agendas for future crews at the stations. In the information systems area, the Mars Society has the opportunity to collaborate with vendors and researchers in seeking to improve the quality of the analog achievable with the Mars Society's research stations.

Introduction

As reported at length elsewhere in these conference proceedings, the Mars Society has been operating two analog research stations.[1,2] On Devon island in the Canadian Arctic is the Flashline Mars Arctic Research Station. In the southern part of Utah in the USA is the Mars Desert Research Station. Other Mars research stations are in the formative stages, with plans for operations in Iceland, Australia, and elsewhere. Whatever their location, the stations are intended to serve as partial simulations or analogs of aspects of conditions and operations that might be encountered or attempted at an actual human crewed base on Mars. Many, but not all, of these would also be applicable at a human crewed lunar base.

Much good research has been conducted thus far at the Mars Society's analog research stations, and much has been learned that appears to be applicable to preparation for activities at an actual crewed base on Mars. Research reports and field reports have been filed with the Mars Society, and reports have been published in these proceedings and elsewhere.

Studying those reports and some conversation with personnel who have served at the analog stations has brought to light an opportunity to improve further the realism of the analog work in the area of information systems gear (equipment and devices and materials). So far, the operations at the analog research stations have given only light attention to the role and characteristics of information systems gear in supporting the research done. To strengthen the attention and take advantage of the opportunity to improve further the realism of the analog work, this paper points out some "do" items and some "don't" items. They are action items. The "do" items are added actions that could enhance the realism of the research work. The "don't" items are actions that could cause impairment of the realism.

In the rest of this paper, seventeen "don't" items are listed in the next section, and then seventeen "do" items are listed in the section after that. For convenience in presentation, the "do" and "don't" sections are annotated bullet lists. Afterward, a brief discussion section gets followed by the section on conclusions.

Seventeen "Don't" Items

- Use No Blank Paper or Cardboard – Blank paper and cardboard have thus far been used in many forms and many ways in the research stations. A few examples of its forms are lab books, notebooks, scratch pads, envelopes, preprinted forms, blank labels, pads and reams of paper, packing containers, packing paper, note cards, file folders, and blank space in reference works such as books. A few examples of the uses of blank paper or cardboard are taking notes in meetings, communicating with other personnel, recording lab activities and findings, posting messages and notes and announcements, making signs, jotting a to-do list, doing computations, labeling specimens, preparing a report or presentation, and doing the day's (sol's) diary. A reminder: Mars has no source of paper until it gets manufactured there!
- Use No Whiteboards or Plastic Sheets – These substitutes for paper typically have been used in group communication settings, for example as transitory signs. While they are much more easily reusable, they are also likely to be redundant with other information systems gear.
- Use No Pens – Without paper, pens' usefulness is greatly reduced, yet the hazards they offers while usually more annoying than serious, still remain. Some of the hazards have been reduced in "space pens," while still preserving the non-erasability that pens offer.
- Use No Pencils and Markers – Without paper, pencils' usefulness is greatly reduced, and the erasability they offer requires erasers and causes cumulative damage to paper. Without whiteboards or plastic sheets, erasable markers' usefulness is greatly reduced except for writing on habitat walls. Non-erasable markers might be justified for marking some kinds of specimens, but other alternatives are more attractive overall.
- Use No Gear with Holes or Porous Surfaces – On Earth, information systems gear is usually made with holes or porous surfaces to keep interior and exterior atmospheric pressures equalized, and to facilitate cooling by air transport. The major exceptions on Earth are nautical gear to help keep the interior dry, and hermetically sealed gear (such as hard disk drives) to preserve a specialized interior atmosphere. While water damage is a risk only within habitats on Mars, dust accumulation is a major risk everywhere on Mars.
- Use No Gear with Exposed Moving Parts (like keys) – Keyboards are a major example. For use in dirty environments on Earth, we have specialized equipments that have no exposed moving parts. On Earth, membrane keyboards are an example and some nautical and hazardous-environment-tolerant equipments are made with no exposed moving parts. Dust is nearly ubiquitous on Mars, as robotic exploration has shown.
- Use No Gear With Fragile or Unprotected Parts – Information systems gear is likely to be dropped once in a while in use, and at anytime may have things dropped on it or collide with it. For a crewed mission working on Mars, the crew personnel are also the information systems repair personnel – and those personnel have more valuable things to do with their time on Mars than try to clean dust out of gear and to repair damaged gear, given the very limited availability of repair tools and equipment.

- Use No Gear with Transparent Surfaces Having a Hardness Less Than Mohs 7 – Removing dust under the best of circumstances from transparent surfaces, such as displays on information systems gear, can lead to scratches. Out on the surface of Mars will undoubtedly be dusty and not be the best of circumstances. To avoid scratches, the transparent surfaces will have to have at least the hardness of quartz, since dust minerals harder than that have not yet been detected on Mars.
- Use No Gear with Fans Without Seals on Rotating Contacting Parts – Cooling of information systems gear, especially the larger sizes, has typically been done by having fans move the atmosphere over or through the gear. Cooling by radiation from enclosed gear has been much less used for many good technical reasons. Dust contamination can cause the early failure of bearings for rotating parts.
- Use No Narrow-Range Temperature Sensitive Gear – Wide temperature swings are possible on Mars from habitat warmth to exterior extreme cold. It will not be possible and sometimes not even desirable always to protect information systems gear from such swings. The gear should give reliable performance whatever the ambient temperature.
- Use No Narrow-Range Ambient-Pressure Sensitive Gear – Ambient atmospheric pressures on Mars may range from in-habitat levels (probably like 3,000 meters above sea level on Earth) to less than 0.5% of typical Earth sea-level atmospheric pressure. Some gear might in transit be exposed to the vacuum of interplanetary space. It will not be possible and sometimes not even desirable always to protect information systems gear from such swings. The gear should give reliable performance whatever the ambient pressure.
- Use No Unprotected Hard-Radiation Sensitive Gear – Mars offers scant protection from the fluctuating stream of radiation from space. To give robust reliable performance, information systems gear must have its radiation tolerance built in, regardless of the size or intended use of the gear.
- Use No Gear with Unsealed Sliding Contacting Parts – Dust contamination is the potential villain. Fortunately, the internal components of information systems gear rarely have unsealed sliding contacting parts. But the human interface in some gear used on Earth includes such sliding contacting parts. One example is the insertion and removal of floppy disks. In addition, the tight tolerance of read-write heads and of head movement in floppy disk read and write operations makes such gear problematical for use in dusty environments.
- Use No Gear Relying Upon Filters Unless the Filters are Both Mechanically Cleanable and Reusable Indefinitely – The usual mechanical method for cleaning filters is to reverse the flow while agitating them. For Mars gear, that process has to be either automated or be very easy and quick for suited-up personnel to do. Of course, this same "Don't" also applies to life support equipment.
- Use No Gear Requiring Lubrication Renewal or Supplementation – Information systems gear typically has small parts that should be neither unlubricated nor over-lubricated. Also, the lubricant required typically is very specialized. For this reason, lubrication failures in information systems gear on Earth typically result in junking the gear and replacing with new gear. This option is very unlikely to be available on Mars for early missions.
- Use No Detergents, Soaps or Other Specialized Cleaning Compounds, Unless the Compounds are Produced On Site – Cleaning of information systems gear has to be done, especially in dusty environments. But specialized cleaning compounds, such as detergents and soaps, are another item to be transported and stored. The need for their use should be avoidable for the information systems gear used, as noted previously for dust. Once greenhouse operations are working well, some cleaning compounds will probably be produced on site. But that will not likely be done on the early Mars missions.
- Use No Single-Function Information Systems Gear – Having and using many types of specialized information systems gear has consequences. It places more demand on the personnel in using them, places more demand on providing appropriate quantities, and potentially raises major interoperability difficulties. The time and effort in making connections among the items of gear becomes increasing difficult as the number and variety in the items of gear increases. This difficulty is clearly tied to the choice of the gear to use in the Mars analog research stations for more realistic simulations – a matter closely tied to the "do" items.

SEVENTEEN "DO" ITEMS

- Use Networked Gear – Networked information systems gear offers major advantages for the personnel. It facilitates communication among the personnel, especially when suited up. It can capture data so that those data can be stored centrally for later use and back up storage. It can enable tracking of personnel during out-of-habitat field work. It could be used to provide routinely some status indications, such as help needed, communication deliberately turned off, low battery state, contact lost due to environmental conditions, etc. The major disadvantage is the need for the base station to support networking.[3]
- Use Small-Cube Lightweight Gear – Storage and transport space and capacity are premium items; hence, minimal demands upon them are valuable characteristics. Also, on portable items of gear, lightweight is appreciated by the using personnel.
- Use Membrane Keypads – These are easy to clean and help keep contaminants out of the internal workings of the gear. While they do not work as well for small or miniaturized gear as they do for larger gear, they are realistic, especially for field use when gloves are worn.

- Use Glove-Compatible Tethered Gear – In out-of-habitat or field use, portable gear can easily be forgotten, mislaid or lost. Tethering is one alternative solution, but can be awkward especially with gloves as when suited up. Tethered gear tends to receive rough treatment, as from bumping against things, and such rough treatment must not produce false or unintended signals from or in the gear. Operating portable information systems gear effectively while wearing gloves, especially when the gear is tethered and conditions are harsh, can be very difficult.
- Use Gear with No User-Maintainable Parts – A good way to provide that characteristic is for the gear not to require any user maintenance other than exterior cleaning. Not having to put time in doing user maintenance of the physical gear leaves more time for the using personnel to use the gear as a tool in getting mission work done.
- Use Gear with Rechargeable Power Packs – Power supply for information systems gear is a recurring concern in practice. Batteries run low and have to be recharged. Rarely should they require replacement, since new batteries will not be available on Mars, and batteries typically tend to be heavy for their cube. Fuel cells of adequate capacity tend to be bulky but do not require light. Solar panels are large and require light to operate. The ideal is the equivalent of an uninterruptible power supply, for whatever the power need, wherever it be located. For portable gear, rechargeable power packs are a currently workable compromise.
- Use Wearable Information Systems Gear – This opportunity has been covered in prior Mars conferences.[4] Since out-of-habitat work requires being suited up, some of the gear could be built into the suit, and some of it could be voice activated. What functionality the gear should provide falls into two categories: desirable for all personnel, and specialized for specific personnel roles, including backup roles. The "what functionality" topic has received scant attention in the analog work thus far.
- Use Gear with Easily Cleanable Surfaces and Connectors – Dust from the environment and debris from the user (such as sweat, skin oils, and flakes of dead skin) top the current list of what has to be cleaned from the surfaces of information systems gear. Any electrical connectors are a special problem area, since they may also be subject to corrosion. Hence, electromagnetic spectrum signal transmission, such infrared or radio, appears to be more reliable and convenient, and less affected by surface cleanliness.
- Use Gear with a Long Mean Time Between Failures (MTBF) with Respect to Mission Duration – Gear will fail – the question is when. For information systems gear, the most acceptable likely time for failure is well after the completion of the entire post-mission phase. But some earlier failures will occur. Hence, backup gear and procedures, and redundancy have to be characteristics of the information systems gear.
- Use Gear Not Requiring Ambient Oxygen or Full Mars Sunlight Even at Higher Latitude – Since these (especially oxygen) are more limited on Mars than on Earth, the analog work has to take care not to require them. Fortunately, nearly all information system gear does not require an oxygen-rich atmosphere, and currently only solar power units require sunlight to be useful.
- Use Gear Usable for a While in Low Light or in the Dark – Information systems gear should be usable at night, in shadows, in caves and during dust storms. Light is usually part of user displays and produced by the gear by using power. More rarely light is used as a power source in information systems gear, as in recharging power packs.
- Use Gear Not Needing Heating, Cooling or Any Atmospheric Pressure – Information systems gear should be usable in the habitat and out in the field, in the full warmth of sunlight and/or human body warmth, and in a cold shadow near "dry ice" before dawn at high latitude.
- Use Shock-Resistant Gear – Gear sent to Mars will have to withstand the shock of landing, and subsequent shocks in use. While stationary gear, such as at a central communications center, may in use get no more than a few bumps, portable gear and wearable gear will get frequent sharp impacts and many minor contacts. The gear should be robust enough to survive and continue to be operational short of suffering major physical impairment, such as being crushed by a falling boulder. Electrical shocks from the power supply should also be taken in stride.
- Use Static-Charge Resistant Gear – Avoiding static charges is especially difficult for wearable gear and portable gear. In their design and construction, such gear should avoid using static charge sensitive components, such as CMOS. In practice, wearable gear and portable gear must be able to continue to operate reliably in the presence of static charges.
- Use High-Capacity Gear – In general, the more capable the gear, the better for Mars work. More storage capacity is better than less; more speed is better than less, and more range is better than less.
- Use Gear That Interoperates in User-Friendly Ways – The complement of information systems gear at a Mars base will consist of more than just one item. Each of the crew will have access to and use more than one item of gear, and will be the primary user of some items of information systems gear. For example, a person doing hydrology research will use different information systems gear in the field to capture and communicate observations, comments and analyses, than in the habitat to organize, integrate and analyze data, and to report findings. How easily and helpfully the gear items in each location work with each other, affects the person's productivity and the quality and quantity of the work accomplished.
- Use Multi-Function Gear – Being able to do more with less weight and less cube makes it easier for the using personnel to get more done in less time with less error. For example, does a person on Mars doing geology work in the field have to carry a two-way communication device, a digital camera, a digital voice recorder, a distance measure, a slope (angle) measure, a location tracker, a physical condition monitor, and a consumables monitor?

The first three items could be made into one. But why not the first six items? And since the last three are needed for all out-of-habitat work of any kind by anyone to provide continuous contact with the central communications base, could not they also be combined with the first three? Then for the geology work, the only specialized information systems gear would be the data capture associated with the distance measure and the slope measure. But are those really that specialized to only geology? And what might be wearable, and what portable? Further work at the Mars analog stations could give helpful direction on the choices, features and capabilities of information systems gear.

Discussion

Fortunately, the limitations on the information systems gear are primarily in the economic area, not the technology area. We are fortunate that the technology hurdles are matters of degree, like the amount of storage or the range of transmission or the fidelity of display. The needs for Mars work extend beyond what is commercially common on store shelves today. Yet the commercial market is changing fairly rapidly. A few years ago, no integration had been made of a cellular telephone and a digital camera. Now that has arrived, and the market focus has shifted to the quality of the image that is transmitted. In testing are new items of information systems gear, with new combinations of features and capabilities.

This gives an opportunity that can benefit both the Mars analog work, and the manufacturers of information systems gear. Collaboration could result in the Mars analog work feeding back experience to the manufacturers from field use of new items of information systems gear, and getting gear at a nominal price to try out that more closely meets the needs of the personnel at the Mars analog stations.

In such collaboration, human factors considerations are of course of critical importance. Always there is a trade-off among many variables, such as usability, robustness, interoperability, weight, power demand, capacity, convenience, cube, etc. Among the more important human factors are user friendliness and safety. For example, how much is the radio frequency radiation exposure to the user for sending and for receiving? Even simple devices can have a potential for misuse. For example, using a carbide-tipped engraver as a marker for specimens avoids the undesirable aspects of ordinary permanent markers, but the engraver could be used as a murder weapon or result in injuries from accidents in use.

Also, while collaboration with information gear manufacturers can enhance the work at the Mars analog stations, it also puts an added burden on the planning for and the management of the stations. The operating plan cannot be thrown together while waiting to transfer airplanes in an airport while en route to take a post in a crew rotation at one of the Mars analog stations. Instead, months of time, including several contact sessions and many exchanges of documents, may be needed not only of the crew commander but also of the individual crew member participants. Trial use of new items of information systems gear usually will be needed in advance before beginning a rotation at a station where the gear is to be tried out. This adds to the challenge and the potential rewards possible from the work done at the Mars analog stations.

Conclusion

The valuable work done at the Mars Society's analog Mars stations has yielded much useful data for the conduct of actual human-crewed Mars missions. So far, the Mars analog stations have given relatively light attention to the area of information systems gear. This paper offers seventeen actions that could be done to enhance the realism of the Mars analog stations as regards information systems gear. To balance those, this paper also offers seventeen actions that should be avoided or minimized to keep up the realism of the simulations possible from the activities at the Mars analog stations. Enhancing the quality of the analog work at the Mars analog stations could be aided by collaborative work with manufacturers of information systems gear. The result could add even more value to the work at the Mars Society's analog research stations.

References
1. Track 3B, Mars Society Third International Convention, 2000.
2. Tracks 2A and 2B, Mars Society Fifth International Convention, 2002.
3. N. Chapin, *Communications Infrastructure to Support Human Activities on Mars, On To Mars*, Apogee Books, Burlington Ontario, Canada, 2002, pp. 159–165.
4. C. E. Carr, *Applications of Wearable Computing to Human Exploration in Extreme Environments*," Track 1D, Mars Society Third International Convention, 2000.

Field Testing a Robotic Assistant for Use in Extravehicular Activities†

Amy Blank, aab193@psu.edu, Akhil Kamat, apk129@psu.edu,
Brendan Knowles, bjk179@psu.edu, Ryan L. Kobrick, rlk193@psu.edu,
Joseph Sapp, jws268@psu.edu and Robert Wilson, rpw133@psu.edu,
The Pennsylvania State University Chapter of the Mars Society, www.psums.org

Abstract

For an astronaut conducting field research, the dangers and constraints imposed by the environment limit what can be accomplished in the time available. To improve the speed and efficiency of field work, the aid of a robotic assistant would be immensely useful. Such an assistant, under development by the Pennsylvania State University Mars Society (PSUMS), was tested at the Mars Desert Research Station (MDRS) during the rotation of Crew 25 (February 29-March 13, 2004). Field testing in Mars-like conditions revealed where and how a robotic assistant would be useful in the field. Future testing will include a comparison of identical Extravehicular Activities (EVAs) conducted with and without robotic assistance. Factors such as safety, speed and ease of EVA completion will be assessed in both situations. Through periodic testing and design revision, the team will be able to determine and implement the most desirable qualities in a robotic assistant.

Introduction

With the renewed desire of many countries to be leaders in space exploration, increased budgets are making manned missions to Mars within two decades a realistic goal. Given the imminence of such missions, the goals of these missions and methods of reaching these goals need to be identified. The extent of the mission objectives will be limited by time constraints and budgetary restrictions. Operating efficiently within these constraints is crucial to ensure that field research produces useful results. With this idea in mind, the use of robotic assistance is vital for efficient operation.

A robotic assistant should have the ability to aid in Extravehicular Activities (EVAs) by performing autonomous tasks as directed by the astronaut. To create a capable assistant, the necessary tasks need to be determined through extensive research. An effective mode of communication between the astronaut and the assistant must also be developed.

The Pennsylvania State University Mars Society (PSUMS) has been developing such a robotic assistant to help an astronaut carry out a large number of tasks quickly and efficiently in the field. The PSUMS rover is a robotic assistant controlled by virtual reality gloves. It features the ability to hold one of several modular tool kits, which can be interchanged to prepare the rover for any particular EVA. The development and features of the PSUMS rover are explained in detail in *A Modular Multi-Function Rover and Control System for EVA*, which can be found in the proceedings from the 2002 International Mars Society Conference[1].

Field testing a robotic assistant such as the PSUMS rover in a Mars-like environment will identify the most beneficial functions the robotic assistant can perform. To determine these functions, a variety of testing methods must be employed under simulated Martian conditions. EVA simulations without robotic assistance reveal which tasks require improved efficiency. Operation of the assistant in the simulated environment further specifies design requirements. Once these requirements are incorporated into an updated design and implemented, the same simulations will be performed with the assistant. To assess the PSUMS rover's ability to aid an astronaut, many factors, such as its effect on EVA efficiency, the simplicity of human-rover interaction, and astronaut safety, will be investigated at this stage. To further evaluate the rover design, various sensors incorporated into the rover design will determine how it is affected by typical operating conditions in the testing environment. The data and observations from the field testing will then be incorporated into another stage of design revisions.

Field Testing of the PSUMS Rover

The first two of the aforementioned testing methods has been put into practice in field testing the PSUMS rover during a two-week Mars exploration simulation at the Mars Desert Research Station (MDRS) by Crew 25 (February 29-March 13, 2004). The crew logs and reports are available from the International Mars Society web site, under the 2003 field season at MDRS.[2] There in the deserts of southern Utah the rover had its first exposure to a realistic Mars-like environment. As such, the field testing revealed many unforeseen issues, which must be taken into account as development progresses.

Rover Testing

At the time of the field testing, the rover consisted of a motorized base that held one module – a soil sampling device. The first difficulty was noted upon assembly of the rover in the habitat. In the extremely space-efficient habitat, finding space to assemble and store the rover was very difficult. Furthermore, simply getting the rover into the airlock and

---

outside the habitat proved to be a major challenge. Thus, the first lesson learned was to make the rover smaller and more portable to deal with the cramped conditions likely to be encountered in a manned Mars mission.

Once outside, the crew had plenty of space to test the rover. However, when traveling a long distance, the rover's slow speed makes carrying it preferable to driving it. Though carrying the rover is normally a simple task, when one is wearing a space suit and a heavy pack, it becomes unwieldy and almost not worth the effort of carrying it along. This observation leads to the idea that a robotic assistant should move quickly enough to keep up with an astronaut, and it should be able to strap onto someone's back in cases where climbing is necessary to reach an EVA site. This in turn requires that the rover also be light enough to make backpacking feasible.

Lastly, the environment itself created some issues. The rough terrain requires a more robust rover design than was anticipated. The frame, steering and suspension were not strong enough to survive repeated impacts and prolonged operation on extremely rough ground, and would have to be strengthened. Furthermore, the sampling device design (shown in Figure 1) in use at the time required flat or nearly flat ground to extract a sample. However, in the Mars-like desert in Utah flat ground is scarce. Though it was anticipated that the crew might have to specifically seek flat ground for sampling sites, the true difficulty of finding such sites was not clear until the field testing. Keeping the terrain in mind, a new system for collecting soil samples would need to be devised.

Intended Uses of a Robotic Assistant

While information about the rover's performance in a realistic environment is crucial, the activities of the crew should also be studied to determine how a rover could best assist astronauts in the field. Though the rover was deemed unprepared to assist in EVAs at this point in its development, the crew was still able to carry out typical EVAs without robotic assistance for the first stage of the field testing. The EVAs conducted by Crew 25 fell into three main categories: biological EVAs, which focused on taking soil samples to check for microbial life; geological EVAs to collect representative rock samples from an area; and general exploration to explore the terrain and identify possible sites for future scientific EVAs. In each of these

Figure 1. Sampling Device Module.

areas, certain difficulties were clearly identified, indicating areas in which a robotic assistant could be invaluable.

Biological Field Work

In collecting soil samples during biological EVAs, numerous tasks must be performed, and many of the more complicated ones could be completed by a robotic assistant. The PSUMS rover, equipped with an effective soil sampling module, would be able to extract and store soil samples on board. Having the rover complete the extraction bypasses the difficulties of transferring soil from a corer to a sample container and sealing that container by hand in a motion-restricting space suit. Storing the samples on board is also easier for the crew than carrying them.

In addition to performing these complex tasks, the rover could also perform simple tasks that are merely very time-consuming. For example, taking a temperature reading from the soil required waiting for the reading to stabilize, sometimes for several minutes. In this case, the rover's ability to multitask would be particularly beneficial, since it could take the temperature reading and extract the soil sample concurrently.

With the robotic assistant available to perform these tasks, the EVA could be conducted with only two crew members instead of the three required otherwise. In addition to putting fewer people at risk and increasing EVA efficiency, this would allow an extra person to remain at the habitat to perform laboratory work or routine maintenance.

Geological Field Work

For sampling rocks during a geological EVA, a robotic assistant would probably not be used to collect the actual sample, but it would be invaluable in recording data about the sampling site. For creating panoramic photographs of an area, a camera on board the rover would provide an excellent alternative to taking a set of photographs by hand. Unlike a camera set up on a tripod, which requires flat ground, the rover's camera could be made self-leveling. In addition, the whole sequence of repeatedly taking a picture and rotating the camera could be automated. A crew member could give a single command to the rover to take a panoramic photograph, and then continue with other tasks while the rover completes the procedure. Again, the increased efficiency frees crew members for other tasks, making the robotic assistant a valuable addition to the team.

General Exploration

For the purpose of general exploration, the role of a robotic assistant will vary depending on the goal of the exploration. For terrain mapping, a GPS module on the rover could keep track of the path taken and note locations at which the crew stops. Photographing capabilities would also be helpful in such situations, in case the crew plans to return to a particular site.

If the crew stops to explore an area at length, a rover could have other purposes. If an interesting site is not easily accessible, further exploration of the area may prove hazardous, or simply troublesome. However, a module for the rover could include a smaller version of itself, either tethered to the main rover or in wireless communication with it. This would simply contain a camera and a light source to give a good view of the area in question. With a view screen incorporated into the onboard part of the module, the astronaut could maneuver the mini-rover into an area as far as the tether or wireless network could reach. This is a much safer and often simpler alternative to sending an astronaut into unknown territory.

Improvements for the Second Generation Rover

Over the next two years, PSUMS will design a new rover to improve and expand upon the current design, based on the lessons learned from the experience of Crew 25. First, the structural issues will be addressed and incorporated into the new rover base. Later, improved modules will be developed to prepare the rover for use in EVA simulations. The goals for this project are to develop a rover with improved performance, human interface, and testability, and to conduct more extensive field tests. With a more capable rover, it is hoped that these field tests will help demonstrate the viability of virtual reality glove control, tool modularity, and the robotic assistant concept in general. In addition, they will provide insight into how to further refine and enhance this and future designs.

Performance

The first rover design showed room for improvement in several areas. The first was the frame, which was constructed of aluminum c-channel. With the addition of a battery that was much heavier than anticipated, the frame was no longer able to adequately support the weight of the onboard systems, so steel bars were fit into the c-channel for reinforcement. The reinforced frame is shown in Figure 2. This change allowed the rover to support all its onboard systems, but increased the weight considerably. Steel was also used for various mountings and ad hoc additions, adding still more weight.

The frame for the new rover will be designed to be both stronger and lighter than the current frame, so the rover is easier to transport and more structurally sound. The most

Figure 2. Steel-Reinforced Frame.

important change will be the use of aluminum tube instead of c-channel. Unlike the old c-channel section, tube is very strong against bending. The increased strength of the new section will eliminate the need for steel reinforcements, thereby reducing the weight. Likewise, the use of aluminum will be favored over steel in all other components of the rover. Where additional strength is needed, titanium will be used. Like the current frame, the new frame will be assembled by TIG welding, which provides very high quality aluminum welds.

The other major issue of the current rover was the design of its steering and suspension. Originally, it was designed to steer using split front and rear axles. Figures 3 and 4 show drawings of this steering system in its conceptual phase. Once constructed and tested, it proved to be overly complicated and unable to support the heavy frame. In particular, the two pivots where the split axles connected to the frame were overloaded due to bending and provided a lot of resistance to steering. Eventually, a single front axle design was used. This worked much better

Figure 3. Old Steering System, Top View.

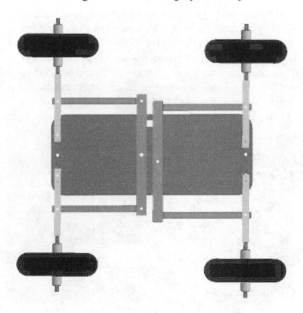

Figure 4. Old Steering System, Bottom View.

because the single central pivot did not undergo any bending. Aside from inflatable tires, the current rover has no suspension, which is unacceptable for use in rough terrain, even at low speeds.

There are many steering options for the new rover including tank style steering, pivoting axles, and pivoting wheels, among others. There are also several suspensions that could be useful, including passive, active, and rocker-bogie. In order to allow for different steering and suspension schemes to be field tested for suitability, the new rover frame will be designed to accept any steering / suspension assembly provided it attaches to the frame in the same way. The first system to be designed will be a simple four-wheel tank style steering assembly. It will feature a passive spring-damper suspension. Additional steering / suspension systems will then be designed to fit the same attachment points as the tank style system. This concept of modular suspension will be utilized during field testing to determine which systems perform the best and if different systems are more suited to some mission profiles than others.

There are several other performance improvements planned for the new rover. Dust proofing enclosures will be used to keep particles out of moving parts and bearings. Power and computing wires will be distributed inside the frame to two module loading bays. This unobtrusive wiring system will reduce the possibility of damaging connections. Finally, lightweight electronics and tool module enclosures will be constructed out of polycarbonate, foamed PVC, or some other rigid, lightweight material.

Human Interface

The primary aspect of the human interface to be improved is the virtual reality glove control system. The current rover can be driven using simple gestures to move forward and backward and steer. The next step is to develop a more comprehensive command set and control structure. The increased set of gestures will allow both manual and automatic control of the rover's motion and the operation of any onboard tool modules. It will also let the user navigate and use an interactive menu, specify the rover's operational state or active feature, and request feedback. Considerable research will need to be done to determine how to structure the gesture set to maximize ease of use and accuracy.

Another improvement will be in the way tool modules are loaded onto the rover. Loading a module onto the current rover involves securing the tool with nuts and bolts and then connecting several wires to the rover's electronics. This process requires a wrench and is not consistent from one module to the next. The new rover will use a standardized module bay. Tools will be locked down using latches in a tool-less process. Also, all power and input / output lines will be run to each module port and consolidated into a single plug. This should make loading and unloading modules much faster and easier.

Testability

In contrast to the current rover design, which provides very limited feedback, the new rover will be designed to provide significant data about the performance of its mechanical, electrical and software systems. In order to provide information about the mechanical performance, various sensors will be embedded in the design. These will include strain gages, vibration sensors, attitude sensors, and optical encoders on the wheels, in addition to others. Data from these sensors will be time-stamped and logged during periods of testing. Voltage and current will be measured across the battery and motors to indicate the draw on the battery and the power used by each motor. Finally, software logs will be kept of the commands sent and the interaction between various programs and the microcontroller to help diagnose problems and improve efficiency.

Damage Prevention
In some cases, the rover may need to operate outside the astronaut's field of vision. For this situation, a camera or cameras on board the rover would provide the astronaut with enough information about the rover's surroundings to reach a destination and avoid obstacles. However, without a view of the rover's position and orientation accidental damage becomes a concern. The same is true of situations in which the rover is acting semi-autonomously and the astronaut's attention is on another task. For such situations, various sensors would report data necessary to evaluate the safety of the rover.

On slanted ground, attitude sensors would report the angle at which the rover rests. The onboard computer would interpret the data to determine whether the rover is likely to tip if it proceeds. Similarly, strain gages and vibration sensors placed on the body of the rover indicate the possibility of structural damage. Together, these sensors would provide the information necessary to determine if a situation is potentially hazardous for the rover. The computer would make decisions on whether to allow a command or not based on this information.

Speed and Position Calculation
Optical encoders or mechanical switches adjacent to the wheels are feasible sensors for determining speed. Incoming data can be interpreted by the rover's onboard computer and reported back to the astronaut to relay the speed of the rover. The readings will not be exact since the computer can only transmit a digital signal in discrete time segments, but they will be enough to determine approximately how fast the rover is moving. From the rover's speed and the time spent at that speed, the distance traveled can be approximated. These data are critical for semi-autonomous functions and remote control using the onboard cameras.

Power Monitoring
Over time, the power a battery can supply decreases due to the nature of a practical battery. By monitoring the voltage of the battery under a small load, the computer can obtain general information about how much power is available for use. With this information, it can shut itself down safely or bring itself and the rover systems safely into a low-power mode when necessary. This prevents problems storing data in the computer's memory due to a power-off when writing. Since the computer will initiate a shutdown sequence before the battery voltage becomes insufficient to power all the onboard systems, the risk of damaging hardware and losing data due to power loss is nearly eliminated.

Software Debugging
Since most software errors are encountered during use rather than by reading through the program, it will be highly beneficial to detect and even correct errors while the rover is in operation. For these purposes, a small LCD screen will be attached to the rover to provide information about the commands received and the status of the modules in use. A keypad will be added to allow interaction with the rover during the diagnostic and debugging processes. This debugging hardware will help reveal unexpected faults and pinpoint the location of these faults within the complex system. In the case that an error cannot be corrected in run time, the module in question can be deactivated to allow the rest of rover to function properly.

Field Testing the Second Generation Rover
When the second rover is ready for field testing, simulated EVAs will be performed to test the overall effectiveness of the robotic assistant concept, the suspension and steering systems, and the glove command concepts. The testing will also reveal new design flaws and additional avenues of development.

Field Testing Methodology
As mentioned earlier, the next stage of field testing will rely on comparative runs. The general concept of a glove-controlled robotic assistant will be evaluated by running identical EVAs with and without the rover. In both cases a time-stamped log of mission events will be recorded. The completion times for different mission objectives such as reaching the target location, completing specified tasks at the target location, and returning to the habitat will be the primary metric of efficiency. Other notable events, such as unforeseen delays or dangerous situations will be recorded and considered as well. In testing this version of the rover, most of the focus will be on the effectiveness of the EVA team at the target location, since the rover's primary function is to cut the workload on site.

Testing of specific systems will also be performed. In order to determine which suspension and steering systems are most suitable for the rover's mission profile, similar EVAs will be performed with different systems attached to the frame. As in previous testing, a mission log will be kept to gauge performance and note difficulties encountered. Additionally, sensor data will be logged to indicate performance in terms of power usage, vehicle speed and vibration.

The glove command system will be tested similarly. Several command structures will be developed, each representing the same commands with different sets of gestures. Investigating different command sets is important because there are many distinct, logical ways to specify tasks like steering, throttling, and system control. By performing field tests with many command sets, the most accurate and intuitive concepts can be selected and developed.

Design and Testing Iterations

As the design for the new rover matures, additional systems will likely benefit from similar testing. However, optimizing each system while also evaluating the need for a robotic assistant may become extremely complicated. To avoid testing every permutation of suspensions, modules, steering systems, and other add-ons, a number of preliminary experiments can be designed to select the optimal systems of each type. Only permutations of the chosen systems need to be tested further. Beyond investigating various components of the rover, repeated field testing will bring to light shortcomings in the rover's overall design and point to improvements which could not be predicted with other tests. Continued iterations of the design revision and field testing processes will further improve the rover, eventually creating a robotic assistant ideally suited for field research on Mars.

Conclusion

Simulations of field research on Mars clearly show that robotic assistance would be of immense value in many situations. An assistant such as the PSUMS rover could increase the safety and efficiency of a variety of Extravehicular Activities by performing repetitive, tedious or dangerous tasks in lieu of an astronaut. The capabilities required of such an assistant are already well defined from the experience of completing simulated EVAs without robotic assistance. Performing the same EVA simulations using a robotic assistant with these capabilities will show the degree to which robotic assistance can be useful. From repeated testing in a Mars-like environment and improvements based on the test results, an effective design can be developed. The ideal rover design is achieved when a human-rover pair acts more efficiently than a pair of humans. As the development of the PSUMS rover progresses, subsequent iterations of field testing and design improvements will bring the rover ever closer to being an ideal robotic assistant.

References

1. McGuire, Stephen, Joel Richter, and Kevin Sloan. *A Modular Multi-Function Rover and Control System for EVA*. Proceedings of the 2002 International Mars Society Conference. http://homepage.mac.com/fcrossman/Marspapers/papers/Sloan_2002.pdf
2. The Mars Society. *Messages and Reports from the MDRS: 2003-2004 Field Season – Daily Reports and Photos.* www.marssociety.org/MDRS/fs03/

Acknowledgments

The authors would like to thank the following individuals and groups for their help and support:

The Pennsylvania State University
The University Park Allocation Committee
The Pennsylvania Space Grant Consortium
PSUMS Advisors: Dr. Lyle N. Long and Dr. David B. Spencer
MDRS Crew 25 – Amy Blank, Dennis Creamer, Daniel Hegeman, Ryan L. Kobrick, Jason Schwier, and Kevin Sloan (Commander and PSUMS Team Lead)
Shannon Rupert, Stacy Sklar and the MDRS Remote Science Team
Paul Graham and the MDRS Engineering Team
Julie Edwards and the MDRS Mission Support Team
The authors would like to especially thank the past and present members of the PSU Mars Society for their dedication and hard work.

//

A Sample Storage System for the Martian Surface†

Joel Donoghue, Sadi Hamut, Christine Kryscio, Kristen Montz and Chad Rowland
Michigan Mars Rover Team, crowland@engin.umich.edu

Abstract

When humans travel to Mars, one of their most time consuming and important activities will be collecting samples from the surface for analysis in the rover or back at the habitat. We have researched the features that are expected from a sample storage system and the capabilities that will be required. This paper will present our design and explain the characteristics that we have included.

Introduction

Much of the science accomplished on the Martian surface will result from collecting samples and analyzing them. How these samples are collected and when each is analyzed can vary tremendously. Our goal is to design a storage system that will work throughout the lifetime of the majority of samples that are collected.

A sample collection mission can be based from many different starting locations including the habitat, a short-range unpressurized rover, or a long-range pressurized rover. This paper is going to focus on a pressurized rover that a crew of three people will use to explore large distances away from the habitat. The rover will be capable of two-week missions and will have some science capabilities aboard.

When the rover crew reaches a site that they are interested in sampling, they will have many different strategies for collecting the sample. The most obvious way is with a human EVA where two crew members will exit the rover in space suits and collect samples similarly to how it is done on Earth. However, to increase the safety of the crew, the rover will also be equipped with a robotic arm, small companion rover, or other robotic sample collection system. The sample storage system that is used must function easily with any sample collection technique that is employed.

Once a sample is taken, it must immediately be placed in a sterile sample container and tagged so that the sample can be identified for later analysis. After a series of samples are taken at a site, they will be returned to the rover for storage. It is important that the samples can be stored from outside the rover because we do not want to risk cross contamination by bringing the samples through the living quarters of the rover.

Analysis of the samples can take place at many different times. The rover crew will analyze some samples during the two-week mission using the rover science lab so that they can immediately identify sites of high interest. Other samples will need to remain protected in storage until they can be returned to the habitat for a more thorough and complete analysis. It is even possible that some samples will be put into storage for a return to Earth with the astronauts.

This paper will describe some basic assumptions and the design process that we used to develop a storage system that can be used for all stages of a sample's life cycle, and will be robust enough to work with many different sample collection techniques.

Design Assumptions

There are a few key design assumptions that were made to narrow the scope of the design and to clarify the range of science that should be possible using this storage system. The basis for many of these assumptions came from the NASA Standards on Toxin and Biohazard Assessment and Sample Curation. These assumptions play a key role in the design of the sample storage system described in this paper.

Human-Sample Separation

Humans and samples need to be kept separate at all costs. Humans should not risk coming into contact with any bacteria or toxins in the samples that could harm them. Also, the scientific value of any biological samples will be ruined with any human-sample interaction.

It is assumed that robotic experiments will test the level of toxicity on Mars, before humans even arrive to help identify the level of separation needed between humans and samples. The safety of the crew is paramount in designing any system that they interact with.

Assumption of Life

Robotic experiments cannot prove that life never existed on Mars because of the extreme difficulty of the task. Therefore, it is necessary to design a sample storage system that will be able to hold biological samples as well as geological samples. One of the biggest goals of a future human mission to Mars will be the search for life.

Specialized Samples

Samples such as ice and permafrost will not be considered in our design because they will need special containers to keep them in their natural state. Long core drills will also not be considered, as they may be too long to store in the rover. Since the majority of samples collected will be surface rock and soil samples, the rare occasions of ice, permafrost and drill core samples will not need to be considered for the design of a simple, standard storage container and system. These samples would need uniquely shaped containers and would probably require separate environmental control.

Minimally Contaminated Sample

A minimum of one sample from each site must be stored in such a way as to minimize any contamination. These samples will be the main source of material used for biological investigations. There is also the chance that these samples could be returned to Earth for further investigations. The minimally contaminated samples require extra effort towards radiation shielding, maintaining Martian pressure, and maintaining Martian temperatures.

Design Requirements

Using information from Earth geologists and biologists, the Apollo mission, and upcoming sample return missions, the basic design criteria were decided upon. The design requirements are a set of rules that should be followed to design an effective sample storage system.

General Requirements

There are a couple of different general requirements that should be considered during the design process. First, the storage system should not allow cross contamination between samples. It should be possible to isolate all samples and be guaranteed of the samples' integrity. One of the most important considerations is the mechanical complexity of the system. Martian dust will get into every groove and piece of the storage system. Moving parts and surfaces that must slide against each other are strongly discouraged. Finally, the storage system that is designed should work for as much of the sample's life cycle as possible. It would be ideal if one system could be used from collection, through storage, and to analysis.

Robust to Collection Technique

As mentioned in the introduction, there will be many different sample collection techniques employed on the Martian surface, including human EVA, robotic arms, and small robotic rovers. To standardize equipment that will be taken to Mars, the sample storage system that is designed should function easily with all different types of collection techniques. This is an extremely important requirement and should be considered in all aspects of the system design.

Size and Number of Samples Taken

The majority of samples collected will be soil and surface rock, with a ratio of approximately 75% soil and 25% rock. It is expected that 20-25 samples will be taken at each major site and that about five major sites will be visited during a two-week rover mission. This leads to a total storage volume of 100-150 samples on the rover. Most research will not require large samples to be taken. A standard storage system should be able to handle material with a diameter between 3-10 cm. The length of the container should be between 8-15 cm.

Geological Sample Storage

Samples that are being used for geological research on Mars will not have storage requirements as strict as those for biological samples. These samples will not require a completely airtight storage container and can be stored at almost any temperature and pressure. The most important aspect of the storage system for these samples is to maintain a separation between samples so that there is no cross contamination. We recommend however, that even the geological samples be stored at Martian temperatures and pressures, using Martian atmosphere as the storage gas.

Biological Sample Storage

It is necessary that the biological samples remain minimally contaminated for accurate testing. These samples need to be stored in ambient Martian atmosphere at the temperature and pressure of the location that the sample was collected. These samples also need as little radiation exposure as possible because the UV rays could alter or destroy any forms of life in the samples.

System Placement

The samples must be accessible from inside the glove box or science area on the rover. This will prevent contamination of the samples and still allow for sample analysis to be done inside the rover during the mission. There should be a door on the outside of the rover to place samples inside the storage system. The door on the outside needs to be accessible to a human standing on the ground, so it will need to be placed at or near the bottom of the rover body. The system design needs to allow for samples to move between the lower door on the outside of the rover and the bottom of the glove box.

There are two options to accomplish this. The first is to place the storage system in the space below the glove box. While space is a premium and the size should be minimized as much as possible, the maximum allowable space for the system in the rover is approximately 100 cm on a side. The other option is to create the storage system as an attachment for the outside of the rover where space is not as important.

Concept Development

Throughout the design process we considered dozens of different designs for the container, ways to place samples in the container, and ways to store the containers. Each design was evaluated using the design requirements that are outlined in the above sections. This part of the paper will describe many of the different designs that were considered and help to explain how we decided upon a final design.

Sample Container

We considered three basic designs for the sample container. We first considered sealable bags because they have been used successfully on the Moon and during Mars analog studies. There are a number of advantages to this system, including efficient use of space and the ability of the bag to easily hold many differently shaped samples. However, we felt that there were a number of problems with this type of container. The biggest was that it would be very difficult for a robotic sample collection system to operate using bags. They are already difficult to use for a gloved human. Also, we questioned the durability of bags and their ability to last over a period of months and during the rough travels that the rover will make.

Next, we considered cylindrical storage containers similar to a test tub. This type of container is often used by Earth-based geologists and biologists to collect samples. They are extremely easy to use and can be stored in a tray for easy transport and access. Again, we identified two problems with a cylindrical storage container. First, when the cylinders are placed into storage, they are not making the most efficient use of space. Also, if a screw-on lid was used, it could be somewhat difficult to operate. Unscrewing the lid would require two hands, or if a robot was doing it, it would have to secure the tube while twisting off the top.

These problems with the cylindrical design led us to consider using a square bodied container. Containers of this design can be easily stacked and arranged to efficiently use all available space. Also, if a screw-on top was used, the square body would automatically secure the container and keep it from twisting.

Container Lid

One of the most important design considerations is the lid of the sample container, because it is the part that will interact with the sample collection system and will keep the samples separated.

The first design that we considered used two door flaps on springs. The top would be able to be pushed inward in order to place the sample inside. The lid would then spring shut. This design would allow the container to be used as a scoop to collect samples ensuring that the same scoop did not touch multiple samples. However, we felt that there was no way to create a seal with this system that would keep the samples from becoming contaminated.

Another idea was to put a membrane with slits in it on the tops of the containers. This type of lid would allow for rock samples to be pushed inside through the slits without risk of them falling out. The major problem with this lid design was the possibility of sample contamination.

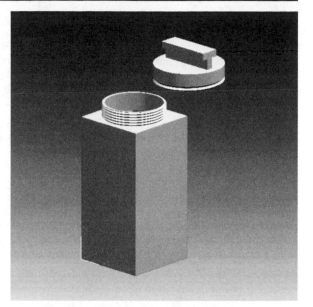

Figure 1. This picture shows a single sample container and lid.

A variety of different lids that would pop on and off were considered, because they would be simple and easy to use by both humans and by a robotic collection system. However, the main concern with this lid was that we could not be certain that the bouncing and jostling of the rover would not pop the lid off.

Finally, we decided to pursue a screw-on lid for the sample container. This type of lid would remain closed and ensure that no cross contamination occurs. The biggest concern was that Martian dust would get into the threads and keep the lid from screwing on. To avoid this problem, we recommend that the threads are somewhat loose and the lid requires only a single twist to close the container. The other concern was how easily it would be for a robotic collection system to grip the lid, screw it off, place a sample inside the container, and then replace the lid. We developed a T-shaped attachment to the lid that would make it easy for a robotic system to grip, twist and lift the lid of the sample container.

Minimally Contaminated

There are two possible approaches to store the minimally contaminated samples. The first is to use a completely different storage container design compared to the normal sample container discussed above. This is what was done on the Moon. The astronauts stored most of the samples in bags, and the minimally contaminated samples for return to Earth were stored in a Special Environmental Sample Container to protect the samples from atmospheric gases. The biggest problem with this approach is that it makes it very difficult to operate using a robotic collection system.

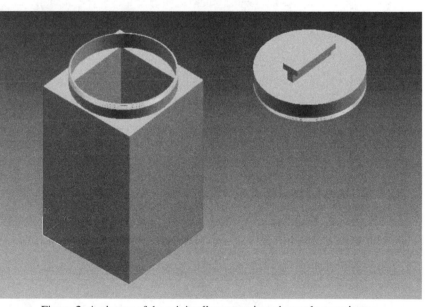

Figure 2. A picture of the minimally contaminated sample container.

The second approach is to place the sample in a regular sample container, and then place that container inside of a "thermos" that would create an airtight seal and make cross contamination impossible. There is also the added benefit of a double seal against contamination. Notice in Figure 2, that the screw on lid for the minimally contaminated sample container uses the same T-shaped attachment as the regular sample container. This makes it easy to integrate the minimally contaminated container into the overall collection system.

Storage of Containers

The containers will need to be stored in some fashion on the rover, on an unpressurized rover, on the space suit during EVA, and on a robotic collection system. We decided that the best way to accomplish this was using a tray that would hold enough sample containers for a single EVA. There would then be a sample tray on the rover for each planned collection mission during the course of the two-week rover excursion.

Each tray should hold approximately twenty sample containers and three minimally contaminated sample containers. A tray can be customized with sample containers of any size necessary. For example, if it is known that a mission will be collecting soil samples, a set of smaller containers can be used than if the mission will be to collect rock samples.

The tray offers the benefit of providing a way to transport the sample containers and at the same time hold them in a position where they are easy to work with. If the containers are being transported on an unpressurized rover, the tray works as a storage compartment for the containers until the destination is reached. The tray also provides a way for a robotic sampling system to hold the containers in place while it collects samples and places them in the containers.

Figure 3. An astronaut with sample containers mounted on the side of his leg.

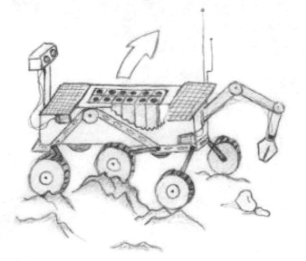

For a human EVA, it would be easiest for the containers to be transported in small compartments along the leg of the astronaut. This provides easy access to the containers and holds the containers in place while the lids are being unscrewed.

Storage of Trays

There are two options for places to store the trays. The first is under the glove box where it will be easy to access the storage area from both the outside of the rover and from the glove box. One idea would be to store the trays in shelves, but this would only allow samples in the very top tray to be accessible from the glove box. Another option would be to place the trays on a carousel. There will be a door on the outside of the rover that would allow trays to be placed in a spot on the bottom of the carousel (due to height differences). There will be a crank on the outside of the glove box that would allow the person to rotate the carousel so the tray with the sample to be studied can be brought up to the top. The tray and sample will then be accessible from inside the glove box. The problem with this design is that it takes up large amounts of space inside the rover and does not use the space efficiently.

The second option is to build the storage system so that it attaches to the outside of the rover in the same area as the glove box. A simple sorting system for the trays could allow trays placed in the bottom to be moved to the top of the storage area where they would be accessible from the glove box.

Figure 4. A robotic companion vehicle with a sample tray mounted on the back of the bot for easy access.

Conclusions and Recommendations

This paper presented a design for a sample storage system for the Martian surface that is robust and will operate easily with both human EVA and with robotic sampling systems. A prototype of this design needs to be created so that it can be tested by humans in analog studies to compare it against current standards for sample storage. It will also be important to test this design using a robotic sampling technique because the sample storage system outlined in this paper was designed specifically to operate easily with a robotic system.

References

- Personal correspondence, Julie Edwards
- *The Mars Surface Reference Mission: A Description of Human and Robotic Surface Activities*, Stephan J. Hoffman, Ph.D., Science Applications International Corporation, December 2001, pgs 52-56
- *Quarantine and Cleaning Issues With Martian Samples*, Timothy Marzullo, Michigan Mars Rover Team, April 14, 2002

//

A Rising Flood of Data[†]

Ned Chapin, InfoSci Inc.
NedChapin@acm.org

Abstract

The successes of the recent robotic missions to Mars have generated a rising flood of data, and greatly enriched our factual knowledge of the red planet. The first successful robotic Mars mission (*Mariner 4*) was nearly 40 years ago and sent back only 21 black-and-white visual images. Taken together, current robotic Mars missions have been sending back during 2004 nearly that much per sol. The current rovers and orbiters can produce visual images in color and in greater resolution than ever before, as well as producing major amounts of non-image data (such as from thermometers, magnetometers, radars, spectrometers, etc.). How much data has been received is summarized in this report.

The incoming data from Mars have been and are being archived for later analysis. But these data are also stimulating some currently done analysis toward two main ends. One is the ongoing direction of the mission. The other is preliminary analyses for status reports on the "science" or "findings" that relate to the formal mission objectives. More detail will be forthcoming for the current missions in the final project reports and in some technical papers that the project science and engineering personnel will tender for publication. What is greatly underfunded is the generation of full and detailed analyses not just of the data from any one mission, but those data in relation to the data from other missions. We have to ask how much knowledge awaits discovery in the rising flood of data from Mars?

Sources of Data

The data received from Mars missions (all robotic up through 2004) are not more than what were designed to be acquired from the mission. Sometimes less data than that has been received because of mishaps to the mission (such as to the *Mars 3* lander) or to the data (such as from transmission equipment failures). In this report, attention is given only to missions specifically targeted at Mars as the prime objective. Thus for example, the data about Mars received from the Hubble telescope are not included in this report, but the data received from *Mars Pathfinder* are included here. Note also that data sent to a Mars lander, orbiter, or spacecraft for the Mars mission, such as from mission support personnel on Earth, are not included since they are outbound data from Earth, not inbound from Mars.

Mission design personnel historically have made the specific selection of what data the mission was to acquire. This human decision process has been constrained by the payload weight chosen, mostly because of the amount of financial funding available. With the weight chosen, the selection has been further constrained by the then-available instrument technology and instrument power demands, operating modes, weights and cubes (payload space required). Fortunately, the forty-year trend in increased reliability, increased capability, and decreased weight for available instruments has yielded an increasing richness in choices available to Mars mission designers, as has been described in the published literature.[1]

Almost all of the available instruments available have had a characteristic in common – their use has required the use somewhere of analog to digital conversion of the data. This has been built into some instruments, but has had to be added to others via analog-to-digital (A-D) converters. For example, a thermometer operates in an analog manner, but the temperature data wanted should be expressed in numbers. Or for example, a camera produces an image. To be transmitted, the captured image has be converted. This usually is like the scan lines used with televisions. A conversion to digital form allows a more reliable transmission of the image data.

Nearly all of the data produced during Mars missions comes during one or more of the following four phases. Up through launch from Earth, data are generated about the status and condition of the spacecraft and its payload. Once in flight, tracking data are generated for the flight duration, as well as data about the status and condition of the spacecraft and its payload. On some missions, some payload instruments are started during the approach to Mars, to begin the mission data acquisition process. Orbital insertion and/or landing generate additional data until landing and/or orbit are achieved. After that, the deployment of the payload instruments generates status and condition data, and is followed by the data generated by those instruments as they are applied to the mission's objectives. This period may have a duration from a few minutes to a few Earth years, and hence is intended to provide the bulk of the data from the Mars mission.

Three kinds of data are environmental data, self-status data, and overhead data. In each of the four mission phases, the relative amount of data is different (see Figure 1; the vertical axis is the percent of the phase data). Each kind of data has relationships with the other two kinds:

- Environmental data. The three major physical parts of a robotic Mars mission are the spacecraft, the orbiter if any, and the lander if any. Each physical part exists and operates in an environment. What the characteristics are of that environment are potentially critical to the success of the mission. Two examples are the density of an atmosphere, and the mineralogy of a rock on Mars. After arrival at Mars, acquiring the environmental data is the main target of the mission. Prior to arrival, the environmental data are relevant in assessing the health of the mission.
- Self-status data. These are data about the internal characteristics of the physical parts of a robotic Mars mission. Four examples are fuel remaining, battery charge rate, velocity of movement, and efficiency of the solar panels. The relationships between the self-status data and the environmental data are different in the different mission phases, and are of particular interest to ground support personnel.
- Overhead data. These are the data that identify and qualify the self-status data and the environmental data. Six examples are date, time, code used, instrument identification, instrument setting (such as filter used and camera pointing direction), and target identification. The consistent presence of these data help the ground support personnel and the associated project personnel in distinguishing any kind of data, such as a battery power drain rate, from any other kind of data, such as a wind speed. The overhead data are critical in distinguishing any particular self-status or environmental data item from among hundreds of thousands of data items acquired during some time period, all expressed in similar appearing codes.

Generic summary of kind of data received by Mars mission phase

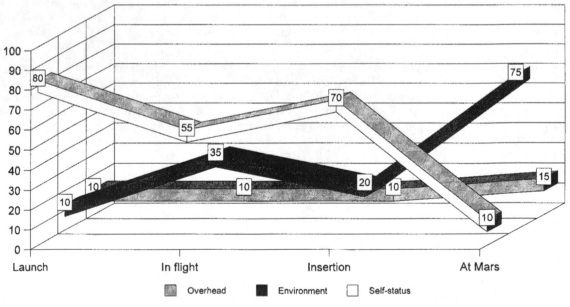

Figure 1.

As an example of the sources of data, Table I summarizes the amount of data received from each of the sources for the *Mars Global Surveyor*. The time period covered is from each instrument's initial deployment to the 2004 September 9 conjunction when Mars and Earth were on opposite sides of the Sun. The instruments in the payload are what vary most from one robotic mission to another. The data sources and their amounts of data will be quite different for crewed Mars missions.[2]

Forms and Formats of Data

Basic to understanding the forms and formats of data is the way data are represented. In this regard, a short reminder may be helpful on the terms used in describing amounts of digital data. The basic unit is the bit (abbreviated as b), also known as a binary digit. A byte of data (abbreviated as B) is eight bits of data, and may be accompanied by a parity bit used for checking for the loss of data in

| Received Data | | Mars Global Surveyor |
Gb	%	(Source Of Data)
2331	53.7	Orbiter camera (~265400 images; 3792 Gb decompressed)
822	18.9	Radio science (RS), 10440 occultations (done from Earth)
466	10.7	Thermal emission spectrometer (TES), 566000000 spectra
390	9.1	Overhead (mostly packet header, transfer frame and parity)
167	3.8	Magnetometer/electron reflectometer (MAG/ER)
92	2.1	Orbiter laser altimeter (LA), 1372000000 shots
73	1.7	Spacecraft engineering telemetry including diagnostics
4341	100.0	Total estimated data received

Table 1. Estimated Data Received by Source from MGS, as of 2004 September 9.

handling or transmission. Practical measures for amounts of data acquired in Mars missions are the kilobit (abbreviated as Kb) for 1,024 bits (but often treated as if it were just one thousand bits), the megabit (abbreviated as Mb) for one million bits, the gigabit (Gb) for one thousand megabits, and the terabit (Tb) for one thousand gigabits. Next up, although we are not there yet with Mars missions, is the petabit (Pb) for one thousand terabits. On Earth, a high quality digital camera picture in full color uses about 4Mb.

For Mars missions, since each Mb of data takes time to transmit between Mars and Earth, data compression techniques are commonly used to save transmission time. Although rarely used in relation to Mars missions, "zip" data are an example of the results of using a compression technique commonly available to people with desktop and laptop computers. In a manner somewhat like data prepared for transmission over the Internet, Mars mission data may be prepared in packets. The preparation of the packets can be combined with the insertion of the overhead data before transmission from Mars to Earth.

Upon receipt on Earth, the first step is to record the data as they arrive. After a transmission is finished, a copy of the received data are decompressed and decoded to make them ready for processing. At a minimum, that processing involves recognizing (via the overhead data) the component parts in the received data, such as fuel consumption, thruster firings, images, etc. The overhead data are used to route copies of the received data to the concerned ground support personnel, and to organize the received data for filing. These files are reformatted for archiving, their contents cataloged, and then copied onto the archiving media (historically, that has normally been magnetic tape).

Upon personnel request, data in the archive may be searched for, and if located, a copy made and provided to the requester either in the archive format, or after being reformatted and/or recoded to meet the terms of the request. In practice, the archived data are the prime source for all data received from a Mars mission other than about activities that are currently going on, such as the in-progress movement of a rover on Mars. This puts the archive in a critical role, and raises concerns about deterioration of the storage media, usefulness of the catalog, and documentation of the archive data formats and codings.

Uses of the Data

During the launch phase, the data are made available in real time for use by the ground support personnel. That practice offers those personnel the most recent self-status and environment data as quickly as they become available as the spacecraft is being prepared for and being launched. During the in flight phase, real time use of the data becomes increasingly impossible, and delayed time use gradually takes over as the spacecraft gets ever closer to Mars. The focus is on guiding the trip to Mars, so the self-status and environment of the spacecraft continue to be the dominant data. If the payload makes it possible, one or more of the payload instruments may be activated in flight and acquire some data for sending back to the project teams on Earth. During the insertion and landing phase, the pattern of data received is like launch, but the specific data are mission dependent. The data received from orbiters are less abundant than and different from those received from landers. Some missions have had both a lander and an orbiter, with different data from each.

For orbiters after insertion into orbit around Mars, the payload instruments are deployed and engage in their data acquisition processes. The acquired data are batched and transmitted to Earth usually no more often than once per orbit. While some orbiters can also serve as data relays for data from landers, in this paper in order to avoid double-counting, relayed data are not counted as data from orbiters. For landers after landing successfully, the instrument and/or rover components are deployed. These generate in use from little (like a wind direction vane) to much (like a rover) self-status data. Some of the self-status data from rovers are dual-use data, such as navigation terrain images with slope and distance indications. In general, the lander instruments, whether static on the lander or mobile on the rover, acquire data either continuously (like an anemometer) or as directed (like a microscope). The acquired data are batched and transmitted to Earth usually no more often than a couple of times per sol, and may be relayed by an orbiter.

Users of the Data

The users of the data received from Mars missions rarely want the data either in the "raw" or in the "ready to archive" form and format. Instead, they want the data in a form and format convenient for their purposes. Putting the data into those forms and formats may involve extensive processing. An example is the production of color images of portions of the surface of Mars. The major users of the data from Mars missions are the ground support personnel running the Mars mission, the principal investigators (PIs) who have projects associated with specific payload instruments, the funding sources (such as NASA or ESA), the news media, the project academic and research personnel, as well as personnel such as graduate students and specialists associated with the Mars Society, and the general public.

The general public and the news media are usually secondary users. That is, they use the results of the processing done by the PIs, the funding sources, and the project academic and research personnel. An example is the visual images of the Martian terrain. The PIs and the project academic and research personnel apply specialized knowledge and technology to process and analyze the data, producing their results in images, graphs, tables, maps, charts and written documents, such as reports and technical papers. The funding sources do similar work with the data, but usually to meet the needs of documenting the mission accomplishments and history, and of the news media. The ground support personnel are concerned with the ongoing direction of the mission, and require data in specialized forms and formats, such as for immediate display on controllers' consoles with comparison prior data.

Opportunity

The only value received from robotic Mars missions is the data that gets back to Earth. And the value of those data depend upon what people do with those data. Hence, a couple of key questions are "How much data have Mars missions sent back to Earth?" and "What use has been made of the received data?"

Firm reliable statistics on how much data have been received from Mars missions are hard to come by for various reasons. Some of the reasons are political, ranging from the personal level to the international level. Some reasons are technical, for both the hardware and the software technology for the transmission and processing of data during the past forty years have changed a lot. Those changes reduce the comparability of data amounts received from different missions. Some reasons are archival, for archive technology and practices have changed, and the storage media used have suffered from varied amounts of integrity degradation. The best that can be done are estimates, such as summarized in Table 2.

Mission Name	Funded By	Active Life	Mission Type	Main Objectives	Data in Gb
Mariner 4	USA - NASA	1965/07/14	flyby	video (21 images)	0.005
Mariner 6	USA - NASA	1969/07/31	flyby	video (76), spectrom.	0.304
Mariner 7	USA - NASA	1969/08/05	flyby	video (126), spectrom.	0.496
Mars 2	USSR	1971/11/17–1972/08/22	orbiter	video (57), radioref.	0.397
Mars 3	USSR	1971/12/02–1972/09/22	orbiter (& lander?)	video images (3)	0.027
Mariner 9	USA - NASA	1971/11/13–1972/10/27	orbiter	map (7329), spectrom.	54.000
Mars 4	USSR	1974/02/10	flyby	video (4), polarim.	0.032
Mars 5	USSR	1974/02/12–1974/03/13	orbiter	video (~60), polarim.	0.050
Mars 6	USSR	1974/03/12–1974/03/15	flyby	atmosphere	0.001
Viking 1	USA - NASA	1976/06/19–1982/11/13	orbiter and lander	exploration (~36,000)	339.890
Viking 2	USA - NASA	1976/08/07–1980/04/11	orbiter and lander	exploration (~16,000)	151.065
Mars Pathfinder	USA -NASA	1997/07/04–1997/09/27	lander and rover	engineering (~17,050)	2.600
Global Surveyor	USA - NASA	1998/04/15–2004 *	orbiter	map (~265,400)	4341.000
Mars Odyssey	USA - NASA	2001/10/24–2004 *	orbiter	map (~160,000)	2053.000
Nozomi	Japan - JAXA	2003/12/14	flyby	atmosphere	0.001
Mars Express	Europe - ESA	2003/12/25–2004 *	orbiter	exploration (~ 154)	414.800
MER-1 Spirit	USA - NASA	2004/01/03–2004 *	rover	past water (~22,300)	32.871
MER-2 Opportunity	USA - NASA	2004/01/24–2004 *	rover	past water (~24,700)	30.397
Estimated total					7420.396

Table 2. Estimated Data Returned by Mars Missions that did send back data about Mars (2004 September 9).
Notes: * Mission still active as of 2004 September 9.

Answers to the "What use has been made of the received data?" question are also hard to come by. Clearly, the ground support personnel have made nearly full use on a transitory basis of all of the data relevant to their concerns with Mars missions. The funding sources have drawn well on self-status data and some environment data from prior missions is preparing subsequent missions. The news media, when they have received data in an acceptable to them form, have not been shy in broadcasting newsworthy accounts on Mars missions to the general public. The PIs (principal investigators) and project academic and research personnel associated with Mars missions have generated technical reports on what have been regarded as advances in human knowledge resulting from their work. These reports have usually been very well done on specifics but of necessity very narrow in their focus and approach. The most general of these have been the work on processing images of the Martian terrain, and some of those results have been outstandingly well done. The general public has been a transitory and cursory user, and not a direct user of the Mars data, except of a relatively few images.

The non-project academic and research personnel have had the least success in using the data received from Mars. The two main roots of the relatively low amount of success appear to have been primarily funding and secondarily access to the data archives. Access to the archives could be done in a manner similar to that used for access to the human genome, but has not been. In the human genome case, the three main elements are specialized access software (usually involving paying license fees), Internet usage, and some authorization. So far, for the Mars archive data, no agency or firm has taken action to provide such access for other than a very limited cadre, such as a doctoral student supported by a NASA grant. Since the Mars archives are different from each other in content, format, form and (often also) media, different specialized access software is needed for each.

The funding situation has also been weak. It has been traditionally difficult to justify spending tax money to search for knowledge. This has been especially true when the likely value of the knowledge cannot in advance be clearly quantified as exceeding the costs, and when the probability of finding that knowledge is hazy. Private funding institutions have thus far very rarely supported the study of Mars, leaving that area mostly to hobbyists and individual philanthropists.

A prime opportunity to discover new knowledge about Mars is to take an integrative approach to the archived Mars data from diverse missions. Mars missions have had some of the characteristics of the group of blind men trying to describe an elephant. Each has sensed some characteristics that the other missions have not, or sensed them differently and provided a different perspective or granularity or location on some characteristics that superficially appeared to be a confirmation. A popular example of one published work based on an integrative approach is the book *Mars* by W.K. Hartmann.[3]

The value from robotic Mars missions comes only from the data they return. But those data are only the vessel for the value. Information about Mars is acquired by processing, analyzing and interpreting the received data. The information extracted from the data of any one mission is less than what may be extracted from an integration of the data of many missions. The ongoing Mars missions are producing a rising flood of data. Integrating those data with archived Mars data offers an opportunity to enrich our understanding of Mars beyond what any one robotic Mars mission can do.

Acknowledgments

For their assistance in locating mission data, the author thanks the Mars-involved personnel at the Jet Propulsion Laboratory (JPL), a NASA facility operated by the California Institute of Technology, and especially Julie A. Cooper, Thomas E. Thorpe, and Guy Webster. The author also thanks Artemis Westenberg and Richard Heidmann of the Mars Society in Europe.

References

1. See the publications of the American Astronautical Society published by Univelt, Inc.: San Diego California.
2. Ned Chapin. *Communications Infrastructure to Support Human Activities on Mars*. On to Mars – Colonizing a New World. Apogee Books Space Series, Collector's Guide Publishing Inc.: Burlington, Ontario, Canada, 2002; pp. 159–165.
3. William K. Hartmann. *A Traveler's Guide to Mars: The Mysterious Landscapes of the Red Planet*. Workman Publishing: New York NY, 2003; 468 pp.

Public Outreach

www.marssociety.org
The Official Mars Society Web Site†

Harold Miller, Mars Society Webmaster
ms@nw.net

The History of the Mars Society Web Site
Before the Mars Society was founded, Dr. Robert Zubrin gave speeches introducing his concept of a manned mission to Mars to groups all over the nation. I met him at one of those lectures, in my home town of Grants Pass, Oregon, early in 1996. After the session, I asked him for the address of his web site. When he told me that he had none, I offered to create one. His site, http://mars.nw.net is still in pretty much it's original form. When the Mars Society was officially started, I created a separate site, just for them. The first domain name was soon registered, and several were added over the next few years. The site and the movement grew.

After the 1999 convention, the site was moved to a Canadian web host, because of the increased traffic to the site. In early 2002 the site was moved again, this time to Interland, the nation's largest web hosting company, and I was asked to again become webmaster. This is where the site resides at this time. The membership database, and several smaller databases are also hosted with Interland.

Today's Web Site

Today (third quarter of 2002) the web site consists of over a gigabyte of data, images, sounds and web pages. Some of the larger multimedia files are served by our own archive server, located in Southern Oregon.

We have links to many services available to our members and the general public. From the main page, users can jump to the NEWMARS Bulletin Board System, the On-Line Chat System, our Mars Store, and links to the Member's Only area. Current events, news and projects are all available for those folks interested in Mars, or the Mars Society.

Projects
- Desert Research Station
- Flashline Arctic Station
- Translife Mission
- Pressurized Rover
- Operation Congress

Join our mailing list

[Subscribe]

Enter your e-mail address

Radio Free Mars

Now broadcasting Music and Audio from the FMARS location

Here's today's image of the Hab and the Weather Sensors donated by Met One Instruments.

July 15, 2002

For the latest breaking news, check out the Flashline Mars Arctic Research Station's web site.

Crew roster:

- **Dr. Robert Zubrin**, Pioneer Astronautics, Astronautical Engineer, USA, Commander
- **Nell Beedle**, Fugro Seafloor Surveys, Geologist, USA
- **K. Mark Caviezel**, Aerospace Engineer & USMC Sergeant, USA
- **Dr. Frank Eckardt**, University of Botswana, Geologist, German citizen living in Botswana.
- **Shannon Hinsa**, Dartmouth Medical School, Microbiologist, USA
- **Dr. Markus Landgraf**, European Space Agency, Physicist, German
- **Emily MacDonald**, Oxford University, Astrophysicist, British

The crew is serving on a 3-week rotation in the Flashline Mars Arctic Research Station on Devon Island. During this time they will employ the Mars-like polar desert of Canada's Devon Island to experiment with techniques for the human exploration of Mars.

A special session reporting devoted to crew reports on the activities of the Flashline Mars Arctic Research Station and Mars Desert Research Station will be held at the 5th International Mars Society Convention, August 8-11 at the University of Colorado at Boulder. Registration for the conference is now open.

Buy it! @The Mars Society

Save up to 40% on some products in the Mars Society mall. A percentage of each purchase goes to support the society's projects.

NEW MARS — A JOURNAL OF THE MARTIAN FRONTIER

Mars Global Surveyor

Mars News

As you scroll down through the first web page, you will see older news items, links to Mars related sites, and links to our ongoing projects. This main page is usually updated weekly, sometimes several times a day. Information, articles, bulletins and news items come from our members, HQ, and items I receive or discover myself.

Communication Aids
Many people cooperate to give the web site its depth and breadth of scope. The web site at www.NewMars.com is run by Adrian Hon. This link provides the Mars Society with a bulletin board service, allowing "conversations" to continue on a variety of subjects. You can follow the "threads" of discussions on many different areas of interest, including the "How," "Why" and "When" of actually getting humans to Mars.

Thanks to folks like Bruce Mackenzie and Randall Severy, we have the ability to hold meetings on line. The Mars Society Chapters Meeting Server allows a group of people to talk (type) in real time, in a controlled environment. The organizer of a meeting can control who is heard and when. With the entire meeting on line there is no need to take minutes, since anyone can log everything to their computer.

The Database
Any organization has a need to keep records and files. The Mars Society is no different, and a lot of those files are kept in secure, on-line storage. Members can access, update and review their own records, and on-line passwords via the Members Only link on our front web page. Headquarters can generate reports and mailing from any location that has internet access. Anyone can join the Mars Society, and pay for it on line, or renew their existing membership. All of these features are possible because of the linkage between the SQL database server and our web forms. This is a new feature, and still has a few kinks in it, but I am sure that it will provide for an easier, faster, more secure method of financial payment in the near future. At this time we can process credit cards, and accept payment through the PayPal service.

Current Statistics
These numbers have been climbing steadily since the beginning of the year.

new) *mars*

New Mars :: New Mars Forums
Website Forums

search   members   help

» Welcome Guest Log In :: Register

New Mars Forums

Forum	Topics	Replies	Last Post Info
New Mars			
Meta New Mars Have your say about this website and forums [OPEN FORUM: Unregistered users may post]	29	180	July 29 2002,17:54 In: World Space Week By: clark
New Mars Articles Comment on New Mars articles	5	40	July 05 2002,00:59 In: The urgency question By: Phobos
New Mars News News about the latest New Mars website developments	1	20	July 20 2002,01:09 In: Forums back online By: Shaun Barrett
Acheron Labs			
Human missions Cost, logistics, funding, technology and timeframe for manned missions to Mars	64	782	Aug. 01 2002,15:23 In: Mars Aeroentry By: Mark S
Interplanetary transportation Present and future methods of spaceship propulsion, e.g. fission, fusion, solar sail	25	311	July 25 2002,16:10 In: Propulsion methods for... By: Canth
Planetary transportation Transportation within Mars; rovers, Mars suits, dirigibles, gliders and more	10	95	July 24 2002,21:44 In: small, high speed bugg... By: ecrasez l infame
Unmanned probes Mars Global Surveyor, ESA Mars Express and future probes	6	36	July 17 2002,18:37 In: Europa By: Byron
Life support systems The problems of producing food, air and water on Mars	13	241	Aug. 01 2002,19:20 In: We need a brainstormin... By: turbo
Life on Mars Discussion of evidence for past and present life on Mars as well as methods of detection	4	28	June 15 2002,13:10 In: New evidence for life on Mars? By: Christina

Average Total Hits per day	Over 26,500
Home Page Hits per day	Over 8,000
Average Visit duration	Over 8 minutes
Total Storage needed	Over 1,000 Megabytes
Database Member Records	Over 5,500

Table 1

The average visit duration is especially satisfying. People are spending more than eight minutes at our web site per visit. The information we are publishing is being read, and maybe the public is giving our Humans to Mars missions serious thought.

Domain Names

The Mars Society has many internet domain names in use or affiliated with our site. Here is a list of the most notable ones:

- MarsSociety.Org
  - Arctic.MarsSociety.org
  - Desert.MarsSociety.org
  - Chapters.MarsSociety.org
  - Home.MarsSociety.org
- MarsSociety.Com
- NewMars.Com
- Mars.NW.Net
- Plus many International Chapter Sites, such as MarsSociety.org.au, MarsSociety.de and planete-mars.com

Future Plans

We have many plans for future projects, upgrades, and improvements. Many will succeed or fail depending on the volunteers, funding and support that is available. You can make a difference.

- Upgrade the SQL / Database support
- Install and connect Web Cam's at the habitat's we support
- Weather information, live from the Arctic or Desert Research Stations
- Automatic News clippings on Mars-related topics
- Our OWN server hardware with high-speed connection to the internet

If one of these projects appeals to you, please contact the webmaster and donate time, money or services to ensure that the Mars Society Web pages will continue to provide interesting, accurate content to our members and the general public.

—#—

How Best to Talk with the Public about Mars, Space Tourism and Space Commerce†

Dr. David M. Livingston
dlivings@davidlivingston, www.davidlivingston.com

Abstract

As host of *The Space Show,* my radio talk show focusing on expanding space commerce, space tourism, and Mars exploration, I frequently receive questions and queries from the program's listeners. The overriding theme of this listener feedback strongly suggests that a significant part of the population does not understand the need for a space program that facilitates the development of space resources, space tourism, a human Mars mission, or even acknowledges the benefits derived from a successful space program. Clearly, space advocates, specific space program proponents, and even NASA do not successfully articulate the case for space development to a wide segment of our population.

This paper addresses the feedback received from listeners since starting *The Space Show* (formerly *Business Without Boundaries*) in June 2001. In addressing the issues raised by the listening audience, recommendations are offered for how best to both talk with and inform the general public about the importance of having an expanded space program, including a new off-Earth economy resulting from a much expanded commercial space industry. Developing a space economy and space resources requires widespread public support and constructive policy making. Obtaining this support and enabling the creation of constructive space policies may largely depend on those in the space community doing a better job in helping the public to understand the benefits of space development. The recommendations contained in this paper can serve as a useful and productive guide to winning public support, as well as influencing policy makers for the benefit of off-Earth commerce and opportunities.

Introduction

There appears to be a major disconnect regarding both the understanding and appreciation by the general US population about enhanced space commercialization programs; space exploration; and private citizens working, living, and playing in outer space. Many factors contribute to this situation, some of which are inherent components of our ability to work and live in outer space, and many of which are largely a result of perceptions which have largely been created by special interest space industry and advocacy groups, governments, and the media. While this paper addresses several of the primary reasons for this lack of connection with the general public, the real focus of this paper is to offer constructive solutions supporting commercial space development and becoming a spacefaring society.

In the course of offering solutions to some of the major problems highlighted in this paper, there must be a discussion of the benefits resulting from developing space resources. Thus, this paper reviews some of the most frequently cited benefits, from a society and cultural viewpoint, as well as from different individual viewpoints. Most of the time, space development advocates and space industry representatives offer benefits for doing something in space, but the benefits are either too general or simply not important to a wide variety of people. Often, the general public simply does not resonate with these stated benefits. This paper provides suggestions for discovering what really is important to people about space development.

Why Space Development?

Space development can mean something different for each person on this planet. Yet there are some facts that cannot be denied regarding space development so far. If these facts are at all useful in predicting future results, space development offers not only advances in science, technology, and other areas that can make positive differences in all our lives, but it also offers hope for a better tomorrow.

The American space program, under the auspices of the National Aeronautic and Space Administration (NASA), has been responsible for many incredible inventions that have benefited people all over the world, not just those in the United States or the wealthy industrial nations. The space program has produced important benefits in the fields of health, medicine, public safety, energy, the environment, resource management, art, recreation, computers, automation, technology, transportation, manufacturing, construction, and much more.[1] Some of these benefits include solar radiation blockers, diabetes pumps, spinal cord rehabilitation protocols, advanced body scanning, cardiovascular advances, global positioning location services, weather analysis and forecasting, environmental and wildlife analysis and protection through pollution control, urban planning, thermal protection for people, equipment, homes, and more.[2] From my experience with my radio program audience, however, unless a person has personal experience with these inventions and their benefits, or even knows they originated from the space program, just saying that wonderful advancements and improvements in medicine and other fields has resulted from the space program is not enough to convince people that space development should be a priority.

In addition to the many benefits mentioned above, an excellent book outlines what I consider to be one of the most important reasons for going into space, producing a powerful benefit for people everywhere on this planet. *The Overview Effect*, by Frank White, documents the "transformative effect" that space and the views of Earth from space have had on the astronauts, cosmonauts, and civilian visitors to space. This transformative effect has the power to dissipate and neutralize border disputes and other similar issues, bringing the world closer together and more peaceful. This was evidenced by the comments made by Congressman Bill Nelson when he rode the Space Shuttle in January 1986. At that time, Nelson said, "If the superpower leaders could be given the opportunity to see the Earth from the perspective from which I saw it – perhaps at a summit meeting in space in the context of the next century – they might realize that we're all in this with a common denominator."[3]

A similar effect was reported by Saudi Arabian Prince Sultan Bin Salman al-Saud when he rode aboard the Space Shuttle in June 1985 and said, "I think the minute I saw the view for the first time was really one of the most memorable moments in my entire life. It really strengthens your convictions. To me, it's an opportunity to prove that there is no conflict being a Muslim, or any other religion. Looking at it from here, the troubles all over the world, and not just the Middle East, look very strange as you see the boundaries and border lines disappearing."[4]

In addition, the space program has proven to be a strong enabling force for world peace and stability. Despite the fierce competition and challenges that were present during the Cold War era between the former USSR and the US, there were no conflicts in space and many treaties and agreements assuring cooperation were created and adhered to by the superpowers as well as the United Nations. Today, former enemies are partners in space development, pursuing space commerce both jointly and competitively.

Since humans first ventured into space, the space development track record is a positive one for mankind. In today's world, beset with so many problems, including terrorism, war, hunger and environmental destruction, space has been a positive arena for humans to be their best. Space seems to be the place where people and nations can work together for positive solutions to our worldly problems.

Even with this enviable track record, space development has not excited people sufficiently to make space programs a priority for government funding and attention. Just understanding these accomplishments is obviously not enough. Based on my experience with *The Space Show* audience, I believe people have to directly connect with the space benefits to be able to personalize them and own them. Then, and only then, will space development be the priority that it should be. Thus, when talking about the benefits of space development and the space program, the discussion cannot be too general or impersonal.

Common Barriers Regarding Space Development

Industry advocates, policy makers, and others working toward becoming spacefaring and expanding space commerce frequently cite barriers that prevent their space objectives from being realized. While the barriers cited are real and often present formidable issues to overcome, barriers are not always the reason why space development does not proceed at a faster rate.

The cost of accessing space is often cited as one of the barriers to furthering space commerce and development. While the cost to orbit a satellite is serious issue, it has not blocked most existing satellite and telecommunications companies from being financially successful. When cost is cited as a barrier, most often the reference is to the cost of launching one pound of material into low-Earth orbit (LEO) at a price of approximately $10,000. This is accurate pricing for some launch vehicles, specifically the Space Shuttle. Such a price far exceeds any possible economic price for a putting one pound into LEO, or even for the more costly launches to higher geostationary orbits. There are many launch vehicle options, however, that launch a pound into orbit at much lower rates for differing payloads and orbits, although still at a high price.

Clearly, the high cost to orbit is not preventing today's commercial space companies from realizing success or profitability. Today's commercial space industry, however, launches only cargo into orbit. Even if the cost to orbit was as low as $3,000 per pound, using an approved hypothetical rocket for carrying humans to space, imagine how costly it would be to take a crew of seven average-size astronauts into orbit, let alone its payload and supplies. When this barrier to space development is presented to the public, the public, on the other hand, sees successful telecommunication, satellite, and rocket companies. This is confusing – how can successful and profitable commercial space companies exist when the cost of getting to and from space is a serious barrier to space commerce and development?

Another common barrier to space development and commerce frequently cited is the lack of investment funding, especially for start-up and entrepreneurial business ventures. Generally speaking, obtaining sufficient investment capital for a space venture is difficult at best. However, just because a venture is not funded or is unable to obtain investment capital from the traditional investment sources, does not mean that the venture should be funded. Many such investments should not be funded when one closely examines the company's business plan and fundamentals.

Space business ventures must compete for financing with terrestrial business ventures. While there may be significant amounts of investment capital available over time, investment capital is still a finite resource. No matter what the economic condition of the country, the supply of capital is not unlimited. Often, the space venture is simply not competitive with a terrestrial investment opportunity, so it is natural for the available investment capital to flow to the investment with the best combination of low risk, high potential return, and a solid array of fundamentals to help the business meet its goals. When a space venture is unable to get funding, one must look to see if the venture should have been funded in the first place.

These two factors – the high cost for space access and the availability of investment capital – are mentioned only to show that while they are barriers to developing space commerce, these obstacles are not completely at fault for delaying and, in some cases, thwarting space commerce. The same case can be made with many of the legal, regulatory, and other barriers to space development.

What's Needed to Facilitate Space Commerce

Facilitating the expansion of space commerce and space development requires positive and constructive public policy and legislation, along with a supportive regulatory environment. While public financing is essential for many of the space programs, private sector investment is also crucial, especially in developing reusable launch vehicle (RLV), lunar, LEO, and other plausible commercial space ventures. In the midst of these requirements, it is important that investors, our elected representatives and policy makers, and the general pubic understand the importance and potential of space.

Furthering space commerce and developing our space resources can happen when we who know, value and understand the many facets and diversity of space are able to lead, educate, persuade, advocate and convince the public and the policy makers of the importance of these objectives. While the engineering, technology, science and costs of space development are important, they pale in comparison to the necessity of reaching these objectives. Space is expensive, risky, mysterious and complex. We are used to space being the complete domain of governments and their agencies, not the domain of people. And we have not personalized the benefits derived from space development and exploration.

The importance of obtaining funding and acquiring investors is a given in launching any new business or industry and the developing commercial space industry is no exception. While the importance of making sure the commercial space venture is properly planned and financially competitive with similar investment opportunities has been mentioned, it is also essential that a space company have a competitive management team and be able to communicate their plans to investors, financiers, and those in charge of funding allocations. Two common mistakes exist in this category. First, the management team is often weak In appropriate business and management experience. Second, management often assumes cutting-edge technology or engineering sells the deal on its own. Except for the occasional space enthusiast who may want to invest or finance a space venture because of his love for space, this approach does not work with most investors and is usually counterproductive.

What is required by management is to assemble an experienced team with the ability to present and market its proposal to the financial community in a highly professional manner, stressing the business components of the venture, showing how the return on investment (ROI) assumptions are plausible, and clearly explaining the market for the goods or services produced. In short, the ability to communicate the essentials to those in the financial community cannot be overstated. An effective, concise and fact-laden business plan executive summary is a good way to prepare for this type of presentation.

When speaking to individuals in the investment community, it is important to ask open-ended questions to find out what the investment attitudes are regarding the type of venture being presented. Understanding what interests the investor, his investment risk and ROI criteria is very important. By understanding the investor's investment strategy, the management team is then able to address these important issues in ways that help secure funding. If the venture to be funded falls outside the investor's criteria, then the management team is better off using its time and resources where it can be more productive. Thus, the ability to listen, question, and to interview those in the investment community to find out their views about space investments and their possible space investment strategy is fundamental to funding commercial space ventures and in obtaining public sector funding of space development projects.

Talking to the Public About Space Commerce, Tourism and Space Development

Among the primary concerns and beliefs held by the general public about space commerce, space tourism, and space development, there is a strong sense that public monies should be spent on projects considered more worthwhile, such as feeding the poor, medical and cancer research, solving our environmental problems, and establishing peace on Earth. Also, many believe that we have so badly messed up Earth, that we should stay out of space because we will undoubtedly do the same to everything we touch in space, including the Moon and Mars. These are difficult beliefs to counter, especially because the holders of these views are not always interested in receiving new, accurate information that might change their viewpoint. Presenting factual evidence to offset these views often results in the hardening of the views by the person holding them. Thus, one has to carefully listen and question the person to discover any underlying beliefs wherein the person might be open to receiving potentially new and view-shifting information.

Space advocates and proponents are well known for stating what the space community believes are the most important benefits for going into space, regardless of what others may think. Such a discussion often centers around the need to save the human race from extinction should there be a life-extinguishing asteroid impact upon Earth, or saving humans from annihilation due to war, environmental meltdown and destruction, disease, or some other destructive event. In addition, conquering the new frontier, opening it to civilization, nurturing the human spirit to explore and reach out, and developing new technologies and equipment that will help us on Earth are all reasons frequently cited for why we should be in space. These reasons, however, do not always connect or resonate with those participating in the discussion.

As with the investment community, it is essential that space advocates and proponents, industry leaders, and policy makers ask questions to find out why space is or is not important to the person or group being addressed. When this approach is undertaken, not only do we find out the key facts underlying the beliefs of others, but we can also discover why the person or group may not have any interest in space. Finding out why there is no interest in space can be just as important and useful as finding out why the interest is there. If we make proper use of this information, we can develop ways of communicating with these people to better explain how and why space may benefit them and their interests.

It is also important that those in the space community avoid talking down to others, regardless of their views or level of knowledge about space matters. Being condescending is counterproductive to winning support for space development and financing.

Going into space, let alone Mars, is costly, risky and complex. To become a spacefaring nation and culture, we need to do it with support and help from many people and groups, not just the few represented by the space community. Therefore, understanding the reasons people have for wanting to be in space or not is a crucial first step to building the alliances and partnerships needed to facilitate space development. Understanding that our reasons or benefits may not matter to others is important. Realizing that we need to make sure that the resulting benefits from our space programs connect on a personal level with those we are addressing is important.

In talking to guests and listeners of *The Space Show*, it has become obvious that selling the negative is less than productive. The better responses come from people who are upbeat and provide positive reasons and explanations for space development. Talk about destroying the Earth, its environment, humanity, and destructive asteroid impacts upon Earth, while resonating with some people, in general does not inspire, motivate, or win people over to being proponents of programs designed to further space commerce, space tourism and space development. Fear and negativity are not good selling points.

Recommendations

Having a positive dialog with people about space development and its benefits is very important. Avoiding space and technical jargon and speaking the same language as the person or group being addressed facilitates a productive exchange of ideas and information. The goal is not to dazzle or impress, but to include those outside the space community in the discussion of why it is important for humanity to be in space.

Rather than making our ideas of the benefits the only important ones, allow the benefits expressed by others to become part of the dialog. At the same time, let people know that the benefits being discussed represent the views of a wide ranging part of society, and that the space community acknowledges the validity and importance of people's views about space development, space settlement, and humanity's place in space.

One crucial recommendation refers back to working with investors and financiers. For the most part, these people are not space advocates or involved in the space community. They invest primarily for profit as part of their businesses. While there are secondary reasons, we must be able to successfully address their primary reason if we want them to play a role in financing in our space ventures. Thus, we need to be able to speak the investment language. We need to know what their hot buttons are and what causes their disinterest in the project. And above all, we must present realistic business plans that are competitive with terrestrial investment opportunities.

Always anticipate the unexpected when talking about space. People may have many reasons for liking or disliking space. We should be able to accept their reasoning even if we disagree. In any event, avoiding placing them in an uncomfortable position and/or putting down their ideas are certainly desirable objectives.

When talking about the benefits derived from developing space, being specific, rather than abstract, academic or general is preferred. Do not assume that we as advocates know all the benefits. Start with a very broad brush and narrow the discussion based on feedback from those involved in the discussion. The goals, to increase support for space development programs and how best to talk to the public about space development, should be held in sharp focus throughout the discussion.

Conclusion

Space advocates and those interested in space development, space commerce, returning to the Moon, or sending humans to Mars need to do a better job in presenting their case to the public, politicians and the media. Bridging the communication gap can best be done by finding ways to personalize the benefits so that people will know why space development or sending humans to Mars is beneficial for them. This personal connection is very important and requires attentive communicating, including listening. Both of these skills are fundamental to being able to speak to a person's interest or lack thereof in space development, and to winning their support for space programs.

Space is expensive and dangerous, especially a manned Mars mission. It is therefore imperative that we have positive, constructive enabling policies, legislation and regulations designed to facilitate our space development and exploration goals. We need to be inclusive, not exclusive, to realize these goals.

I believe it is also important to recognize that space development is no longer the domain of the government and NASA. Expanding space commerce and becoming spacefaring is now a global effort, involving many nations that have divergent and competitive interests with the US. We need to find ways to work with other countries and their space programs to make space an addition to all our lives. Thinking globally about space development, settlement and project financing will certainly be helpful in humanity's quest to become truly spacefaring.

References

1. Baker, David. 2000. *Inventions From Outer Space. Everyday Uses for Technology,* p. 5. New York: Random House.
2. Ibid, (multiple pages throughout the book).
3. White, Frank. 1998. *The Overview Effect: Space Exploration and Human Evolution.* 2nd ed. p.265. Washington DC: American Institute of Aeronautics.
4. Ibid, pp. 45, 256-7, 259.

---//---

A Discussion of the Practical Advantages of Designating a Woman to be the First Person on Mars†

Bob McNally

The Purpose of This Paper

This paper is not intended to prove that having a woman step on Mars first is the best thing to do. It is intended to provoke a discussion and reflection on whether who is first on Mars should be seen as a mission important resource, affecting the future exploration and settlement of Mars. In addition, we'll ask should this resource be employed to the benefit of Mars exploration, and future human presence on Mars. Since there are some cogent reasons why choosing a woman for that role may be a very good choice, we will look at that in terms of asking: how do we optimize support for a continued human presence on Mars?

Generating the will, the spirit, the enthusiasm, the courage for going to Mars, in vast numbers of people, is perhaps the biggest obstacle we face. Support for going to Mars is by no means universal. Many people argue that resources devoted to going to Mars would be better spent improving conditions here on Earth. Generating support for continuing Mars exploration is a necessary and legitimate goal of the first mission, and well worth being an important priority in mission planning.

The Importance of Point of Contact

I spent two weeks at the Utah Mars Desert Research Station this winter. We had a lot of press visits in our rotation and, to our surprise, with each press group we were asked "what do you eat?" as one of their first questions. At first that boggled our minds . . . we are running around in simulated space suits, simulating research on Mars, searching for signs of fossil life and driving ATVs and keeping the Hab running and they ask what do we eat? Who cares what we eat! But! — they knew what they were asking. They were looking for a "point of contact" with the story for their readership. "What do you eat?" is a real point of contact question. "What are you doing with that Magnetic Susceptibility Meter?" is not a point of contact question.

Point of Contact is a key to generating public support. The outreach challenge is to get the largest number of people possible to see Mars as something that contacts and affects them, and even directly benefits them. They need to be shown that going to Mars can contribute to the quality of their life, be in their self interest; then support will be widespread. The importance of outreach cannot be overstated. We achieved the Moon and then abandoned it because outreach ultimately failed. If we are going to Mars to stay, we have to have deep political support that will stay the course in the face of adversity, not just in America, but in the world. If going to Mars will take public support, staying for centuries will take public enthusiasm and commitment (and it better be global). We cannot gain that support without outreach, generating continuing points of contact, and demonstrating to people that it makes a difference in their lives that we are going to Mars.

Designing for Point of Contact

Designing boosters, spacecraft, life support, pressure suits, rovers, orbiters and habitats; as well as crew factors, the search for life, geology, exploration, and even future politics on Mars – all of these are fascinating, stimulating, creative and exciting. The "fun stuff," and quite necessary. However, for many Mars enthusiasts, political outreach and generating public support are not seen as "fun stuff." But generating public support is the *sin qua non* of Mars exploration. It is the most important job we have to accomplish if we intend going to Mars and staying there. Fortunately, it is really possible for all the "fun stuff" to contribute to outreach, and this is a particular opportunity I want us to look at in this talk.

There are many aspects of mission planning that can include generating public support as part of the design equation. As we dream and plan rovers and Mars bases, we can ask ourselves over and over:

- What is the point of contact with the public here?
- How can the solution to this design problem also contribute to public support?
- If there are three colors that are technically equivalent for the pressure suits, which one feels most exciting to viewers?

Even at the most mundane level we can ask these questions:

- We are drilling holes in this support beam to make it lighter, can we arrange them in an intriguing pattern that charms and fascinates people?
- Can we have a contest to name the vessels?
- Can we paint the rovers in bright colors and patterns?

I do not mean that *On To Mars* is the next reality TV show. I mean that point of contact, and creating support and enthusiasm are necessary design criteria along with the weight, durability, and performance of a component or system. They are needed to sustain momentum of support, and are fundamental aspects no less than the mechanical performance of the systems.

This may be a new mode of thinking for planners and engineers, but it is a necessary new skill we have got to master. Any mission planning aspect that can have strong impact on support and enthusiasm for going to Mars should be examined carefully, to optimize efficiency and performance in the generation of support. Remember that Congress (as an expression of the public sentiment) got tired of sending people to the Moon. Point of contact excitement was lost. The Vikings had settlements in North America long before Columbus, but they stopped sending longboats to Vinland. We would all be speaking Danish or Norwegian today if the Vikings had sustained their colonies in the New World. The mere fact of going to Mars once is no guarantee that we will be there a century later. A primary goal of the first mission must be generating deep support for future missions. For many people it is not self-evident that we can and should settle Mars. Our job is to keep morale high, until the human presence on Mars becomes self sustaining.

Is "First on Mars" a Critical Element?

Let us consider whether who is the first to step on Mars is a critical mission aspect which can significantly affect the generation of support for the first and subsequent missions, and for the eventual exploration and settlement of Mars.

The first person on Mars will become one of the most widely known personages in history. Great symbolic importance will accrue to this position, despite the fact that this person did not invent Mars, or create the mission that got us there. The first person on Mars will be a representative of all Earth. They will be the primary point of contact for billions of people on Earth. This can translate into increased support if we plan wisely for that outcome.

Historically, we have given the role of "first ashore" to the mission commander. The great explorers organized their missions, generated funding, selected the teams, and planned the equipment. They knew that saying "I claim this land in the name of the Queen" as they strode ashore through the foam was extremely important public relations that affected their future opportunity to explore again. Now, explorations are massive industrial team efforts. The position of first ashore should still serve the mission purposes of generating future support, but the mission commander is not generically the person to perform that function. If the position of first person on Mars were simply honorary, it is an honor that can be extended to the commander of the expedition as a traditional prerogative. But if the role of first person serves a mission important purpose, then it should be assigned based on who can best meet the needs of that mission requirement.

Therefore, two critical questions to ask are:

- Can the determination of who is first on Mars really have an affect on generating and sustaining support for going to Mars?
- How can that determination be made in a way that maximizes support?

Where Does Support Come From?

Let's look at where support comes from:

- If going to Mars is seen as a primarily scientific mission, it will gain support from people who see that as a worthwhile priority.
- If we look at going to Mars as an engineering challenge, a technological advancement, it will achieve support from people who are stimulated by that prospect.
- If going to Mars is seen as an adventure and an exploration, and first steps out into the universe, yet another group of people will value that aspect most highly, and extend their support.

These are all points of contact for different interest groups. To the extent that going to Mars is seen as all of the above ideas, support from multiple groups of people can be expected, more than from any one purpose alone.

Going further, if going to Mars can be seen in a way that inspires, unifies and connects people here on Earth, that speaks to the improvement of life here on Earth, people who could care less about science, engineering, or exploration will have a point of contact.

We cannot ignore these people if we want to be on Mars a century from now. Anything about going to Mars that speaks to life here on Earth, that raises the quality of life (or culture) on Earth, that makes people excited or enthusiastic about life in general, not just Mars, is an asset in terms of garnering support. Maximum support for the exploration and settlement of Mars is gained by maximizing the points of contact values for many diverse groups of people.

Think about it this way: if we want to get to Mars, the population of Earth is the primary crew, and we better take care of their morale, because, without their support – no mission. The team that actually goes is the second crew. The morale of the first crew, the crew that stays behind, is just as critical to the mission as the morale of the crew that flies.

How Does First on Mars Relate to Support?

Things that catch and hold the human imagination have great and subtle power; witness our fascination with the *Titanic*, with vampires, with sharks, with dinosaurs . . . no one will ever go broke writing another book or making another movie about any of those things. People will keep ponying up the money for each encounter with those subjects. Mars is powerful in a like way, and has a similar hold on the human imagination. Other papers at these conventions have addressed the deep fascination people have with Mars. This fascination is a reality; how do we translate it to further the success of the mission? One way is by paying attention to who steps on the planet first; that is a key element in the story of Mars.

The positions on the crew will be chosen for mission related reasons, pilots, doctors, geologists, biologists, engineers . . . why not have the role of first person on Mars also serve a mission related purpose?

We may anticipate there will be four, six or eight in the first landing crew. There will be competent men and competent women aplenty to make the final crew selections from. The hardest job will be distinguishing among the top candidates, not finding enough who are qualified. Everyone on the mission will be multi-capable, cross-trained in numerous disciplines.

Does it serve the mission best to reward the mission commander with the plum of "First on Mars," or should that be considered a job, and the person selected for it who best supports the mission's success. Do we need the strongest, fastest person to step down first? The smartest? The best geologist? Or should it be the person whose taking that role most creates support for the mission as a whole?

The role of first on Mars has no real function beyond its symbolic aspect. So optimize for that aspect. It is a courtesy position, but one with great potential consequences politically, so extend that courtesy where it will genuinely help the whole long mission of settling Mars. Consider First on Mars as one of the duty descriptions, and select for who best to perform that duty.

Again, we know there will be great historic attention given to the first person on Mars. The first person on Mars will be known across the world, will be a spearhead for the "Humans to Mars" movement, and will be called upon to speak and lecture and inspire. This will be critical to the long-term success of Mars. We can see that who is first on Mars can have an effect on generating support; the next question is how can the selection be made to optimize that support?

How to Optimize for Broad Support

We might assume, based on familiar experience, that the likely first person on Mars will be male, white and American. But if there were important mission advantages accruing to the choice of a non-white, or non-male, or non-American being first on Mars, then it would be hard to argue that a white American Male should be first. Only one person can be first, so what choice would optimize the chances of long-term success in terms of support and the opportunity to return to Mars?

Unlike our landing on the Moon, landing on Mars is not motivated by Cold War competition or conflict. It will require cooperation to succeed – one country may achieve landing and even some exploration, but settling and terraforming will require global resources, and global commitment. How do we best demonstrate that this is a human project (not just an American project) and gain broad human support?

I say choose one of the women on the crew to be first on Mars, and you accomplish a masterstroke of demonstrating what kind of cooperative, inclusive future we intend to make on Mars, and on Earth. We only have to give the job to the most articulate of the several competent women in the crew.

Designating a woman to be first would send a powerful, all inclusive message about who is going to Mars – humanity is going to Mars; not men; not Americans alone, but humanity.

Choosing a woman represents including all the diverse parts of humanity better than choosing a man does. Think about the opportunity for inclusive, ecumenical, cosmopolitan generation of support if we say "a qualified woman is just as valuable as a qualified man, and we'll put our money where our mouth is."

Symbolism has deep point of contact overtones. Symbolically Mars has meant war and masculinity for many cultures. Historically, women have represented life and peace. There is an intense and persuasive symbolism in having a woman of life and peace be first on to the male planet of war. This is a magnificent point of contact potential, if we want broad human support for our presence on Mars.

Women are the majority of our species. A Gallup poll in 1989 asked men and women "if our investment in space research is worthwhile, or would it be better spent on domestic programs?" In this poll, men supported space 51% to 43%, and women opposed 59% to 35%. It is clear that we have to work hard to win the support of women. We cannot sustain a long commitment without them. Designating a woman to be first is a powerful way of reaching for the support of women, and demonstrating that we understand that we need women's support. I believe women will warm to this project in a new way. Not because a woman per se will be first on Mars, but because respect and recognition will be clearly demonstrated. Human beings warm when treated with respect and recognition.

Broader International Support

In addition, the generation of support will not be confined to women. People of many races and nationalities, who may have reasoned, self-interest-based objections to going to Mars, can be warmed by the larger human commitment to equality that is expressed by having a woman be first.

Those who criticize going to Mars as a First World stunt, as a drain of resources that could be helping the developing world, can be given a point of contact by the broader human commitment that designating a woman first on Mars makes. It says, "it's all of us, not some of us." When the first crew places a plaque about coming to Mars for all "humankind," there will be reason to believe that it is true. Imagine the Earth-changing, all-in-the-same-boat, committed-to, cooperative transformational effect of going to Mars being seen without the glare of nationalism, sexism and business-as-usual shining in our eyes. That is point of contact working to maximize broad public support.

It is unlikely that men will cool to going to Mars because of this (well, maybe a few). There will still be a first "man" on Mars, for those who care. The plan has to be presented as a straight forward, intentional optimization of support resources, not a stunt, not patronizing, not tokenism, but a calculated, real-politick decision to further mission objectives. Even for the most macho among us, there is still the huge courageous task of exploration and colonization.

Generations of children will start out with a fundamental notion of equality of the sexes that we have had to struggle to learn. In one stroke, the playing field is shown to be more leveled. Better than the first woman president, because that is just one country. The first person on Mars will also be a first citizen of earth; she will belong to all of Earth, and represent all of Earth.

Conclusion

Columbus could discover and explore the New World, but he could not settle it by himself, and he could not explore it all. It took the whole world to explore the whole world. Mars is a whole world, and will require all of Earth's support to explore and settle. Columbus's voyage that brought the Americas to the European's attention was deemed an economic endeavor, but spawned changes in nearly every sector of European society, and eventually global society. It is this aspect of going to Mars that will most change life on Earth. We hardly know what we are biting off, but it will change us by what it requires. Designating a woman to be first makes total sense when we look at what it is going to take to be on Mars for the long haul.

Summary

In summary, I am saying that:

1. Going to Mars will require strong public support; staying on Mars will require broad public commitment.
2. Points of contact are how people connect to and bond with complex projects.
3. Points of contact and broad public support are more important for Mars exploration than any other single planning concern.

4. People are moved by experiencing things making a difference to their and their children's lives.

5. We can and must design for maximum support, without sacrificing mission success, safety, or scientific objectives. Many aspects of Mars mission planning can include points of contact in their purpose criteria.

6. The role of first person is a huge symbolic opportunity, and should be seen as a support-enhancing resource, an asset to be spent wisely, not simply as a traditional honor for the mission commander.

7. Designating a woman to be first maximizes our effort to achieve long-term success for the presence of humans on Mars.

8. Designating a woman to be first sends a powerful, all inclusive message about who is going to Mars — all humanity is going to Mars.

//

Miscellaneous Mars Papers

//

The Real Reason for Exploring Mars†

Marvin Hilton
skymar@tcac.net

Here we are, sitting in an outer arm of our Milky Way galaxy. We don't know really where we are. We do not know our origin. We do not know whether or not we are alone in the universe. The real and very important reason for exploring Mars is to find some answers to these questions. Or at least to take some steps in finding some of these answers.

I have the impression that the majority of people do not like to talk about these questions. I have had people tell me that, "we have no business asking these kind of questions. It is not for us to know. Only God knows those answers. These are unanswerable questions."

Some people may feel that we are somehow violating the heavens when we explore beyond the Earth. The Christian Bible speaks to this in Genesis. "Go to, let us build us a city, and a tower, whose top may reach unto heaven; . . . And the Lord came down to see the city and the tower, . . . now nothing will be restrained from them, which they have imagined to do. Go to, let us go down, and there confound their language, that they may not understand one another's speech, and they left off to build the city."[1]

People may also be uncomfortable with these questions because we cannot comprehend the dimensions of the universe. We are about 30,000 light years from the center of our Milky Way galaxy and the nearest other galaxy is about 2.4 million light years away. The latest estimate is 100,000 galaxies in the known universe and each galaxy contains on the order of several hundred billion stars.[2] Kenneth Patton, a noted theologian, said, "We are not able emotionally to cope with the dimensions of the universe."[3]

The majority of people can now conveniently forget the vast universe that lies beyond the Earth. According to the International Dark Sky Association, two thirds of the people in North America cannot see the glow of the Milky Way due to light pollution. Recently I was in a Tucson, Arizona hotel lobby when a man ran in jabbering very rapidly and loudly. He had returned from a drive outside the city into the southwestern desert. Being from New York City, he had never before seen the glow of the Milky Way and the brilliant stars shining undimmed by city lights. Although he was at least 30 years old, the wondrous new sight was a revelation for him.

Most modern cities have automatic street lights that are switched on as their part of the Earth rotates into the darkness. And of course we turn on our car lights, yard lights and interior lights until we seldom get a glimpse of the beautiful stars and the glow of the Milky Way.

In my hometown of Fayetteville, Arkansas, hundreds of man-hours are used each autumn to string thousands of Christmas lights. The downtown area is made to glow with these lights in order to celebrate Christmas and attract tourist dollars. I would love to have a celebration by turning out every light in town for at least one night. Then the whole town could celebrate the night of stars and wonder at the glow of the Milky Way. That also would be a good way to celebrate Christmas, considering that the first Christmas had a bright star in the heavens.

These questions about our origin and our place in the universe seem to create a dichotomy within our minds. They seem to reside in the back of our minds most of the time, but we are often uncomfortable speaking about them. I believe that facing the hard truth about our limited understanding of the universe will launch the human family into a new renaissance and a maturing of the human family. Exploring Mars to answer these questions will be another quantum step in becoming more comfortable with the real universe. Most importantly, each of us and the entire human family will have a greater regard for ourselves. Our status will increase in our own human eyes. Our self-esteem will increase both individually and collectively.

How will exploring Mars increase our self-esteem? Steven Weinberg, a Nobel prize winning physicist speaks to this when he says, "The effort to understand the universe is one of the very few things that lifts human life a little above the level of farce, and gives it some of the grace of tragedy."[4]

Nathaniel Brandon, in *The Psychology of Self-Esteem*, speaks to this from a psychological viewpoint. He says, ". . . an unbreached determination to use one's mind to the fullest extent of one's ability, and a refusal ever to evade one's knowledge or act against it, is the only valid criterion of virtue and the only possible basis of authentic self-esteem."[5]

Continually ignoring these questions about our origin and place in the universe is a form of evading our knowledge. That is also why, in a sense, it is dishonest for us not to explore Mars. To not explore Mars is to avoid looking at the evidence on the planet next door and to ignore our truth. Imagine how the discovery of one small fossil on Mars would affect our perception of our place in the universe.

Writings abound proclaiming that humans have an almost instinctual urge to explore to see what is over the next horizon. Victor Frankel, who wrote *Man's Search for Meaning*, said, "I know that we cannot stop going into space. It belongs to the very foundations of human existence."

Giordano Bruno, an Italian sixteenth century renaissance humanist, taught Copernicus's new astronomy that the Earth revolved around the sun, instead of the sun revolving around the Earth. He also said that the universe was filled with an infinite number of stars and that the stars were like our Sun, with planets revolving around them. He was imprisoned by the Roman authorities and questioned on charges of blasphemy, immoral conduct and heresy. On February 17, 1600, he was led from his prison cell, stripped of his clothes, and led naked through the streets of Rome. He was staked to a pyre in the Square of Flowers and an Inquisitor, with torch in hand, demanded that Bruno recant. Bruno refused to recant, the pyre was lit and some say that one of the most profound minds in history was burned alive.[6]

Galileo, with his improved telescope, saw that there were craters and mountains on the Moon, and that Jupiter had moons revolving around it. His observations confirmed that Copernicus' theory of the solar system was correct. The Church also threatened Galileo with torture and death for these teachings if he did not recant. Galileo, perhaps knowing what happened to his contemporary Bruno, recanted in 1632.

The Catholic Church of today should be given due credit for its progress. The Church apologized to Galileo in 1993, 360 years after he recanted his teachings that the Earth revolves around the sun.

All of this reminds me of the writings of the astronomer Chet Raymo in 1985:

> "When we are in the temples, then who will hear the voice crying in the wilderness? Who will hear the reed shaken by the wind? Who will watch the galaxy rise above the eastern hedge and see a river infinitely deep and crystal clear, a river flowing from the spring that is creation to the ocean that is time? . . . Antares is a lamp, burning and shining; rejoice in its light."[7]

The glow of the Milky Way galaxy always evokes within me a feeling of wonder and mystery which beckons to explore beyond the Earth. And the red glow of Mars is like a beacon to step into the world next door.

In the early 1900s there were three men who were among the first to do something other than dream and speculate about what lies beyond the Earth. Konstantin Tsiolkovsky, a Russian; Herman Oberth, a Romanian; and Robert Goddard, an American, were among the first to conceive of and design liquid fueled rockets.[8] They did this work with the idea of actually walking on the Moon and planets. They worked mostly alone and were sometimes ridiculed, but their work was utilized by others, such as Wernher Von Braun.

In 1920, at the age of only eight, Wernher Von Braun's mother gave him a telescope. He immediately ran outside into the streets of Berlin and looked up at the Moon with his new telescope. He exclaimed, "Oh, how near it is. I am going to go up there some day!" Not long after that, as a very young man, he was designing and building the German V-2 rockets for Hitler's army. He later made a joke, "I aimed for the Moon, but I missed and hit London instead."

In 1945 Von Braun was brought to the United States as a prisoner of war. In the Southwestern United States desert he was seen to slip under the fence at night and stand alone and stare up at the stars. Not very long after that, he was made the director of the US Space and Rocket Center in Huntsville, Alabama. There he conceived and designed the Saturn Five rocket that sent men to the Moon in 1969.

Although Von Braun worked on the V-2 rockets for Hitler's army, he seemed to want only to explore beyond the Earth and reach out to the stars. After the Apollo Moon project, he attempted to persuade Congress to fund an expedition to Mars. But the war with Hitler and the race to the Moon had been won. Not enough people were interested in space travel only for the sake of exploration.

Four hundred years before men walked on the Moon, men were being ridiculed and severely punished for merely speaking about what lies beyond the Earth. But we are now the first generation to physically thrust ourselves beyond the Earth and walk on another world. We are having serious scientific debate about the evidence for life on another planet. Unmanned spacecraft are exploring throughout our solar system. Over 100 planets in other solar systems have been discovered. Scientists continue to embark on projects to listen for signals from other civilizations in other solar systems and other galaxies. The basic questions of our origin, our destiny and whether or not we are alone in the universe are beginning to be answered with real scientific evidence.

It is obvious that sending an expedition to Mars would increase human knowledge in nearly every discipline. Ongoing Mars exploration projects are already considering factors in human relations, nutrition, theology, governmental systems, communications and on to an endless list. Much the same could be said of the Apollo Moon project and, to a lesser extent, of the German V-2 rocket project. But both of these projects were competitions, sometimes deadly competitions, between entire nations of humans.

To explore Mars for the real reason, to discover our origin and whether or not we are alone in the universe, would be a great unifying endeavor. Fred Hoyle realized this in 1948 when he said, "Once a photograph of the Earth taken from the outside is available, a new idea as powerful as any in history will be let loose."[9] A mere two decades later men walked on the Moon, but we were more enthralled with the photo of ourselves, our planet taken from the outside. Apollo 14 astronaut Edgar Mitchell said that seeing the Earth from space produced, "[what] I prefer to call instant global consciousness."[10]

To explore Mars in order to discover our place in the universe addresses those seldom spoken, but often thought of questions of our origin and whether or not we are alone in the universe. I believe it is a universal human thought, but only spoken by a few brave minds. That is why the largest TV audience in history, at that time, watched the first human step on the Moon. That is why every human on Earth who has access to a TV will watch the first human step into the red dust of Mars. Images of a marble-sized Earth beamed from the vicinity of Mars and the sight of a crew living far into space will intrigue billions on the Earth for months.

When we start to explore Mars for the real reason, we will begin to lift human life to a new level of authenticity. It will represent a new renaissance, a growing up of the human family. A human family that will leave far behind the one that burnt Bruno at the stake and knocked down the twin towers of New York. Then we will become more comfortable with the real universe, not the one that we imagine. Our status will increase in our own human eyes. That can lead only to success for the entire human family.

The real and very important reason for exploring Mars is to discover who we are and our place in the universe.

References
1. Genesis, Chapter 11, versus 1-9
2. Stephen Hawking, A Brief History of Time, (1988) p. 37.
3. Kenneth Patton, A Religion for One World (1964).
4. Steven Weinberg, The First Three Minutes, (1993).
5. Nathaniel Branden, The Psychology of Self-Esteem (July 1979) p. 114.
6. Dorothea Waley Singer, Bruno: His Life and Thought (1950).
7. Chet Raymo, The Soul of the Night (1985).
8. Carsbie C. Adams, Space Flight (1958).
9. Peter Russell, The Global Brain (1983).
10. Ibid.

//

The Prospects for Space Commerce in the Aftermath of 9/11†

Dr. David M. Livingston
dlivings@davidlivingston, www.davidlivingston.com

Introduction
The first humans landed on the Moon with Apollo 11 on July 20, 1969. Our last lunar landing was with Apollo 17 on December 11, 1972. In the thirty years that have past since men walked on the Moon, we have seen the commercial space industry, consisting mainly of telecommunication satellites and expendable launch vehicles (ELVs), grow into a multibillion dollar industry. Yet, today's commercial space industry is a limited industry. If the commercial space industry is to develop beyond today's level of space commerce, it needs to expand by developing New Space Industries (NSIs) and emerging space business ventures. Is this expansion plausible in the aftermath of 9/11, a near global recession, and our war on terrorism; and if so, over what time period?

This paper discusses some of the more plausible NSIs and what is needed for the overall success of new and emerging commercial space ventures. In addition, an attempt is made to gauge the full impact of the aftermath of 9/11, the recession, and the war on terrorism upon a newly developing and expanding commercial space industry. The effect of current launch costs upon the developing space industry is also considered, including whether the payload consists of cargo or humans to orbit, and the payload destination. Business planning, including the financing of space ventures, is also discussed.

Policy considerations play an important role in developing our space economy. As such, some of the conflicts of interest within both the established aerospace industry and NASA that have the potential to adversely impact commercial space development are discussed. Sometimes the commercial space industry can be its own worst enemy, through improper planning, damaging rhetoric, and conflicts of interest. By noting these problems we can learn how best to avoid them in the future and move forward in expanding the commercial space industry.

The paper concludes with a discussion of the awesome responsibility we have in seeding the off-Earth environment with our people, culture, business practices, and our economic principles and values. What we do today in space has the potential to impact those of us here on Earth, as well as those who will choose to work, live and play in space from this time forward, through future generations. Therefore, my Code of Ethics for Off-Earth Commerce[1] is cited as one possible tool for ethical behavior and for assuring corporate, business, and personal responsibility for our future in a spacefaring world.

Resolving the problems associated with transporting people to and from space in a cost-effective space transportation vehicle – most likely the reusable launch vehicle (RLV) – is crucial to the successful development of the commercial space industry. Suggestions are set forth that have the potential for making a positive impact on the RLV, space tourism, and commercial space industries. Public and private sector cooperation is essential. Financing NSIs and emerging space business ventures is fundamental to developing an expanded commercial space industry.

Today's Commercial Space Industry

Lest anyone doubt the profitability of businesses commercially operating in space today, we need look no further than the commercial satellite industry. This industry has a thirty-eight year track record of commercial space operations dating back to April 1965, when the Early Bird satellite was successfully launched. Since then, commercial space ventures have grown and profited to an impressive degree.

To gauge the growth and profitability of the commercial space business, we need reliable data. Fortunately, such data exist. KPMG Peat Marwick in collaboration with SpaceVest, the Center for Wireless Telecommunications, and Space Publications, has published an informational journal, *1997 Outlook: State of the Space Industry*. This comprehensive report analyzed the commercial space industry of 1996 with industry projections for the year 2000. The report stated that worldwide revenues from space commerce totaled $77 billion in 1996 and projected annual revenues of $121 billion by the year 2000.[2] In 1996, for the first time, revenues received by private launch companies were greater from commercial organizations than from the federal government, with 53 percent of the revenues coming from the commercial side and 47 percent coming from the government side.[3] Also in 1996, the space industry as a whole employed an estimated 835,900 employees.[4]

In addition to the KMPG *Outlook* report, two other industry reports are useful in demonstrating the financial success of this industry, although they don't evaluate and analyze the industry in the exactly the same way. Merrill Lynch, in its annual satellite industry review, *Global Satellite Marketplace 99*, projected the industry to increase from an estimated $36 billion in 1998 to $171 billion by the year 2008. This represented a 17.5 percent annual growth rate.[5]

C.E. Unterberg, Towbin, a noted financial company with offices in New York and San Francisco, produces *The Satellite Book* with quarterly updates. According to the second quarter 1999 issue, the commercial satellite industry was estimated to grow from a $54.8 billion industry in 1998 to an estimated $116.3 billion in the year 2003.[6]

Clearly there is a profitable commercial space industry operating, both worldwide and in the United States today. Putting today's commercial space industry into perspective with leading businesses at the time these reports were made, however, gives a sense of the relative size of this newly developing industry and the potential awaiting its continued development. For example, it is interesting to compare the entire launch industry segment of the total commercial space industry with General Motors, the largest revenue-earning corporation in the world in 1999. General Motors reported total international sales of $176.6 billion for 1999, with a corresponding total net income for the same year of $6 billion.[7] In 1997, Arianespace studied the size of the launch market and determined that its growth would be flat over the next ten years, generating gross revenues totaling $34 billion.[8] This is slightly more than $3 billion per year and includes both the commercial launch demand and the demand for government research launches. One corporation alone, General Motors, generated almost 59 times the revenues of the total launch industry!

New Space Industries (NSIs)

NSIs have the potential to increase revenue streams and profits dramatically for the space industry. The most commonly mentioned NSIs include reusable launch vehicles (RLVs), space transportation systems, fast package express, space tourism, space business and theme parks, space manufacturing, microgravity research and development, remote sensing, communication, space solar power, entertainment, space rescue services, space mining / resource development, waste disposal, space servicing and transfer, space utilities, and orbital debris removal. In addition to this known list, I like to include one's imagination as an NSI, because once a space-based economic infrastructure exists and NSIs are being developed, new ventures will come into being that are not even in today's consciousness.

From the list of possible NSIs, those with stronger market certainty and needing less in the way of new technology or costly engineering are more likely to be realized. Some of these near-term NSIs are the RLVs, space transportation systems, space tourism, fast package express, orbital debris removal, remote sensing, communications, entertainment, and space rescue. Most, if not all, of these plausible space businesses are ideal for the private sector to develop and exploit. They can also be built upon common and shared space infrastructure.

It is worth noting that NASA and the White House are considering the development of nuclear power for space travel. Identified as Project Prometheus, this government program would enhance access to space, especially to locations outside low-Earth orbit (LEO). Nuclear powered space transportation vehicles would be able to operate more efficiently, with larger payloads than today's chemical rockets and the travel time to destinations would be greatly reduced. For example, a mission to Mars, which might take six months or more using one of today's chemical rockets, might only take six weeks to make the same trip using nuclear power.[9] Should Project Prometheus come into being over the coming decade, it has the potential to revolutionize and accelerate the timetable for NSI development, just as Apollo revolutionized the space program thirty years ago.

Cost-Effective Space Access

More than ever, for NSIs to come into existence in the aftermath of 9/11, the recession, and the war on terrorism, there needs to be cost-effective space access, constructive public and private sector policies, comprehensive and strong business planning, risk mitigation, market validation and exploitation, capital acquisition, and proper timing. There must be a focus on fundamentals and essential economic components of the space business venture. Thus, we need to start by taking a look at our ability to both frequently and economically access space.

At the present time, there is no cost-effective space access for either cargo or humans. Today, we access space using expendable launch vehicles (ELVs) or evolved expendable launch vehicles (EELVs) for satellites of all types. We require the Space Shuttle in the United States and the Soyuz in Russia for men and women going to space. Although ELV and EELV costs can be high, ranging upwards of more than $10,000 per pound to orbit depending on the launch vehicle used (the Space Shuttle) and the orbit objective, most satellite and telecommunication companies have been able to operate profitably, grow, and expand their markets. However, transporting people to and from space is a completely different matter.

The US Space Shuttle costs upwards of $500 million per launch, sometimes approaching $1 billion per launch. While the Soyuz is considerably less costly at a price of about $100 million per launch, the Soyuz only holds three people and the number of flights it can make per year is limited as is the number of annual flights using the Space Shuttle. For human space flight there simply is no affordable space transportation system; thus, all commercial ventures requiring the presence of people in space are thwarted until the economics of space access dramatically change, along with the ability to go to space frequently with large numbers of people.

Many companies are working toward designing and developing RLVs, which will allow for cost-effective human space access. These companies are in various stages of their business development program, attempting to raise capital, working on engineering and technology requirements, and competing to be the first with a passenger certified RLV. NASA and the large players in the aerospace industry are also working on second-generation launch vehicle designs, but at a much slower pace and using a more research- and new-technology-oriented approach. Regardless of how a new launch vehicle comes into being, the fact is that today, there is no cost-effective launch vehicle of any type. Having such a vehicle is absolutely essential to developing and expanding space commerce. In the aftermath of 9/11, many companies, especially the start-up space transportation companies, have been adversely impacted in their ability to attract and secure capital. Still, for businesses with solid planning, market confirmation, and a cost-effective marketing program which allows for market development with a proven management team, the opportunities for success are remain strong. Regardless, it is important to understand that an RLV or new space transportation system remains the key to the future economic success of space development, and that such a vehicle makes possible space tourism and the other plausible near-term NSIs.

The newly emerging space businesses must also show they are capable of mitigating their risks, capitalizing on the right timing, and working within the present policy and regulatory environments. Having an experienced and proven management team is a crucial component of success and is highly valued by investors and financiers. While modifying or changing space policies and regulations to enhance a business opportunity can be a good thing, it will probably be costly and time consuming.

Public and Private Sector Issues of Concern

There are several policy issues that concern commercial space development. These issues include competition with the government, government financial and tax incentives, aerospace industry issues, determining the proper role for NASA, the Space Launch Initiative (SLI),[10] and comprehensive business planning. Examples of this concern stem from the following: the National Transportation Space Policy of 1994, the National Space Policy of 1996, the NASA Strategic Plan, the NASA Administrator's Strategic Outlook, various US laws and regulations, the lack of both private- and public-sector financing, the X-33 failure,[11] the Iridium bankruptcy[12] and other constellation satellite system problems, misperceptions and damaging rhetoric from leading industry and NASA executives, conflicts of interest within the large aerospace industry as well as the government, and the SLI.

Damaging Rhetoric

It is important to highlight the damaging rhetoric, as this may be the most insidious of all space barriers. This is because it directly and adversely impacts people's minds, and the result is very difficult to overcome. Changing a law, policy or regulation has a process to it. Changing a point of view, a deeply entrenched belief system, or altering one's perception based on new information can be quite challenging. The examples of damaging rhetoric that can skew a person's mind, thinking, and ultimately their behavior come from both the leaders of the aerospace industry as well as NASA.

Addressing Congress on May 21, 1999, Peter B. Teets, then-President and CEO of Lockheed-Martin, when referring to their VentureStar RLV, stated that the project was unsuccessful in "attracting Wall Street investors and would need some form of added government funding or loan backing. Wall Street has spoken. They have picked the status quo – they will finance systems with existing technology. They will not finance VentureStar."[13]

Another example comes from a comment made by Daniel Goldin, the former NASA Administrator, on July 12, 1999. Goldin was reported saying that US companies and investors won't finance costly new launch vehicle programs without further reducing the technical and financial risks. *Space News* quoted Goldin as saying that "NASA will probably have to retire the technical risk. There isn't one corporate executive in their right mind that would take on a multibillion-dollar investment that won't have a payoff until 10 years from now. In the space community, we have space in our heart. When you're in corporate America, you've got to meet the numbers."[14]

Yet another example comes from a Boeing study of commercial space tourism showing that the costs of developing a space tourism capability were currently impractical, according to Vice Chairman Harry C. Stonecipher. Boeing found that development of a two-stage commercial vehicle to provide 50 passengers with short orbital flights would cost at least $16 billion. Tickets would have to cost $150,000 each, and the vehicle would have to fly at least 800 times per year for the project just to break even. Stonecipher and other Boeing managers said they believe the $16 billion figure is itself seriously understated because it does not deal with costs associated with regulatory issues or other expenses.[15]

What all of these comments failed to mention was important. With VentureStar, for example, the engineering was widely considered by experts to be flawed from the beginning, with no chance of VentureStar succeeding. Also, its selection in the competition over other companies with already successfully tested technology and a prototype vehicle was an issue. Yet for Wall Street financiers, not knowing the history or the specifics of the matter, hearing directly from the Lockheed CEO was enough for them to agree with his analysis. The same for the Goldin comment. Wall Street believed (and still believes) that if NASA or Lockheed can't build it, it simply can't be built. Goldin's comments did not tell the entire story. From the perspective of those who were knowledgeable about the issue he was discussing, it was easy to see how his comments could be construed as misleading.

As for the Boeing space tourism comments, this too needs to have the light of day shined upon it. While Boeing has this study, it would obviously be from the Boeing perspective. It is important to understand that Boeing is a very large and profitable maker of ELVs and the newer Delta IV, an EELV. Along with Lockheed, Boeing benefits from a multibillion-dollar contract to operate the Space Shuttle and to improve and upgrade it. At the present time, it is not in the financial interest of Boeing or its stockholders to develop a new launch vehicle that would eat into the company's existing launch vehicle and shuttle operating stream of revenues. Hearing or reading these comments without having a full understanding of the complete picture, however, one might easily think that commercial space tourism is simply not plausible in a reasonable time period. It is easy to see how an understanding like this could result from the very authoritative Boeing report and statement, but the Boeing perspective needs to be considered. Thus, the rhetoric can be damaging and people's minds can easily be adversely influenced by one-sided statements that do not address all sides of an issue. In a perfect world, such statements would be accompanied by an explanation of the legitimate and very real pressures that are exerted on the people or the company making the statement. Were this the case, any statements made would reflect the complete picture regarding a specific issue.

Recommendations for Expanding Space Commerce

It is relatively easy to look at the way the commercial space industry is structured today, including all of the components that have an impact on it, and point out weaknesses, problems, barriers and troublesome policies. The challenge, however, is coming up with practical, viable and cost-effective remedies to these difficulties, and in getting support from those involved in space commerce that can make a difference. With this in mind, I have suggested a series of recommendations that have the potential to move the commercial space industry forward.

A long-term recommendation is for young people to begin to consider, plan and choose the commercial space industry as a career. Real opportunities await those who can take advantage of the developing commercial space industry. While this particular recommendation will take several years to develop, since education and skill building are not achieved overnight, such actions undertaken now by students, teachers, space advocates, policy makers and space-oriented businessmen and women, will help pave the way for commercial space development during the coming years.

A second recommendation is to realize that political activity on the part of space advocates and commercial space promoters is important. Our nation thrives on political activity and effectively communicating with our elected representatives and policy makers is an important part of our political, social and economic way of life. Even if the immediate response is tepid, we must push forward with our goals and our focus in the political arena. By doing so we can accomplish much over a shorter time frame than if we did not advocate in the political arena. The key is to make sure that our efforts are productive and that we understand how specific barriers in the form of policies, regulations and laws actually interfere with space business opportunities, and how these can be changed. Many of the existing space advocate organizations have very effective political action components in their organizations, so learning how to do this does not mean reinventing the wheel.

Third, applying constructive pressure to accomplish positive results is essential in space development. Challenging the contradictory and limiting policies, regulations, laws and even international treaties is important. It is through these challenges and the creation of an effective strategy for dealing with these challenges that solutions to the problems will be achieved.

Fourth, it is important that space businessmen and women understand that their business plans and concepts must compete with terrestrial businesses. This puts added burdens on the space venture to project and competitively perform, mainly because the associated risks for the space venture may be higher than those for similar terrestrial ventures. It also demonstrates to financiers and investors that the space venture management team understands the financial and risk markets. Thinking that the business is special because it involves space is not sufficient for attracting capital and other business support.

Fifth, when talking about space commerce or related fields, it is extremely important to personalize the benefits for the person or group being addressed. Just to say that the space program is responsible for pace makers or Velcro is not enough unless one has had or is having a valuable experience or relationship with someone using these items. Finding out what may be important to the person or group being addressed, and seeing how that relates to space can be transforming as it makes a direct and personal connection with those in the audience. When this connection is made and when a firm relationship is established, new advocates for the space program are created and while they may not be very active in promoting space, they no longer are part of the barrier side of the issue.

Misperceptions are common in the space field and are largely a result of damaging rhetoric from agencies and businesses that seem to have authority in space matters, as has already been pointed out. A sixth recommendation, then, is to support quality space educational programs and political action to counter the rhetoric and the misperceptions, something that all of us can readily do. The results from such positive action taken to counter the rhetoric and existing misperceptions can also be realized over a reasonable period of time.

Seventh, it is strongly urged that, within the space community itself, infighting and conflicts of interest be eliminated, or exposed and dealt with in an open and honest fashion. The infighting among those with competing agendas is damaging and counterproductive. While competition is desirable and can produce positive results and benefits for everyone, it must not be destructive. Conflicts of interest exist widely in the commercial space environment, especially in relationships with NASA, the aerospace industry, and even with some of the smaller, entrepreneurial start-up space companies. It is not possible to avoid conflicts of interest, but they should be stated up front so that people know the perspective and pressures on a particular company, agenda, or course of action. One such example of this latter point is the development of a reusable launch vehicle by private industry. Boeing and Lockheed are logical builders for this type of vehicle, yet they are only involved in long-term research and are working on the problem within the framework of NASA programs. This position is clearly understandable when one is aware of the contracts to maintain and eventually remodel the Space Shuttle to keep it flying for another 15 years, and their sale of both ELVs and EELVs. It would not be prudent for these companies to expedite an R&D program that would reduce or even replace a company profit center. It may even be counter to their fiduciary responsibility for their investors, given that a small number of RLVs can theoretically replace the Space Shuttle and all the ELVs and EELVs launched each year.

Eighth, the commercial space industry needs to ethically and professionally manage its business ventures, space resources, and overall space development. Across the board on a worldwide basis, there is an ethical consciousness about what and how we do things here on Earth as well as in space. As expanded space commerce comes closer to reality, ethical concerns will increase. The space industry, through professional management and a demonstrated concern for ethical issues and principles, can win friends and supporters during this crucial initial stage of commercial space development. This support will prove vital in helping achieve constructive policies and legislation. If people perceive that space businesses and the related space development are following the greed and unethical models of business that seem so prevalent as we move into this new century, then they will make sure their concerns and restrictive goals for space development are heard by those responsible for the very policies and regulations needing change. By demonstrating ethical and professional management of their plans for space exploitation, much of the policy, regulatory and legal barriers can be avoided or at least minimized. By working within an ethical framework, or by adopting my Code of Ethics for Off-Earth Development, today's active space entrepreneurs and businessmen and women can certainly facilitate the commercial space industry.

Concluding Thoughts

Space commerce and all that this field entails will expand not only within the United States, but also throughout the rest of the world. Eventually, we will evolve into a spacefaring world. The only real issue is the timing of when this will happen. To a large degree, we can influence the timing by our actions of today. It is also important to understand that it is an awesome responsibility to export our way of life, our culture, our economy, and even our presence off Earth. To start life on a different world is probably as big a step forward as is cloning humans. It behooves us all to make sure we give our future space brethren the best foundation possible. We the people, not the government, and not a special few, should decide what qualities and characteristics of ourselves, our culture and values we take with us off Earth and use for seeding future space settlements with our future generations. Those of us concerned about space development, advocating an economic infrastructure for space development and an expansion into a commercially driven space program, must come to know that it is we who control and shape space development and our future in space, not events such as 9/11, or the policies and programs flowing forth from this or related events.

References

1. www.davidlivingston.com/presentations.htm.
2. KPMG Peat Marwick, *1997 Outlook: State of the Space Industry* (KPMG Peat Marwick, SpaceVest, Space Publications, and Center for Wireless Telecommunications, 1997), 9.
3. Ibid.
4. Ibid.
5. Thomas W. Watts and William W. Pitkin, Jr., *Global Satellite Marketplace 99*, (New York: Merrill Lynch, Pierce, Fenner and Smith, Inc., 1999), 15.
6. J. Armand Mussey, William B. F. Kidd and Patrick Fuhrmann, *The Satellite Book*, vol. 1, no. 2, (New York: C.E. Unterberg, Towbin, 1999), 7.
7. General Motors 1999 Annual Report, "Financial Highlights," www.generalmotors.com/company/investors/ar1999/fh/index.htm, (21 June 2000), 2.
8. Patrick Collins and H. Taniguchi, "The Promise of Reusable Launch Vehicles for SPS," presentation for SPS 97, Space Transportation Systems Group, Tokyo, www.spacefuture.com/the_promise_of_reusable_launch_vehicles_for_sps.shtml&terms=SST, (19 September 1998), 3.
9. www.nuclearspace.com/a_project_prometheus.htm
10. As defined by NASA, the Space Launch Initiative is a program designed to develop a lower cost, safer, privately operated space transportation capability to replace the Space Shuttle early next decade.
11. The X-33 was an attempt to develop a prototype for a second-generation reusable launch vehicle, VentureStar, that was fraught with problems and eventually abandoned after the government development funds were exhausted.
12. Iridium was a company created by Motorola to put dozens of satellites into LEO to provide worldwide cell phone-like communications throughout the world. The satellites were expensive to launch and maintain, the phones were heavy, and the cost to users was excessive. Motorola declared Iridium bankrupt and the few satellites launched are now owned and operated by the US government.
13. Spacedaily.com, June 26, 1999.
14. *Space News*, July 12, 1999, p. 1.
15. *Aviation Week and Space Technology*, 'On Orbit' section, February 4, 2002, Edited by Bruce A. Smith.

The Mobile Agents Integrated Field Test –
Mars Desert Research Station April 2003†

William J. Clancey, Maarten Sierhuis, Rick Alena, Sekou Crawford, John Dowding,
Jeff Graham, Charis Kaskiris, Kim S. Tyree and Ron van Hoof
NASA / Ames Research Center and Florida Institute for Human and Machine Cognition
William.J.Clancey@nasa.gov

Abstract

The Mobile Agents model-based distributed architecture which integrates diverse components in a system for lunar and planetary surface operations, was extensively tested in a two-week field "technology retreat" at the Mars Society's Mars Desert Research Station (MDRS) during April 2003. More than twenty scientists and engineers from three NASA centers and two universities refined and tested the system through a series of incremental scenarios. Agent software, implemented in run-time Brahms, processed GPS, health data, and voice commands – monitoring, controlling and logging science data throughout simulated EVAs with two geologists. Predefined EVA plans, modified on the fly by voice command, enabled the Mobile Agents system to provide navigation and timing advice. Communications were maintained over five wireless nodes distributed over hills and into canyons for 5 km; data, including photographs and status was transmitted automatically to the desktop at mission control in Houston. This paper describes the system configurations, communication protocols, scenarios and test results.

Background

The Mobile Agents project anticipates exploration of Mars, in which a crew of six people are living in a habitat for many months. One long-term objective is to automate the role of CapCom in Apollo, in which a person on Earth (in Houston) monitored and managed the navigation, schedule, and data collection during lunar traverses (Clancey, in press b). Because of the communication time delay, this function cannot be performed from Earth during Mars exploration, and other crew members will often be too busy with maintenance, scientific analysis or reporting to attend to every second of a four to seven hour Extravehicular Activity (EVA).

This project is a collaboration across NASA centers and other organizations:

- Brahms Project Group (NASA-Ames: W.J. Clancey, Principal Investigator; M. Sierhuis, Project Manager; R. van Hoof, lead programmer; C. Kaskiris, modeler)
- RIALIST Voice Commanding Group (RIACS: John Dowding)
- MEX Vehicle & Wireless Communications Group (Ames: Rick Alena, John Ossenfort, Charles Lee)
- EVA Robotic Assistant Group (NASA-JSC: Jeff Graham, Kim S. Tyree (née Shillcutt), Rob Hirsh, Nathan Howard)
- Space Suit Biovest (Stanford: Sekou Crawford, in collaboration with Joseph Kosmo, JSC)
- NASA-Glenn Research Center/NREN satellite communications (Marc Seibert).

We have previously described how the Brahms simulation system (Clancey et al. 1998; Sierhuis 2001) has been adapted to provide both a tool for specifying multi-agent systems and an implementation architecture for run-time agent interactions on mobile platforms (Clancey, et al., 2003). We have described how Brahms is used to model and control system interactions, and outlined two preliminary field tests at Johnson Space Center and Meteor Crater (September 2002). We presented a summary of advantages and limits of the Brahms architecture for multi-agent applications. We have emphasized that building a practical system in a difficult terrain prioritizes issues of network robustness and diminishes, at least initially, theoretical questions about agent competitiveness and cooperation.

Mobile Agents Configuration

In an AI system, computational models make "intelligent" operation possible. The models in the Mobile Agent architecture include:

- Agents representing people in the simulation system (used for testing the design protocols).
- Models of devices (e.g., camera).
- Dynamic location model, including each agent and object (in terms of "areas" such as a habitat, and then specified by the LAT / LONG coordination system).
- Network connectivity model, distributed in design of Comm and Proxy agents (which relate external devices and agents to a local platform).
- EVA Activity Plan: Sequence of activities specifying start and stop coordinate locations, a duration, and thresholds allowed.
- Language model: Word models and mapping of phrases to commands (with agent, location, object parameters).

- Command semantics, distributed in agent interactions, constituting a work flow for communicating, accessing, and storing data (e.g., photographs, coordinates, biosensors).
- Alarms, representing data thresholds (e.g., expected length of an activity within an EVA) and actions (e.g., where to communicate this information).
- ERA plans, locations, and control procedures (e.g., to take a photograph of a location).

In preparation for the April 2003 test, the project team developed three Brahms models to run respectively on laptops located on the EVA Robotic Assistant (ERA; Burridge & Graham 2001; Shillcutt et al. 2002), on an ATV, in space suit backpacks (for two astronauts), and in the Mars Desert Research Station (MDRS) habitat ("HabCom"). The HabCom model is entirely new, to monitor the EVA activity and biosensors.

Tests were performed at Ames in early March, first by wiring the laptops, and then with the wireless data network (MEX/KaoS) linking all components. The biosensors are wired to an iPaq PDA worn by the astronaut, which transmitted by bluetooth to a Minibook computer attached to the top of backpack. A GPS unit, camera and headphone-microphone were all connected to the Minibook. The ERA Brahms controlled the ERA's camera through an API to the ERA's computer.

Changes were made after September 2002 (Meteor Crater) to handle key problems:

- **Agent architecture:** Cope with a brittle wireless network, with methods for handling lack of communication, as well as means for people to monitor agent status.
  - Used NetPhone to allow communications between support personnel at EVA site and HabCom
  - Implemented AgentMonitor display to allow HabCom to view entire state of every agent on the network (running on remote platforms)
  - Separated low-level sensor processing on iPaq from interpretation and data archiving on AstroBrahms
  - Implemented a rudimentary "forgetting" operation in Brahms, so modeler can deliberately delete beliefs that represent transient states (e.g., communications from other agents after they have been processed).

- **Hardware:** Sensors to indicate remaining power and provide warnings – short-term solution is running remote monitoring of Minibooks from laptop at ATV; bandwidth interference and microphone sensitivity – resolved by using two wired headsets for voice communications and recognition system; space suit dome greatly reduced noise; improved discipline for configuring connectors, by using dedicated kits; did not augment ERA capabilities.

- **Logistics:** Eliminated the pressurized space suit to focus on MAA infrastructure; in situ tests were carefully staged over two weeks to meet objectives; MDRS provided a permanent field shelter for working in raining, cold, windy conditions (also sometimes dry and hot).

Comparison to recommendations in the Flairs 2003 report will show substantial progress was made. With the ability to test the system more thoroughly, yet more challenges were discovered.

The following is a summary of functionality implemented for April 2003 (* indicates new functions, mostly handled by the HabCom personal agent):

- **Location Tracking**
  - GPS mounted on *backpacks & ATVs
  - Flexible logging interval
  - *Sent to Remote Science Team (RST) at variable intervals
  - Naming from predetermined vocabulary
- **Science Data Logging**
  - Sample bags associated with locations
  - *Voice annotations associated with sample bags & locations
  - *Photographs logged by time & place
  - *Stored at Hab & transmitted to RST
- ***Biosensor Logging**
  - Transmitted via iPaq to backpack
  - Logged at intervals & interpreted
- ***Activity Tracking**
  - Indicate start from predetermined list
  - Modeled by location & duration, sequence
  - Alerting for exceeding thresholds
- **ERA Commanding**
  - Move, follow, take a picture

For each of these functions, language was developed to support natural spoken phrasings in the dialog system. The essential expressions are:

- **start tracking my location every** <N> **seconds**
- **start tracking my biosensors every** <N> **seconds**
- **start** <activity name> **activity**
- **call this location** <location name>
- **where is** [<location name> | Boudreaux]?
- **Boudreaux take a picture of** [tracked astronaut | <location name>]
- **record/play** {a} **voice note** {associated with <location name> | sample bag #}
- **create sample bag** #
- **associate** {this} [**voice note|sample bag** {#}] **with** <location name>
- **how much time is left** {until next activity}?
- **upload [all | one] image**{s}

Figure 1 shows that personal Agents (PAs) are communicating locally with external systems via "communications agents," providing an Application Programming Interface (API) to read data and control devices (e.g., camera). PAs (all implemented in Brahms on different computers) communicate via a wireless network (with repeaters) using KAoS "agent registration" system.

PEOPLE	Microphone & Headphone	PERSONAL AGENTS	Comms Agents	External Systems
Astro1	⇓voice◊	**Agent**_{Astro1}	⇓API◊	GPS, ERA Biosensors Camera
| radio |		| wireless network |		
HabCom	⇓voice◊	**Agent**_{HabCom}	⇓API◊	Email (to RST)
| radio |		| wireless network |		
Astro2	⇓voice◊	**Agent**_{Astro2}	⇓API◊	GPS, ERA Biosensors Camera

Figure 1. Simplified schematic of MA configuration April 2003: Astronauts and HabCom communicate via radio (not all links are shown); people speak to their Personal Agents (PAs) on local computers using a microphone and receive feedback on headphone.

Six scenarios were designed, ranging from a simple walk around the Hab to a full-day drive onto a plateau. Actual testing involved incremental, preplanned stages:

1. Wired test in the lower deck of MDRS to confirm communications protocols and peripheral connections.
2. Wireless test inside the Hab (without full suits, emphasizing communications and biosensors).
3. Test standing on the front porch of MDRS (allows use of GPS for first time).
4. Full walk (over one hour) around MDRS, following a script to test basic functionality; all systems running except ERA.
5. "Pedestrian EVA" with ERA, walking from porch to a dry wash about 100 meters south, gathering samples, taking photos, commanding ERA to take a photo, and return.
6. "Lith Canyon EVA" involving two repeaters, ATV providing gateway to a LAN in the canyon, and a hour or more of scripted sampling and photography.

We arrived Saturday, March 30, and completed setup of all equipment in and around MDRS on Sunday. First tests began Monday morning. The scenarios (already pruned to eliminate three EVAs to remote sites), were accomplished very gradually (Table 1). The day count subtracts one for Sunday April 6, a rest day. Nearly 5 of 11 work days were devoted to model modifications and testing. Functionality errors stem from incomplete end-to-end simulation (caused by inadequate resources and planning).

The astronaut-geologists were provided with scripts to indicate the sequence of activities, including locations, and requested or optional commands to test. The astronauts could skip or repeat activities by indicating what they were doing (subject to predefined location and timing constraints). This was most useful for the more realistic exploration at Lith Canyon, where Astro2 improvised a voice annotation (Table 2).

Field Test Milestone	Day	New Problems Discovered
Wired test inside MDRS	3	Heat burned power regulator; voice commands misinterpreted
Wireless test inside	4	Alarms & email bogging down system
Test porch and walk inside	5	Many command processing bugs/incomplete code
Front Porch with GPS & Walk around MDRS	6	Initialization logic: No GPS inside, can't start biosensors until headset/helmet on
Repeat MDRS Walk with ERA	7	Basic command workflow still being tested first time
Pedestrian EVA	8	Multipaths from hills, lose network
Pedestrian w/ERA	9	Unreliable ERA navigation
Lith Canyon EVA	11	System tolerant to node reboot & network dropout, but topography-specific limits

Table 1. Field test work day on which milestone was accomplished with problems discovered.

ACTIVITY PLAN	GEOLOGISTS' SCRIPT
<EVA PREP> (-,-,-)	1. Drive in EVEREST with backpacks, helmets, suits, all equipment 2. Start Minibooks & GPS 3. Don suit with boots, gaiters, radios & headsets 4. Put on backpack & helmet & connect cables
Checking equipment (20, 5, 20)	1. "Start CHECKING EQUIPMENT activity" 2. "Start tracking my location every 60 seconds" 3. "Start tracking my biosensors every 5 minutes" 4. "Start WALKING TO TOP OF CANYON activity"
Walking to top of canyon (10, 0, 10)	{Astronaut 2 improvised a voice note during the walk}
Sample fossils (10, 5, 0)	1. "Start SAMPLE FOSSILS activity" 2. Sample bag, voice annotation, association, photo 3. "Start WALK TO HEAD OF CANYON activity"
Walk to head of canyon (10, 0, 10)	< Walk carefully down the hill and proceed to the head of the canyon to the south (your left) >

Table 2. Lith Canyon Scenario script (first half) provided to Astronauts.
Activity plan key: (Duration, Duration threshold in minutes, Distance threshold in meters).

Figure 2 shows the topography configuration for Lith Canyon; it required nearly two days to deploy communication equipment in preparation for this EVA.

Lith Canyon Relay Test

The Lith Canyon site involves broken ledges and steep cliffs 5 km from MDRS. The topography caused several serious problems:

• A wireless "shadow" occurred at the head of the canyon (as expected), causing the computerized backpacks to drop out of the network linking the astronauts back to MDRS. (Communications were properly queued and handled when possible.)
• The astronauts were unwilling to pass over a meter-high drop off in the canyon, requiring them to change the plan and walk around.
• The ERA was unable to follow the astronauts into the canyon because of the terrain, and even along the ledge had to be directly teleoperated with a person standing nearby.

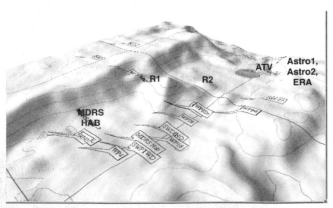

Figure 2. Topographic layout of MDRS (Hab), Cisco repeaters, ATV LAN gateway,and astronauts with ERA in Lith Canyon (about 5 km from MDRS).

The Lith Canyon field test was a major accomplishment for the Mobile Agents project. The geologists' backpack computers running Brahms were wirelessly networked to another computer on an ATV 75 m away on a ledge across the canyon, and from there to a laptop running in MDRS more than 5 km away (Figure 2). The EVA lasted about 1.5 hours, as batteries allowed.

Table 3 shows communications that were generated by personal agents during the canyon test. In addition, one command was given to the ERA PA by the Astro2 PA (Astro1 asked the ERA to take a picture; however it was not executed.)

An *explicit confirmation* mode was added to the dialog system, requiring each command to be confirmed by the astronaut prior to execution (e.g., "Do you want to start 'walk to head of canyon' activity?"). Most commands are forwarded to the MDRS computer (HabCom) and then responses returned to the astronaut agents, causing the dialog system to generate a verbal feedback (e.g., "Activity 'walk to head of

Communication	Astro1	Astro2	HabCom
broadcast_alert	2	0	4
create_location	1	1	0
download_images	3	1	0
GPS	3	2	0
GPS_start_logging	1	2	0
initialization	2	4	4
location_with_samplebag	1	1	0
newData	0	0	9
new_activity_started	2	1	7
SampleBag	3	1	0
start_specified_activity	2	0	2
storeData	27	13	0
TakePicture	0	1	0
voice_annotation_with_location	1	1	0
voice_annotation_with_samplebag	1	1	0
VoiceAnnotation	7	1	0

Table 3. Lith Canyon Communications Processed by Personal Agents.

canyon' activity started"). This confirmation and feedback protocol was implemented through trial and error during the first week of the field test. It ensures that the correct command is being processed and that it has been executed. Although seeming obvious now, use of the system in context was required to discover what kind of feedback was required.

Analysis And Conclusions

Field test results can be summarized according to lessons learned about the hardware, the agent architecture, and logistics of setting up the system and carrying out the scenarios:

- **Hardware**
  - Technology required for field science is strongly topography driven.
  - A robot should be capable of working in terrain that geologists explore (e.g., tractors / spider legs).
  - Need automated antenna and video tracking.
  - Need faster computers all around.
  - Adapt ERA's differential GPS for astronauts.
- **Agent Architecture**
  - Copes well with loss of signal; but Astro PAs must take over some HabCom monitoring functions.
  - Need assertions to verify end-to-end functionality (e.g., has photo arrived at RST?).
  - Must integrate speech output, astronaut voice, and HabCom for recording.
  - Astronauts work in parallel, lack voice loop; they must coordinate to avoid work redundancy.
- **Logistics**
  - Need formal data network specification (GPS, computer, radio, biosensors).
  - Need written specs / deliverables (not just design documents).
  - Need field backups for all computers.

Ideally, mobile agents should run for the duration of an entire mission – perhaps several years. However, agents' interactions with their environment results in an ever expanding belief set (memory set). Not managing memory results in serious performance degradation of the agents' reasoning state network. An interim solution, implemented during the field test, is to retract transitory communication beliefs (i.e., that a request has been received). A more systematic, built-in model of forgetting is required.

In this early stage of development, the HabCom person was responsible for monitoring the various Brahms systems. He listened and responded to: the voice loop, information spoken by his PA (such as alerts), and field radios (or IP phones) to verify system responses. Because all three require a different focus, he frequently missed problems and requests. This discovery exemplifies our approach of "empirical requirements analysis." One may design clever agent algorithms and architectures, but in practice one will find that simple services are needed that were never considered back in the home lab. In particular, we are frequently discovering new tools required to monitor and verify the system's operation during this developmental process. This phase can be expected to last many years, and involves viewing the agents as assisting in the research and ongoing redesign of the system.

Brahms was not designed from the ground up to be a distributed, highly available mission support system. Persistence and failure recovery are some aspects that require near-term redesign. Additional requirements may need to be imposed on the Brahms language. For example, better methods are required to allow agents to recover from system failures, multi-task, inspect their own state and adjust themselves in different contexts. Methods must be improved for handling large volumes of data that may need to be stored, but only sampled and interpreted periodically.

We conclude that the following aspects of the MA design and field test worked especially well and were crucial to our progress:

- **Human-Centered Design (Technology Pull)**
  - Authentic work scenarios (real geologists doing real work)
  - Analog simulations during MDRS5 (Clancey 2002a) plus historical analysis of Apollo CapCom (Clancey in press b)
  - On-site requirements analysis (MDRS5) —> Voice command design —> Simulation in Brahms —> Distributed Implementation
- **Technology Retreat Facility**
  - Attractive, isolated, evocative setting
  - Utilities augmented: Broadband ISP, LAN, toilets
  - Nearby inexpensive motels & restaurants
  - Resident handyman
- **Management Structure**
  - Commander plus subsystem point-of-contacts
  - Realistic 9 am–5 p.m. schedule
  - Required 9 am briefing; replanning at end of day
  - Arrive Friday, start Monday am, Sunday off, cleanup second Saturday (11 work days)

As suggested by the previous tests, we conclude that a multi-agent simulation with scenario-based formal specification accelerates cross-institution collaboration for integrating sensors, automation, procedures and communications.

We plan to return to MDRS in April 2004 with these objectives:

- Complete the Pedestrian and Lith Canyon scenarios.
- Extend navigation & schedule monitoring.
- Develop a medical agent to interpret biosigns.
- Develop a mission console for HabCom to log alerts.
- Provide HabCom with a map of locations for all people and agents, including their activities.
- Add science database, with RST access for shared annotation.
- Plan and track multi-day EVAs with the RST.

Discussion Of Related Work

Here we compare the MAA with some other multi-agent frameworks that focus on team coordination. To begin, the Teamcore agent framework is similar to the MAA in the sense that it allows new or existing agent programs that have no preexisting coordination capability to work together. While in the MAA the PAs provide mobile agents with the capabilities to work together performing team tasks, in Teamcore this function is provided by "proxies." In the MAA proxy refers to agents that represent PAs in agent environments running on different machines. Proxy agents in MAA thus represent the PAs in other agent systems, whereas Teamcore proxy agents represent agents that can be written in different languages with a Teamcore wrapper to allow these agents to communicate using the KQML agent communication language .

Another similarity between Teamcore and MAA is the need to facilitate human collaboration. Our objective is to allow an EVA crew to work together (including the crew in the habitat and remote science teams on Earth, as well as with robots and other science tools and devices). The Teamcore infrastructure has been applied to a computer simulation of the evacuation of civilians from a threatened location and to assist the Teamcore's research team in their routine coordination (e.g., scheduling team meetings). In this second application, Teamcore is also addressing the mobility issue of people by integrating GPS devices and PDAs. Differences between the MAA and Teamcore include: 1) MAA allows the human user to dynamically create new locations; 2) agents in the MAA are mobile, therefore the architecture and agent design have to deal with the fragility of a wireless communication network; and 3) MAA provides assistance to people coordinating with robots, external devices and medical monitoring.

RETSINA is a multi-agent system (MAS) infrastructure that allows developers to implement large distributed societies of software agents . RETSINA is an agent infrastructure, not an agent language. Research with RETSINA focuses on understanding what is needed to provide a "domain independent and reusable substratum on which MAS systems, services, components, live, communicate, interact and interoperate, while the single agent infrastructure is the generic part of an agent that enables it to be part of a multi-agent society" . We also intend for the MAA to be an MAS infrastructure that allows independent agents (people, software agents, robots and devices) to be part of a large multi-agent society, or even multiple societies – including EVA and habitat agents, and RST societies (science, medical, engineering, mission control teams).

One key difference between RETSINA and Mobile Agents is the domain. The vision of the RETSINA team is the internet's computational world, populated with agent societies. In the current MAA, agents are not directly connected to the internet, but run on a dedicated wide-area wireless communication network. Connection to the internet is currently provided by a single agent that can email to members of an RST. Sycara, et al. (2003) provide a definition of the MAS infrastructure service layers. We aim to provide many of these services in the MAA as well. To this extent we are hoping to use existing systems, and perhaps in the future we will use some of RETSINA's capabilities within the MAA. Currently, some of the MAA infrastructure is provided by KAoS , such as the communication infrastructure, name to location mapping, and some of the multi-agent management services. We are also currently integrating Brahms with the policy capability of KAoS to handle agent interactions when communication breakdowns require local agents to assume responsibilities, and then relinquish them when the network is reestablished.

Acknowledgments
Funding is provided in part by the NASA / Ames Intelligent Systems Program, Human-Centered Computing area, managed by Mike Shafto. Our team includes a dozen other Brahms, ERA, and communications specialists. For related information (including daily field reports from MDRS16), see http://bill.clancey.name, http://www.marssociety.org, and http://www.agentisolutions.com.

References
- Alena, R., Ossenfort, J., Lee, C., Walker, E., Stone, T., in press. Design of Hybrid Mobile Communication Networks for Planetary Exploration. *IEEE Aerospace Conference 2004*.
- Bradshaw, J. M., Dutfield, S., Benoit, P., and Woolley, J. D. 1997. KAoS: Toward an industrial-strength generic agent architecture. In J. M. Bradshaw (ed.), *Software Agents*, 375-418. Menlo Park, CA: AAAI Press.
- Bradshaw, J. M. et al. 2003. Representation and Reasoning for DAML-Based Policy and Domain Services in KAoS and Nomads. *AAMAS '03*, Melbourne, Australia.
- Clancey, W. J., Sachs, P., Sierhuis, M., and van Hoof, R. 1998. Brahms: Simulating practice for work systems design. *Int. J. Human-Computer Studies*, 49: 831-865.
- Clancey, W. J. 2002a. A Closed Mars Analog Simulation: the Approach of Crew 5 at the Mars Desert Research Station. *Mars Society Annual Conference*, Denver.
- Clancey, W. J. 2002b. Simulating activities: Relating motives, deliberation, and attentive coordination. *Cognitive Systems Research*, 3(3) 471-499.
- Clancey, W. J., Sierhuis, M., van Hoof, R., and Kaskiris, C. 2003. Advantages of Brahms for Specifying and Implementing a Multiagent Human-Robotic Exploration System. *Proc. FLAIRS-2003*, 7-11.
- Clancey, W. J. in press a. Roles for Agent Assistants in Field Science: Understanding Personal Projects and Collaboration. *IEEE Transactions on Systems, Man and Cybernetics*. Issue on Human-Robot Interaction.
- Clancey, W. J. in press b. Automating CapCom: Pragmatic Operations and Technology Research for Human Exploration of Mars. In C. Cockell (ed.), *Martian Expedition Planning*. AAS Publishers.
- EVA Robotic Assistant. URL http://vesuvius.jsc.nasa.gov/er_er/html/era/era.html.
- Finin, T., Y. Labrou, and Mayfield, J. 1997. KQML as an agent communication language. In J. M. Bradshaw (ed.), *Software Agents*, 291-316. Menlo Park, CA: AAAI Press.
- Pynadath, D. N., and Tambe, M. 2003. An Automated Teamwork Infrastructure for Heterogeneous Software Agents and Humans. *Autonomous Agents and Multi-Agent Systems*, 7, 71-100.
- Shillcutt, K., Burridge, R., Graham, J. 2002. Boudreaux the Robot (a.k.a. EVA Robotic Assistant). *AAAI Fall Symp. on Human-Robot Interaction*, Tech Rpt FS-02-03, 92-96. Menlo Park, CA: AAAI Press.
- Sierhuis, M. 2001. *Modeling and simulating work practice*. Ph.D. thesis, Social Science and Informatics (SWI), University of Amsterdam, The Netherlands, ISBN 90-6464-849-2.
- Sycara, K., Paolucci, M., Velsen, M. v., and Giampapa, J. 2003. The RETSINA MAS Infrastructure. *Autonomous Agents and Multi-Agent Systems*, 7, 29-48.

//

Mars and the Republic†

Matt Pearson

Introduction

Much of the speculation regarding colonization of Mars has at its core the idea of a Martian Republic. This is not because Mars is particularly suited to a republican, or more loosely "democratic" system, or that careful examination of the facts has led to the conclusion that such a system is the best option, but simply that those expounding their opinions on the matter are almost invariably citizens of republics. Discussions of the matter are entered into with a base assumption derived from the very recent experience of only a fraction of humanity, under conditions radically different from those the Martian colonists will face. At present, the republican system is primarily the government of choice for North America and Western Europe. Several nations, particularly in Eastern Europe and South America, possess the trappings with varying degrees of sincerity; many more nations do not even bother to feign it. At present republics, while not altogether uncommon, are by no means universal. Historically they are rare indeed. Go back even as little as a century and their number drops dramatically.

A Martian Republic is certainly an ideal to strive for and can potentially convey on human civilization the same sort of benefits as the American republic during its frontier period. Mars can act as a proving ground for new ideas, a cradle for invention and a societal safety valve all at once, but a republic is not assured. It will require a great deal of preparation and careful cultivation if it is to succeed, and to assume its inevitability is to willfully blind ourselves to the perils of the course we have set ourselves upon.

The Birth of Republics

Contrary to modern Western bias, republics are not a form of government that people naturally tend to organize into. Their formation requires certain conditions, to endure requires a minor miracle. As is often the case when discussing Mars colonization (and despite the false assumptions such analogies can lead to), there are lessons to be learned from the history of the Americas.

Britain's American colonies (which we will focus on exclusively, for clarity) were uniquely suited to the formation of a new and lasting republic when they declared their independence. That collective act of defiance was the culmination of centuries of cultural development in the colonies. The individual colonies were able to unite not so much because of the compelling arguments of the benefits of doing so, but because of their common culture, common language and common historical experience. The specifics of that cultural and historical base were essential for the republic's development; without it the Constitution of the United States of America would likely be nothing more than a figment of political theory.

The American colonies, while subjects of the Crown, were essentially on their own from the very beginning. The colonies governed themselves for the most part, though not always in a representative manner. Even when they had orders from their mother country, those orders could often be ignored to a degree. The distance was so great and the means of travel so slow that even garrisoned soldiers charged with enforcing the English laws acted locally and independently of direct control from London. Their orders originated in America, not in England. America was a world unto itself, cut off from the mother country. Centuries of practical independence laid the cultural groundwork for formal independence. By the time of the revolution, the people had a long history of self-reliance and self-governing.

America had ample resources in an easily accessible form. Everything the colonists needed was there for the taking – food, water, building materials – all were present in quantities far in excess of what was required. The American colonies were an ideal medium for the formation of a new republic. Isolation from the mother country provided the rational motivation for independence. The availability of resources provided the means to make independence practical, but only the centuries of cultural evolution in the presence of those qualities allowed for the formation and survival of a free republic.

Neither will be the case on Mars. Unlike the American frontier, Mars will not be a wide open field of seemingly endless resources stretching for as far as the eye can see, and communications delays will be measured in minutes, not months. The Martian environment makes even the harshest regions of Earth seem a paradise by comparison. The American frontier had to be tamed, the Martian frontier is colder and harder, and must be broken. The task need not take centuries, a generation or two of grappling with the Martian environment, cut off from significant material help from Earth, may be enough to create the necessary cultural framework, but it cannot be bypassed entirely if the Martian Republic is ever to exist.

It cannot be stressed enough that simply living under an established republic is not sufficient. Just because the first Martian colonists will likely be Americans or Europeans who have lived under and been firmly indoctrinated in the ideals of representative government does not mean that those colonists will be well suited to building a new one. While Westerners are firmly rooted in democratic ideals, we are certainly not self-reliant to anywhere near the degree of the American pioneers. All the nations that have the means to take part in the colonization of Mars are well developed with high standards of living and a rather extensive welfare apparatus. Average American or European citizens, even those with the technical skills to make colonization physically possible, are not up to the task. In essence, what is required is a tech-savvy mountain man – a "Daniel Boone" in a space suit. If there is to be a Martian Republic its founders cannot come from Earth, they must have lived their lives on Mars. They must develop a Martian culture that allows for the growth of a Martian Republic, it cannot exist in a societal vacuum. Republics require more than elections. The people must believe in it, they must have faith in the integrity of the system, and they must have a spirit of liberty and self-reliance. A constitution cannot provide the essential elements needed for a republic to function.

Martian Society in a Vacuum

While it seems unfortunate at first that a republic cannot be established on Mars out of the can, it is good to be discouraged from trying at this stage. The types of decisions that the government of a new Mars colony will face will fall into two basic categories. The first is organizational questions such as size of work details, use of the habitable space, what type of time keeping system to use and a host of other minor details, and variations from the already established plan that will never be resolved to everyone's satisfaction by debate. The second category comprises decisions that need to be made very quickly, most likely in cases where something has gone horribly wrong. In neither of these cases is a system based primarily on popular opinion desirable; in the first case it is simply inefficient and in the latter it is a sure way to significantly reduce the colony's odds of survival. While some degree of popular representation is needed, if only from a cultural standpoint, the government at this early stage must have the authority to act without consulting. This requires a small and centralized authority.

Building a new nation from the most basic raw materials is something that must be planned meticulously in advance and then strongly guided when all the well-laid plans begin to fall apart. Nothing as complex as establishing a colony on Mars will go entirely according to design, and when things begin to go wrong a mechanism must be in place for quickly deciding on a solution and implementing it. With such requirements, the introduction of something as crude as a popular vote is a recipe for disaster. On the other hand, simply establishing an unquestionable hierarchy won't do either. What is required is a system which provides the needed command structure while still allowing for change instituted from below, a system of government well suited to an environment where it may become necessary to very quickly scrap the old way of doing something and replace it with a method more suited to the moment. The colonization of Mars will require a pragmatic and adaptable form of government (ad-hocracy, if you will) that allows for a great deal of individual initiative down to the lowest levels while having the authority to change drastically and quickly should the need arise. The future Martians will need to find a balance between seemingly contradictory requirements. In addition there are likely social trends that will have to be considered and, if possible, utilized.

The establishment of a Mars colony will require workers in several widely varied fields, all carrying out their tasks simultaneously. Such an endeavor would not need dedicated management personnel, and therefore even those who do have a management function would be rooted and active in one of the fields needed. The combination of the natural division of workers by trade and the lack of a rigid and ossified management bureaucracy would likely have profound effects on the project. Work on every aspect of the colony could proceed unhindered by interference from uninformed meddling, resulting in more efficient work and likely in superior results. Unlike on Earth, Martian colonists will be in a very real sense working for themselves – the more infrastructure they erect, the more secure and comfortable their lives will become. Their successes will be rewarded with a steady and highly visible increase in the standard of living conditions for the entire colony and they will be acutely aware that failure will in all probability result in Mars' swift return to uninhabited status. The Martian colonial labor force will be well motivated.

During the building of a colony those colonists will live in an environment the likes of which humans have never fully experienced – modern needs coupled with fairly primitive societal forms. There will be no outside social pressures or regulators, and labor will be specialized and therefore, to a large degree, segregated. As every colonist will know their job, be self-motivated to do it to the best of their abilities, and generally be free from interference, those who work in a given field will likely tend to form fairly closed groups. While this may be good from a standpoint of building the colony's infrastructure, it results in a social structure without any direct Terran equivalent. While collectivism or corporatism are the most obvious parallels, such a society is really more like a hive. The "hive" model offers distinct advantages during the building of a colony, but the societal framework it creates does not readily lend itself to the establishment of a republic. If left to develop freely, the result of Mars' colonization could well be a rigid caste society constantly reinforced by a hard and unforgiving environment.

The prospects for a Martian Republic developing on its own seem slim, especially taking into account the human factor. Fledgling republics are fragile things and humans are selfish creatures. Those with authority rarely relinquish it voluntarily if they can avoid it. If an approach combining the natural tendencies listed above with a legal framework intended to nurture republican ideals can be executed, the odds for success increase dramatically.

A transitional form of government is needed, something that can effectively carry out the task of building the colony's infrastructure, while at the same time helping to create the conditions needed for the republic to be formed. Such a transitional government may take many forms, but it should be kept in mind that the more it approaches the status of a complete governmental system, the less likely it will be that the colonists will elect to entirely replace it. If something works, we tend not to want to change it. This tendency will be reinforced by the cold and hard reality of the Martian environment.

Outline for a Martian Transitional Government
As every colonist will have a specific function, and their numbers will initially be small, organizing the workers of various fields or "trades" is a reasonable basis for representation. As they will tend to group together on their own and the overall numbers will be small, organizing the colonists in this way will be fairly simple. Each grouping, or Guild as they will be referred to for this discussion, will be solely responsible for its role in the colony. This gives us the starting point for a government and is based on what the colonists will likely do anyway, it is a simple and stable start. But something more is required if the "hive" is to be avoided and the groundwork for a republic is to be laid. Some central authority is required, both to direct the development of the colony toward a republic, and to take control in the event of a catastrophe. If its intervention is required, it must be quick, decisive, and free to take

whatever action is needed. The primary purpose of the Executive is to have in place an authority to direct the colony if things do not go as planned.

The makeup of the Executive body could take many forms, though a single person or a body of three is likely the most effective. Two makes gridlock likely when swift action is required, and anything more than three takes on the complexity and inefficiency of a committee. Either case destroys the Executive's ability to perform its function should the need arise. For a number of reasons, both practical and political, an Executive Triumvirate may well be the preferred option, from the perspective of efficiency versus excessive authoritarianism.

The Executive must be well informed about any action it may have to take, and to that end each guild should elect one or more of its members to serve on an "Advisory Council." This not only ensures that the Executive has the necessary "big picture" perspective, but also forces the guilds to have regular contact with one another rather than focusing solely on their own work. This forced contact may prove to be the most important function of the transitional government. Should the need arise to implement new laws during the colony's early development, the advisory council can also act as a ready-made legislative body. Such an arrangement is ideally suited to the conditions because rather than having technical matters decided by people with little or no understanding of the issue in question, at least some of the guild representatives will have first-hand experience with any issue that arises.

This results in a Martian colony with a small, efficient government possessing the knowledge to perform its function. The only real policy should be the survival and development of the colony. Ideally the colonial government shouldn't have to do anything at all, but if it must then it will require the power to act. Although the Executive could in theory have the power to rule by decree, if the guilds took care of their own affairs the Executive would be irrelevant unless a situation arose in which its intervention was absolutely necessary for the colony's survival.

This system, while simple and effective for building a colony, does have the potential to become an authoritarian monstrosity, particularly given the Martian conditions that will tend to reinforce any strong authority already in place, coupled with the lack of immediate alternatives. Living in a pressurized dome is likely to stifle the desire to question authority, given that the entire population will be dependent on the smooth and orderly functioning of society to a degree unheard of on Earth. Clearly, the transitional government requires some guidelines. First, the inclusion of some derivative of the Bill of Rights of the US Constitution would be a good start.

Aside from a brief list of the things the government cannot do, some method of expiration must also be put in place, otherwise it will not be the transitional system it is intended as. A regular referendum on whether the system is still required or to begin drafting the constitution of a Martian Republic is the simplest method. The absolute minimum time frame for a referendum of this sort should be no less than twenty years. The republic cannot exist until the colony's infrastructure is well established and the colonists themselves will require at least that long to begin building the cultural foundation needed for the republic to endure. By the very fact of the ad-hocracy's simplicity and incompleteness, it will need to be discarded as soon as the colony outgrows it, though to be truly ready a time frame of 50-100 years is likely. While it may seem excessive, half a century is a fraction of the time it took for the United States to emerge.

Rights, Laws and Freedom

While the transitional government will most likely not form the basis of the Martian government once the colony is firmly established, it will leave a lasting legacy. The principles and institutions of the transitional system will heavily influence those of the Martian Republic, therefore the transitional government must not only physically pave the way and secure the material well-being of the colony, it must lay a sound philosophical foundation for the future Martian state. It is an awesome responsibility, and reason must prevail over emotion. The Martians would do well to keep a few simple principles in mind.

Dr. Robert Zubrin has written a list of the "Rights of Mars" in his book *Entering Space: Creating a Spacefaring Civilization*." This list can be used to illustrate some important points. He wisely begins with derivatives of the Bill of Rights of the US Constitution. These principles are proven and are the basis for a free state. Additions to these must be carefully considered. While well intentioned, the careless dispensing of "rights" can result in a great deal of harm.

Of particular note are numbers 12 and 17, "The Right to Access to Means of Mass Communication" and "The Right to Free Education," respectively. The first nine rights are things that government cannot do *to* you. These two are things which government must do *for* you. They are not "rights," but entitlements. Rights make the people free, entitlements make them dependents. If the people of Mars want access to means of mass communication, they would be much better off building it for themselves than waiting for the Martian government to provide it to them. A colony of people who believe they are entitled to services simply by virtue of their own existence will never build a free republic. A people who know they are entitled to nothing but an opportunity to improve their conditions will strive to do just that, and Mars will belong to them.

Many other additional "rights" are simply superfluous if the first nine are taken at face value. Still others, such as "The Right to Self Government by Direct Voting" are aspects of a particular governmental system rather than an individual right. Even in the United States, the so-called "right to vote" is merely a collection of laws guaranteeing the opportunity to vote. If it were truly a right, citizens could just vote; there would be no need to register. It is an aspect of the American system of government, but it is not a "right" in the proper sense. Rights are simple things, generally things easily done by individuals. Some rights exist to limit what the State can do to its citizens, but when a state apparatus is required for the execution of the so-called "right," it isn't a right at all. For example, a right to due process simply prevents the state from arbitrarily imprisoning its citizens, a "right to trial by jury" refers to a specific form of due process and does not constitute a right in the proper sense of the word.

Whatever form a Martian Bill of Rights may take, laws are needed to give order to the new Martian society. It should be noted that, contrary to popular misconception, laws do not prevent crime. Laws define crime, and therefore, in a sense, laws create crime. To keep the transitional government small and allow the colony plenty of room to grow into its own path, the laws should be as few and as simple as possible. A good guideline would be something to the effect of:

> If an action brings no quantifiable harm to the colony or its citizens, nor infringes upon the rights of others, it cannot be criminal.

Using this as a guide, any actions that are truly damaging (such as murder, rape, theft, vandalism, etc.) are clearly criminal. Many of today's laws do not meet such requirements of criminality and can therefore be discarded. Such a legal code retains all the authority to punish true offenses while cutting back significantly on the opportunities for abuse by the government, an important consideration in an environment such as Mars, where any established authority will tend to be reinforced by the inhospitable conditions and interdependence of the colony.

The establishment of property rights is of vital importance for the future of Martian colonization as well, whatever form it may take. At first there will be no privately owned land as the colony's habitable space will be extremely limited resulting in all interior space being publicly owned to some degree, a sort of default communism. As mining operations and other activities on the planet's surface become practical there will be a need for establishing property rights over Martian land. Allowing the Martian government to simply grant title should be avoided, the implication being that the Martian government owns the entire planet. Ownership based on possession and use is the simplest method. An individual or corporate entity which begins mining operations gains ownership of the land. Simply declaring intent to mine is insufficient. This simple approach prevents corporations from claiming vast areas of Mars on speculation while still allowing individuals to homestead.

Once property rights are established they must be secure, the more freely the landholders can profit by improvements of their land the greater the incentive to develop further. Any activities that infringe on property rights, whether from other colonial interests or from government detract from the growth of the colony and must be avoided as much as possible. There will likely be intense competition over particularly lucrative lands as soon as anything of value is discovered, but this only serves to build the colony and Martian civilization more quickly and should be encouraged.

The transitional government outlined here is not what many want for Mars. It does not fulfill idealistic visions of a utopian future nor does it take a bold and dramatic jump toward a new, Martian philosophy. However, unlike some more idealistic proposals, it can be implemented in the earliest stages of a colony, and most importantly, it is workable. While the building of the Martian Republic is a project for another generation, the colonization of Mars need not be. Yet, while the task can be accomplished, attempting it will take us in new directions, likely some that we did not intend. We must be prepared, as the American pioneers before us were, to face the challenges and overcome them. At times those who remain on Earth may not approve of what emerges, but the endeavor is not about them.

If there is to be a Martian Republic we must first overcome the Martian environment. We may at times be forced to adopt very undemocratic methods to do so. There are many perils, but the path to freedom has always been beset with hardship, and it is never guaranteed or secure. Employing non-republican methods in order to build a republic is not without risks, but the risks are worth taking. If we blindly succumb to the modern prejudices of democracy we will almost surely fail. With eyes and minds open, on to Mars.

//

CD-ROM Table of Contents

The included Bonus CD-ROM includes the following material: